Statistics and Computing

Series Editors:
John Chambers
David J. Hand
Wolfgang K. Härdle

Statistics and Computing

Brusco/Stahl: Branch and Bound Applications in Combinatorial Data Analysis
Chambers: Software for Data Analysis: Programming with R
Dalgaard: Introductory Statistics with R, 2nd ed.
Gentle: Elements of Computational Statistics
Gentle: Numerical Linear Algebra for Applications in Statistics
Gentle: Random Number Generation and Monte Carlo Methods, 2nd ed.
Härdle/Klinke/Turlach: XploRe: An Interactive Statistical Computing Environment
Hörmann/Leydold/Derflinger: Automatic Nonuniform Random Variate Generation
Krause/Olson: The Basics of S-PLUS, 4th ed.
Lange: Numerical Analysis for Statisticians
Lemmon/Schafer: Developing Statistical Software in Fortran 95
Loader: Local Regression and Likelihood
Marasinghe/Kennedy: SAS for Data Analysis: Intermediate Statistical Methods
Muenchen: R for SAS and SPSS Users
Ó Ruanaidh/Fitzgerald: Numerical Bayesian Methods Applied to Signal Processing
Pannatier: VARIOWIN: Software for Spatial Data Analysis in 2D
Pinheiro/Bates: Mixed-Effects Models in S and S-PLUS
Unwin/Theus/Hofmann: Graphics of Large Datasets: Visualizing a Million
Venables/Ripley: Modern Applied Statistics with S, 4th ed.
Venables/Ripley: S Programming
Wilkinson: The Grammar of Graphics, 2nd ed.

Robert A. Muenchen

R for SAS and SPSS Users

 Springer

Robert A. Muenchen
University of Tennessee
Knoxville, TN, USA
muenchen.bob@gmail.com

S-PLUS® is a registered trademark of the Insightful Corporation.

SAS® is a registered trademark of SAS Institute.

SPSS® is a registered trademark of SPSS Inc.

Stata® is a registered trademark of Statacorp, Inc.

MATLAB® is a registered trademark of The Mathworks, Inc.

Windows Vista® and Windows XP® are registered trademarks of Microsoft, Inc.

Macintosh® and Mac OS® are registered trademarks of Apple, Inc.

ISBN: 978-0-387-09417-5   e-ISBN: 978-0-387-09418-2
DOI 10.1007/978-0-387-09418-2

Library of Congress Control Number: 2008931588

© Springer Science+Business Media, LLC 2009

Printed on acid-free paper

springer.com

Preface

While SAS and SPSS have many things in common, R is *very* different. My goal in writing this book is to help you translate what you know about SAS or SPSS into a working knowledge of R as quickly and easily as possible. I point out how they differ using terminology with which you are familiar, and show you which add-on packages will provide results most like those from SAS or SPSS. I provide many example programs done in SAS, SPSS, and R so that you can see how they compare topic by topic.

When finished, you should be able to use R to:

- Read data from various types of text files and SAS/SPSS datasets.
- Manage your data through transformations or recodes, as well as splitting, merging and restructuring data sets.
- Create publication quality graphs including bar, histogram, pie, line, scatter, regression, box, error bar, and interaction plots.
- Perform the basic types of analyses to measure strength of association and group differences, and be able to know where to turn to cover much more complex methods.

Who Is This Book For?

This book is, perhaps obviously, for people who already know a statistics package. While aimed at SAS and SPSS users, many statistics packages share their main attributes, especially the use of a rectangular dataset as their only data structure. I expect users of most other statistics packages could benefit from this book. An audience I did not expect to serve is R users wanting to learn SAS or SPSS. I have heard from quite a few of them who have said that by explaining the differences, it helped them learn in the reverse order I had anticipated. However, I explain none of the SAS or SPSS programs, only the R ones, and how the packages differ, so it is not ideal for that purpose.

Who Is This Book Not For?

I make no effort to teach statistics or graphics. Although I briefly state the goal and assumptions of each analysis, I do not cover their formulas or derivations. We have more than enough to discuss without tackling those topics too. This is also not a book about writing R functions; it is about using the thousands that already exist. We will write only a few very short functions. If you want to learn more about writing functions, I recommend John Chamber's *Software for Data Analysis: Programming with R* [1]. However, if you know SAS or SPSS, reading this book should be a nice leisurely step to take before diving into a book like that.

Practice Datasets and Programs

All the programs, datasets and files that we use in this book are available for download at http://RforSASandSPSSusers.com. Also check that site for updates and/or corrections to this book.

Acknowledgments

I am very grateful for the many people who have helped make this book possible, including the developers of the S language upon which R is based, Rick Becker, John Chambers, and Allan Wilks; the people who started R itself, Ross Ihaka and Robert Gentleman; the many other R developers for providing such wonderful tools for free and all the r-help participants who have kindly answered so many questions. Virtually all the examples I present here are modestly tweaked versions of countless posts to the r-help discussion list, as well as a few SAS-L and SPSSX-L posts. All I add is the selection, organization, explanation, and comparison to similar SAS and SPSS programs.

I am especially grateful to the people who provided advice, caught typos, and suggested improvements including: Patrick Burns, Peter Flom, Martin Gregory, Ralph O'Brien, Charilaos Skiadas, Phil Spector, Michael Wexler, and seven anonymous reviewers who provided pages of invaluable advice to a neophyte.

A special thanks goes to Hadley Wickham, who provided much guidance on his ggplot2 graphics package. Thanks to Gabor Grothendieck, Lauri Nikkinen, and Marc Schwarz and for the r-help discussion that led to Sect. 14.15. Thanks to Gabor Grothendieck also for a detailed discussion that lead to Sect. 14.4. Thanks to Patrick Burns for his assistance with the glossary of terms in Appendix A.

My thanks also go to these people who helped compile *Appendix B: A Comparison of SAS and SPSS Products with R Packages and Functions* including: Thomas E. Adams, Jonathan Baron, Roger Bivand, Jason Burke, Patrick Burns, David L. Cassell, Chao Gai, Dennis Fisher, Bob Green, Frank E. Harrell Jr., Max Kuhn, Paul Murrell, Charilaos Skiadas, Antony Unwin, and Tobias Verbeke. Thanks to Henrique Dallazuanna for the code to count packages presented in that appendix.

I also thank SPSS Inc., especially Jon Peck for his helpful review of this book and his SPSS expertise which appears in the programs for extracting the first/last observation per group and generating data.

Finally, I am grateful to my wife, Carla Foust and sons Alexander and Conor, who put up with many lost weekends as I wrote this book.

About the Author

Robert A. Muenchen is a consulting statistician with 28 years of experience. He is currently the manager of the Statistical Consulting Center at the University of Tennessee. He holds a B.A. in Psychology and an M.S. in Statistics. Bob has conducted research for a variety of public and private organizations and has assisted on more than 1,000 graduate theses and dissertations. He has coauthored over 40 articles published in scientific journals and conference proceedings.

Bob has served on the advisory boards of SPSS Inc., the Statistical Graphics Corporation and PC Week Magazine. His suggested improvements have been incorporated into SAS, SPSS, JMP, STATGRAPHICS and several R packages.

His research interests include statistical computing, data graphics and visualization, text analysis, data mining, psychometrics and resampling.

Contents

1	**Introduction**	1
	1.1 Why Learn R?	1
	1.2 Is R Accurate?	2
	1.3 What About Tech Support?	3
2	**The Five Main Parts of SAS and SPSS**	5
3	**Programming Conventions**	7
4	**Typographic Conventions**	9
5	**Installing and Updating R**	11
	5.1 Installing Add-on Packages	11
	5.2 Loading an Add-on Package	13
	5.3 Updating Your Installation	15
	5.4 Uninstalling R	16
	5.5 Choosing Repositories	17
	5.6 Accessing Data in Packages	18
6	**Running R**	21
	6.1 Running R Interactively on Windows	21
	6.2 Running R Interactively on Macintosh	23
	6.3 Running R Interactively on Linux or UNIX	25
	6.4 Running Programs that Include Other Programs	27
	6.5 Running R in Batch Mode	27
	6.6 Running R from SPSS	28
	6.7 Graphical User Interfaces	32
	6.7.1 R Commander	33
	6.7.2 Rattle for Data Mining	34
	6.7.3 JGR Java GUI for R	36

7   Help and Documentation   41
   7.1   Help Files   41
   7.2   Starting Help   41
   7.3   Help Examples   42
   7.4   Help for Functions that Call Other Functions.   44
   7.5   Help for Packages.   44
   7.6   Help for Datasets   45
   7.7   Books and Manuals   45
   7.8   E-mail Lists.   45
   7.9   Searching the Web   46
   7.10   Vignettes.   47

8   Programming Language Basics   49
   8.1   Simple Calculations   50
   8.2   Data Structures.   51
       8.2.1   Vectors.   51
       8.2.2   Factors.   53
       8.2.3   Data Frames   55
       8.2.4   Matrices.   57
       8.2.5   Arrays   59
       8.2.6   Lists   59
   8.3   Saving Your Work So Far   60
   8.4   Comments to Document Your Programs   62
   8.5   Controlling Functions (Procedures).   62
       8.5.1   Controlling Functions with Arguments   62
       8.5.2   Controlling Functions with Formulas.   64
       8.5.3   Controlling Functions with an Object's Class.   65
       8.5.4   Controlling Functions with Extractor
               Functions – ODS, OMS   67
       8.5.5   How Much Output Is There?   69
       8.5.6   Writing Your Own Functions (Macros)   73

9   Data Acquisition   77
   9.1   The R Data Editor   77
   9.2   Reading Delimited Text Files.   79
   9.3   Reading Text Data Within a Program (Datalines, Cards,
       Begin Data...)   84
   9.4   Reading Data from the Keyboard   86
   9.5   Reading Fixed-Width Text Files, One Record per Case   87
       9.5.1   Macro Substitution   90
   9.6   Reading Fixed-Width Text Files, Two or More Records per
       Case   92
   9.7   Importing Data from SAS   95
   9.8   Importing Data from SPSS   96

9.9   Exporting Data.................................... 97
     9.9.1   Viewing an External Text File.................. 98

10 Selecting Variables – Var, Variables = ... 103
10.1   Selecting Variables in SAS and SPSS.................. 103
10.2   Selecting All Variables 104
10.3   Selecting Variables by Index Number 104
10.4   Selecting Variables by Column Name 107
10.5   Selecting Variables Using Logic...................... 108
10.6   Selecting Variables by String Search (varname: or
     varname1-varnameN)............................... 110
10.7   Selecting Variables Using $ Notation.................. 112
10.8   Selecting Variables by Simple Name: `attach` and `with` .. 113
10.9   Selecting Variables with the `subset` Function (varname1-
     varnameN) 114
10.10 Selecting Variables by List 115
10.11 Generating Indexes A to Z from Two Variable Names.... 115
10.12 Saving Selected Variables to a New Dataset 116
10.13 Example Programs for Variable Selection 116

11 Selecting Observations – Where, If, Select If, Filter 123
11.1   Selecting Observations in SAS and SPSS............... 123
11.2   Selecting All Observations 124
11.3   Selecting Observations by Index Number 124
11.4   Selecting Observations by Row Name 127
11.5   Selecting Observations Using Logic................... 128
11.6   Selecting Observations by String Search 132
11.7   Selecting Observations with the `subset` Function 133
11.8   Generating Indexes from A to Z from Two Row Names .. 134
11.9   Variable Selection Methods with No Counterpart for
     Selecting Observations 135
11.10 Saving Selected Observations to a New Data Frame...... 135
11.11 Example Programs for Selecting Observations 135

12 Selecting Both Variables and Observations 141

13 Converting Data Structures 143
13.1   Converting from Logical to Index and Back 146

14 Data Management 147
14.1   Transforming Variables 147
14.2   Procedures or Functions? The `apply` Function Decides... 152
     14.2.1   Applying the `mean` Function.................. 152
     14.2.2   Finding N or NVALID 155
14.3   Conditional Transformations......................... 158

14.4   Multiple Conditional Transformations   162
14.5   Missing Values   165
        14.5.1   Substituting Means for Missing Values.   166
        14.5.2   Finding Complete Observations   167
        14.5.3   When "99" Has Meaning.   168
14.6   Renaming Variables (. . . and Observations).   171
14.7   Renaming Variables – Advanced Examples.   174
        14.7.1   Renaming by Index   174
        14.7.2   Renaming by Column Name.   175
        14.7.3   Renaming Many Sequentially Numbered Variable
                 Names   176
        14.7.4   Renaming Observations   177
14.8   Recoding Variables.   180
        14.8.1   Recoding a Few Variables.   180
        14.8.2   Recoding Many Variables   181
14.9   Keeping and Dropping Variables.   185
14.10   Stacking/Concatenating/Adding Datasets   186
14.11   Joining/Merging Data Frames   190
14.12   Creating Summarized or Aggregated Datasets   194
        14.12.1   The `aggregate` Function   195
        14.12.2   The `tapply` Function.   196
        14.12.3   Merging Aggregates with Original Data   198
        14.12.4   Tabular Aggregation   200
        14.12.5   The `reshape` Package   201
14.13   By or Split File Processing   204
        14.13.1   Comparing Summarization Methods   208
        14.13.2   Example Programs for By or Split File Processing   208
14.14   Removing Duplicate Observations.   210
14.15   Selecting First or Last Observations per Group.   213
14.16   Reshaping Variables to Observations and Back   217
14.17   Sorting Data Frames   221

15 Value Labels or Formats (and Measurement Level)   225
15.1   Character Factors.   226
15.2   Numeric Factors.   227
15.3   Making Factors of Many Variables   229
15.4   Converting Factors into Numeric or Character Variables. . .   232
15.5   Dropping Factor Levels   233

16 Variable Labels   239

17 Generating Data.   245
17.1   Generating Numeric Sequences   245
17.2   Generating Factors.   246
17.3   Generating Repetitious Patterns (not factors)   247

17.4 Generating Integer Measures 248
17.5 Generating Continuous Measures 249
17.6 Generating a Data Frame. 251

18 How R Stores Data 259

19 Managing Your Files and Workspace 261
19.1 Loading and Listing Objects 261
19.2 Understanding Your Search Path 264
19.3 Attaching Data Frames 264
19.4 Attaching Files 266
19.5 Removing Objects from Your Workspace 267
19.6 Minimizing Your Workspace. 268
19.7 Setting Your Working Directory 268
19.8 Saving Your Workspace. 269
19.9 Saving Your Programs and Output 271
19.10 Saving Your History (Journal). 271

20 Graphics Overview 273
20.1 SAS/GRAPH 273
20.2 SPSS Graphics 274
20.3 R Graphics 274
20.4 The Grammar of Graphics. 275
20.5 Other Graphics Packages 276
20.6 Graphics Procedures Versus Graphics Systems 277
20.7 Graphics Devices 277
20.8 Practice Data: Mydata100 278

21 Traditional Graphics 281
21.1 Barplots 281
21.1.1 Barplots of Counts. 281
21.1.2 Barplots for Subgroups of Counts. 285
21.1.3 Barplots of Means 286
21.2 Adding Titles, Labels, Colors, and Legends. 288
21.3 Graphics Parameters and Multiple Plots on a Page. 290
21.4 Pie Charts 292
21.5 Dotcharts 293
21.6 Histograms 293
21.6.1 Basic Histograms. 294
21.6.2 Histograms Overlaid 297
21.7 Normal QQ Plots 299
21.8 Strip Charts. 301
21.9 Scatterplots. 303
21.9.1 Scatterplots with Jitter. 304
21.9.2 Scatterplots with Large Datasets. 305

21.9.3   Scatterplots with Lines   307
21.9.4   Scatterplots with Linear Fit by Group   308
21.9.5   Scatterplots by Group or Level (Coplots)   309
21.9.6   Scatterplots with Confidence Ellipse   311
21.9.7   Scatterplots with Confidence and Prediction
          Intervals   312
21.9.8   Plotting Labels Instead of Points   316
21.9.9   Scatterplot Matrices   318
21.10   Dual Axes Plots   320
21.11   Boxplots   322
21.12   Error Bar and Interaction Plots   324
21.13   Adding Equations and Symbols to Graphs   324
21.14   Summary of Graphics Elements and Parameters   325
21.15   Plot Demonstrating Many Modifications   328
21.16   Example Traditional Graphics Programs   330

22   **Graphics with ggplot2 (GPL)**   341
22.1   Overview qplot and ggplot   342
22.2   Bar Charts   344
22.3   Pie Charts   347
22.4   Bar Charts with Subgroups   348
22.5   Plots by Group or Level   349
22.6   Pre-summarized Data   351
22.7   Dotcharts   352
22.8   Adding Titles and Labels   353
22.9   Histograms   354
22.10   Normal QQ Plots   359
22.11   Strip Plots   360
22.12   Scatterplots   361
22.13   Scatterplots with Jitter   363
         22.13.1   Scatterplots with Large Datasets   364
22.14   Scatterplots with Fit Lines   367
22.15   Scatterplots with Reference Lines   368
22.16   Plotting Labels Instead of Points   371
22.17   Changing Plot Symbols by Group   372
22.18   Adding Linear Fits by Group   373
22.19   Scatterplots Faceted by Groups   374
22.20   Scatterplot Matrix   374
22.21   Boxplots   376
22.22   Error Barplots   380
22.23   Logarithmic Axes   381
22.24   Aspect Ratio   382
22.25   Multiple Plots on a Page   382
22.26   Saving ggplot2 Graphs to a File   385

22.27   An Example Specifying All Defaults   385
22.28   Summary of Graphic Elements and Parameters   386

23   Statistics   403
23.1   Scientific Notation   403
23.2   Descriptive Statistics.   404
23.3   Cross-Tabulation   408
23.4   Correlation   413
23.5   Linear Regression.   417
        23.5.1   Plotting Diagnostics   420
        23.5.2   Comparing Models   421
        23.5.3   Making Predictions with New Data   422
23.6   t-Test – Independent Groups   422
23.7   Equality of Variance.   424
23.8   *t*-Test – Paired or Repeated Measures   424
23.9   Wilcoxon Mann–Whitney Rank Sum Test – Independent
        Groups   425
23.10   Wilcoxon Signed-Rank Test – Paired Groups   426
23.11   Analysis of Variance.   427
23.12   Sums of Squares   431
23.13   Kruskal–Wallis Test   432

24   Conclusion   441

Appendix A A Glossary of R Jargon   443

Appendix B A Comparison of SAS and SPSS Products with R Packages
        and Functions   449

Appendix C Automating Your Settings   453

Appendix D A comparison of the major attributes of
        SAS and SPSS to R   457

Bibliography   459

Index   463

Chapter 1
Introduction

The availability of R [5] has dramatically changed the landscape of research software. It provides a powerful common language for data analysis and graphics that is freely available to all. SPSS users can now call R functions within their programs, dramatically expanding the capability of SPSS.

For each aspect of R we discuss, we will compare and contrast it with SAS and SPSS. Many of the topics end with example programs that do almost identical things in all three. The R programs often display more variations on each theme than do the SAS or SPSS examples, making the R programs longer.

SAS and SPSS are so similar to each other that moving from one to the other is straightforward. R, however, is very different, making the transition confusing at first. I hope to ease that confusion by focusing on the similarities and differences in this book. When we examine a particular analysis, say comparing two groups with a *t*-test, someone who knows SAS or SPSS will have very little trouble figuring out what R is doing. However, the basics of the R language are very different, so that is where we will spend most of our time.

I introduce topics in a carefully chosen order so it is best to read from beginning to end the first time through, even if you think you do not need to know a particular topic. Later you can skip directly to the section you need. I include a fair amount of redundancy on key topics to help teach those topics and to make it easier to read just one section as a future reference. The glossary in Appendix A defines R concepts in terms that SAS or SPSS users will understand and provides parallel definitions using R terminology.

1.1 Why Learn R?

If you already know SAS or SPSS, why should you bother to learn R? Both SAS and SPSS are excellent statistics packages. I use them both almost daily. If they meet your needs, and you do not mind paying for them, there is little point in learning another package. However, R offers a lot:

R.A. Muenchen, *R for SAS and SPSS Users*, DOI: 10.1007/978-0-387-09418-2_1,

- R offers more analytical methods. There are now well over 1000 add-on packages available for R, and R can download and install them directly from the Internet. It takes most statistics packages at least 5 years to add a major new analytic method. Statisticians who develop new methods often work in R, so R users often get to use new methods immediately.
- You can use R while knowing very little about it. You can do all your data management with any software you prefer, and learn just enough R to import a file and run the procedure you need. If you are an SPSS user, you can run R programs from within SPSS programs, allowing you to do much of your work in a familiar environment while avoiding the cost of the various add-on modules for SPSS.
- R is far more flexible in the type of data it can analyze. While SAS and SPSS require you to store your data in rectangular datasets, R offers a rich variety of data structures that are much more flexible. You can perform analyses that include variables from different data structures easily without having to merge them.
- R's language is more powerful than SAS or SPSS. R developers write most of their analytic methods using the R language; SAS and SPSS developers do not use their own languages to develop their procedures.
- R's procedures, which it calls *functions*, are open for you to see and modify.
- Functions that you write in R are automatically on an equal footing with those that come with the software. The ability to write your own completely integrated procedures in SAS or SPSS requires using a different language such as C or Python, and in the case of SAS, a developer's kit.
- R's graphics are extremely flexible and are of publication quality. They are flexible enough to overlay data from different datasets, even at different levels of aggregation.
- R runs on almost any computer, including Windows, Macintosh, Linux, and UNIX.
- R has full matrix capabilities that are quite similar to MATLAB, and it even offers a MATLAB emulation package [6]. For a comparison of R and MATLAB, see http://wiki.r-project.org/rwiki/doku.php? id = getting-started:translations:octave2r.
- R is free.

1.2 Is R Accurate?

When people first learn of R, one of their first questions is "Can a package written by volunteers be as accurate as one written by a large corporation?" People envision a lone programmer competing against a large corporate team. Having worked closely with several software companies over the years, I can assure you that this is not the case. A particular procedure is usually written by one programmer even at SAS Institute and SPSS, Inc. The testing process is then carried out by a few people within the company and then more thoroughly by a group of "beta-testers" who are volunteers from outside the company.

It is to their credit that SAS Institute and SPSS Inc. post databases of known bugs on their websites, and they usually fix problems quickly. R also has open discussions of its known bugs and R's developers fix them quickly too. However, software of this complexity will never be completely free of errors, regardless of its source.

The most comprehensive study to date [4] compared nine statistics packages on the accuracy of their univariate statistics, analysis of variance, linear regression, and non-linear regression. The accuracy of R was comparable to SAS and SPSS and by the time the article was published, R's accuracy had already improved [5].

1.3 What About Tech Support?

When you buy software from SAS or SPSS, you can call or e-mail for tech support that is quick, polite, and accurate. Their knowledgeable consultants have helped me out of many a jam. With R, you do not get a number to call, but you do get direct access to the people who wrote the program via e-mail. Since they are scattered around the world, you can usually get an answer to your question in well under an hour, regardless of when you post your question. The main difference is that the SAS or SPSS consultants will typically provide a single solution that they consider best, while the r-help list responders will often provide several ways to solve your problem. You learn more that way, but the solutions can differ quite a bit in level of difficulty. However, by the time you finish this book that should not be a problem. For details on the various R e-mail support lists, see Chap. 7.

Chapter 2
The Five Main Parts of SAS and SPSS

While SAS and SPSS offer hundreds of functions and procedures, they fall into five main categories:

1. Data input and management statements that help you read, transform, and organize your data.
2. Statistical and graphical procedures to help you analyze data.
3. An output management system (OMS) to help you extract output from statistical procedures for processing in other procedures or to let you customize printed output. SAS calls theirs the Output Delivery System, (ODS), SPSS calls theirs the Output Management System, (OMS).
4. A macro language to help you use sets of the above commands repeatedly.
5. A matrix language to add new algorithms (SAS/IML and SPSS Matrix).

SAS and SPSS handle each of these five areas with different systems that follow different rules. For simplicity's sake, introductory training in SAS or SPSS typically focuses on only the first two topics. Perhaps, the majority of users never learn the more advanced topics. However, R performs these five functions in a way that completely integrates them all. The integration of these five areas gives R a significant advantage in power and is the reason that most R developers write procedures using the R language.

Since SAS and SPSS procedures tend to print all of their output at once, a relatively small percent of their users take advantage of their output management. Virtually all R users use output management. That is partly because R shows you only the pieces of output you request and partly because R's output management is easier to use. For example, you can create and store a regression linear model using the `lm` function.

```
myModel <- lm(y~x)
```

You can then get several diagnostic plots with the `plot` function.

```
plot(myModel)
```

You can compare two models using the `anova` function.

```
anova(myModel1,myModel2)
```

R.A. Muenchen, *R for SAS and SPSS Users*, DOI: 10.1007/978-0-387-09418-2_2,
© Springer Science+Business Media, LLC 2009

That is a very flexible approach! It requires far fewer commands than SAS or SPSS and it requires almost no knowledge of how the model is stored. The `plot` and `anova` functions have a built-in ability to work with models and other data structures.

The price R pays for this output management advantage is that the output to most procedures is sparse and does not appear as publication quality within R itself. It appears in a monospace font without a word processor style table structure or even tabs between columns. Variable labels are not a part of the core system, so if you want clarifying labels, you add them in other steps. You can use functions from add-on packages to write out HTML, ODF, or TEX files to use in word processing tools. SPSS and, more recently SAS, make output that is of publication quality by default, but not as easy to use as input for further analyses.

On the topic of matrix languages, SAS and SPSS offer them in a form that differs sharply from their main languages. For example, the way you select variables in the main SAS product bears no relation to how you select them in SAS/IML. In R, the matrix capabilities are completely integrated and follow the same rules.

Chapter 3
Programming Conventions

Although R has many ways to generate practice data and has a variety of example datasets, we will use a tiny practice dataset that is easy to enter in various ways and print repeatedly to enable you to see how we changed it.

You can download the practice datasets and program files from http://Rfor SASandSPSSusers.com The example programs are set to look for their matching data files in the directory (folder) named *myRfolder*, but that is easy to change to whatever location you prefer. Each program begins by loading the data as if it were a new session. That is not required if you already have the data loaded, but it makes it easier to ensure that previous programming does not interfere with the example. It also allows each program to run on its own.

Each example program in this book begins with a comment stating its purpose and the name of the file it is stored in. For example, the programs for selecting variables each begin with a comment like the one below.

```
# R Program for Selecting Variables.
# SelectingVars.R
```

The filename in the practice files will always match, so the three files for this topic are SelectingVars.sas, SelectingVars.sps, SelectingVars.R.

The R data objects in this book are each available in a single file. Their name is the same as that used in the book, with the extension, ".Rdata". For example, our most widely used data object, mydata, is stored in *mydata.Rdata*. Also, all the objects we create, data and functions, are stored in *myWorkspace.Rdata*.

R.A. Muenchen, *R for SAS and SPSS Users*, DOI: 10.1007/978-0-387-09418-2_3,
© Springer Science+Business Media, LLC 2009

Chapter 4
Typographic Conventions

All programming code and R package and function names are written in: `this courier font`.

Names of other documents and menus appear in *this italic font*.

Menus appear in the form *File> Save as*, which means "choose *Save as* from the *File* menu".

When learning a new language, it can be hard to tell the commands from the names you can choose (e.g., variable or dataset names). To help differentiate, I CAPITALIZE statements in SAS and SPSS and use lower case for names that you can choose. However, R is case-sensitive, so I have to use the exact case that the program requires. Therefore, to help differentiate, I use the common prefix "my" in names like mydata or mySubset.

When examples include both input and output, I leave in place the symbols that R uses for input. That helps you identify which is which. R uses ">" to prompt you to input a new line and "+" to prompt you to enter a continued line. So the first three lines below are the input I submitted, while the last line is the mean that R wrote out. I also add spacing in some places to improve legibility.

```
> q1 <- c(1, 2, 2, 3,
+            4, 5, 5, 5, 4)

> mean(q1)

[1] 3.4444
```

R.A. Muenchen, *R for SAS and SPSS Users*, DOI: 10.1007/978-0-387-09418-2_4,
© Springer Science+Business Media, LLC 2009

Chapter 5
Installing and Updating R

When you purchase SAS or SPSS, they sell you a "binary" version. That is one that the company has compiled for you from the "source code" version they wrote using languages such as C, FORTRAN, or Java. You usually install everything you purchased at once and do not give it a second thought. Instead, R is modular. The main installation provides R and a popular set of add-on modules called *packages*. You can install other packages later when you need them. With over 1000 to choose from, it is a rare individual who needs to install them all.

To download R itself, go to the Comprehensive R Archive Networks (CRAN) at http://cran.r-project.org/.

Choose your operating system under the web page heading, *Download and Install R*. The binary versions install quickly and easily. Binary versions exist for many operating systems including Windows, Mac OS X, and popular versions of Linux such as Ubuntu, RedHat, Suse, and others that use either the RPM or APT installers.

Since R is an Open Source project, there are also source code versions of R for experienced programmers who prefer to compile their own copy. Using that version, you can modify R in any way you like. Although R's developers write most of the analytic procedures using the R language, they use other languages such as C and FORTRAN to write the most fundamental R commands.

Each version of R installs into its own directory (folder), so there is no problem having multiple versions installed on your computer. You can then install your favorite packages for the new release.

5.1 Installing Add-on Packages

While the main installation of R contains many useful functions, many additional packages are available on the Internet. The main site for additional packages is at the CRAN website under *Packages*. That is the best place to read about and choose packages to install, but you usually do not need to download them from there yourself. R automates the download and installation

R.A. Muenchen, *R for SAS and SPSS Users*, DOI: 10.1007/978-0-387-09418-2_5,   11 process. A comparison of SAS and SPSS add-ons to R packages is presented in Appendix B.

On the R version for Microsoft Windows, you can choose *Packages> Install package(s)* from the menus. It will ask you to choose a CRAN site or "mirror" that is close to you (Fig. 5.1, left). Then it will ask which package you wish to install (right). Choose one and click OK.

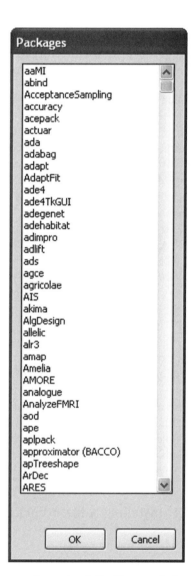

Fig. 5.1 When installing software, you first choose a mirror site (*left*). Then a window appears, where you choose the package you need (*right*). You install a particular package only once, but you must load it in every R session in which you need it

If you prefer to use a function instead of the menus, you can use the `install.packages` function. For example, to download and install Frank Harrell's `Hmisc` package [6], start R and enter the command:

```
install.packages("Hmisc", dependencies=TRUE)
```

The argument `dependencies = TRUE` tells R to install any packages that this package "depends" upon, and those that it "suggests" as useful. R will then prompt you to choose the closest mirror site and the package you need.

If you do not have administrative privileges on your computer, you can install packages to a directory to which you have write access. For instructions, see the FAQ at http://www.r-project.org/.

5.2 Loading an Add-on Package

Once installed, a package is on your computer's hard drive, but not quite ready to use. Each time you start R, you also have to load the package from the library. You can see what packages that are installed and ready to load with the `library` function.

```
library()
```

That causes the window in Fig. 5.2 to appear showing the packages that I have installed.

You can then load a package you need with the menu selection, *Packages> Load packages*. It will show you the names of all packages that you installed but have not yet loaded. You can then choose one from the list.

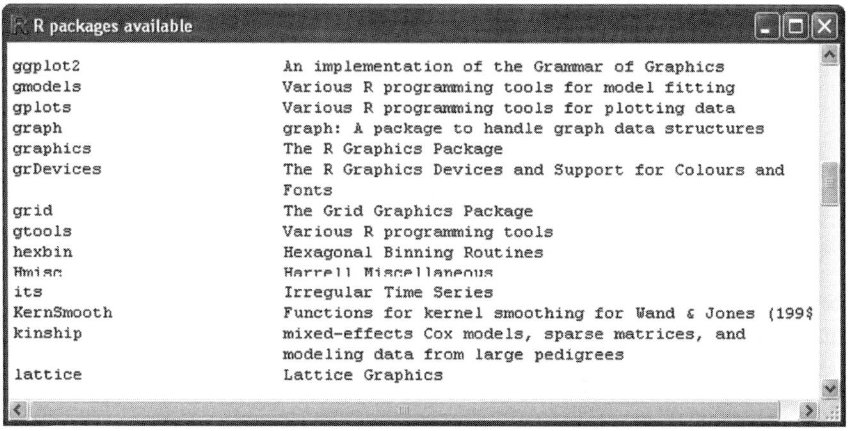

R packages available	
ggplot2	An implementation of the Grammar of Graphics
gmodels	Various R programming tools for model fitting
gplots	Various R programming tools for plotting data
graph	graph: A package to handle graph data structures
graphics	The R Graphics Package
grDevices	The R Graphics Devices and Support for Colours and Fonts
grid	The Grid Graphics Package
gtools	Various R programming tools
hexbin	Hexagonal Binning Routines
Hmisc	Harrell Miscellaneous
its	Irregular Time Series
KernSmooth	Functions for kernel smoothing for Wand & Jones (199$
kinship	mixed-effects Cox models, sparse matrices, and modeling data from large pedigrees
lattice	Lattice Graphics

Fig. 5.2 The `library` function shows you the packages you have installed and are ready to load

Alternatively, you can use the library function. Here I am loading the Hmisc package [6]. Since the Linux version lacks menus, this function is the only way to load packages.

```
library ("Hmisc")
```

Many packages load without any messages; you will just see the ">" prompt again. When trying to load a package, you may see the error message below. It means you have either mistyped the package name (remember capitalization is important) or you have not installed the package before trying to load it. In this case, the package name is typed accurately, so I have not yet installed it.

```
>library ("prettyR")

Error in library (prettyR) : there is no package
called 'prettyR'
```

To see what packages you have loaded, use the search function. We will discuss this function in detail in Chap. 19.

```
> search ()
[1] ".GlobalEnv"        "package:Hmisc"
[3] "package:stats"     "package:graphics"
[5] "package:grDevices" "package:utils"
[7] "package:datasets"  "package:methods"
[9] "Autoloads"         "package:base"
```

Occasionally two packages will have functions with the same name. That can be very confusing until you realize what is happening. For example, the Hmisc and prettyR [8] packages both have a describe function that does similar things. In such a case, the package you load last will mask the function(s) in the package you loaded earlier. For example, I loaded the Hmisc package first, and now I am loading the prettyR package (having installed it in the meantime!). The following message results.

```
> library ("prettyR")

Attaching package: 'prettyR'
        The following object (s) are masked from package: Hmisc:
         describe
```

You can avoid such conflicts by detaching each package as soon as you are done using it by using the detach function. For example, the following command will detach the prettyR package.

```
detach ("package:prettyR")
```

One approach that avoids conflicts is to load a package from the library right before using it, and then detach it immediately as in the following example.

```
> attach ("Hmisc")
> describe (mydata)
> ...(output would appear here)...
> detach ("package:Hmisc")
```

If your favorite packages do not conflict with one anther, you can have R load them each time you start R by putting the commands in a file named *.Rprofile*. That file can automate your settings just like the autoexec.sas file for SAS. For details, see Appendix C.

5.3 Updating Your Installation

Keeping your add-on packages current is very easy. You simply use the update.packages function.

```
> update.packages()

graph :
Version 1.15.6 installed in C:/PROGRA~1/R/R-26~1.1/
  library
Version 1.16.1 available at http://rh-mirror.linux.
  iastate.edu/CRAN
Update(y/N/c)? y
```

R will ask you if you want to update each package. If you enter "y" it will do it and show you the following. This message, repeated for each package, tells you what file it is getting from the mirror you requested (Iowa State) and where it placed the file.

```
trying URL 'http://rh-mirror.linux.iastate.edu/CRAN/
  bin/windows/contrib/2.6/graph_1.16.1.zip'
Content type 'application/zip' length 870777 bytes (850 Kb)
opened URL
downloaded 850 Kb
```

This next message tells you that the file was checked for errors (its sums were checked) and it says where it stored the file. As long as you see no error messages, the update is complete.

```
package 'graph' successfully unpacked and MD5 sums checked

The downloaded packages are in
        C:\Documents and Settings\muenchen\Local
            Settings\Temp\Rtmpgf4C4B\downloaded_packages
updating HTML package descriptions
```

Moving to a whole new version of R is not as easy. First, you download and install the new version just like you did the first one. Multiple versions can co-exist on the same computer. You can even run them at the same time if you wanted to compare results across versions. When you install a new version of R, you also have to install any add-on packages again. You can do that in a step-by-step fashion as we discussed above. An easier way is to define a character variable like "myPackages" that contains the names of the packages you use. Here is an example that does this for all the packages we use in this book.

```
myPackages <- c("car","foreign","hexbin",
   "ggplot2","gmodels","gplots","Hmisc",
   "reshape","Rcmdr")

install.packages(myPackages, dependencies=TRUE)
```

We will discuss the details of the c function used above and how to run statements like this later.

You can automate the creation of myPackages by placing that line in a special file named .Rprofile. Similar to the SAS autoexec.sas file, R will execute the functions stored in this file every time it starts. Putting it there will ensure that myPackages is defined every time you start R. As you find new packages to install, you can add to the definition of myPackages. Then installing all of them when a new version of R comes out is easy. Of course you do *not* want to place the install.packages line into your .Rprofile! There is no point in installing package every time you start R. For details, see Appendix C.

5.4 Uninstalling R

When you get a new version of any software package, it is good to keep the old one around for a while in case any bugs show up in the new one. Once you are confident that you will no longer need an older version of R, you can remove it.

On Microsoft Windows, R does not have an uninstaller accessible from the usual Windows *Add or Remove Programs* control panel. Instead, you can choose *Start> Programs> R, Uninstall R 2.7.0*. That menu choice runs the uninstall program, unins000. That program will remove R and any packages you have installed. That file is located in the folder c:\program files\R\R x.xx\.

To uninstall R on the Macintosh, simply drag the application to the trash.

Linux users can delete /usr/local/lib/R.

Although it is rarely necessary to uninstall a single package, you can do so with the uninstall.packages function. First though, you must make sure it is not in use by detaching it. For example, to remove just the Hmisc package, use

```
detach("package:Hmisc")#If it is loaded.

remove.packages("Hmisc")
```

5.5 Choosing Repositories

While most R packages are stored at the CRAN site, there are other repositories. If the *Packages* window does not list the one you need, you may need to choose another repository. Several repositories are associated with the Bioconductor project. As they say at their main website, http://www.bioconductor.org/, "Bio-Conductor is an open source and open development software project for the analysis and comprehension of genomic data [8]." Another repository is at the Omegahat Project for Statistical Computing, http://www.omegahat.org/ [9].

To choose your repositories, choose *Packages> Select repositories...*, or enter the following command and the *Repositories* selection window will appear (Fig. 5.3). Note that two CRAN repositories are selected by default. Your operating system's mouse commands work as usual to make contiguous or non-contiguous selections. On Microsoft Windows, that is Shift-click and Ctrl-click, respectively.

```
> setRepositories()
```

If you are working without a widowing system, R will prompt you to enter the number(s) of the repositories you need.

```
--- Please select repositories for use in this
    session ---
1: + CRAN
2: + CRAN(extras)
3:   Omegahat
4:   BioC software
5:   BioC annotation
6:   BioC experiment
7:   BioC extra
Enter one or more numbers separated by spaces
1: 1,2,4
```

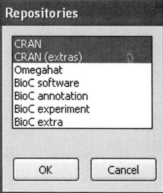

Fig. 5.3 Selecting repositories will determine which add-on packages R will offer to install

5.6 Accessing Data in Packages

You can get a list of datasets available in each loaded package with the data function.

```
data()
```

The window listing the default datasets will appear (Fig. 5.4).

You can use these practice datasets directly. For example, to look at the top of the CO2 file (capital letters C and O, not zero!), you can use the head function.

```
> head(CO2)

  Plant    Type  Treatment conc uptake
1   Qn1 Quebec nonchilled   95   16.0
2   Qn1 Quebec nonchilled  175   30.4
3   Qn1 Quebec nonchilled  250   34.8
4   Qn1 Quebec nonchilled  350   37.2
5   Qn1 Quebec nonchilled  500   35.3
6   Qn1 Quebec nonchilled  675   39.2
```

The similar tail function shows you the bottom few observations.

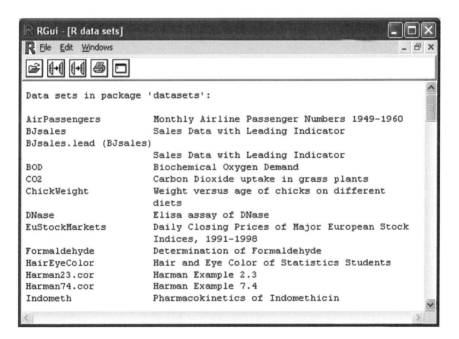

Fig. 5.4 The data function displays all the practice datasets for the add-on packages you have loaded

```
R data sets                                                    [_][□][X]
Mroz                      U.S. Women's Labor-Force Participation
OBrienKaiser              O'Brien and Kaiser's Repeated-Measures Data
Ornstein                  Interlocking Directorates Among Major Canadian
                          Firms
Pottery                   Chemical Composition of Pottery
Prestige                  Prestige of Canadian Occupations
Quartet                   Four Regression Datasets
Robey                     Fertility and Contraception
SLID                      Survey of Labour and Income Dynamics
Sahlins                   Agricultural Production in Mazulu Village
Soils                     Soil Compositions of Physical and Chemical
                          Characteristics
States                    Education and Related Statistics for the U.S.
                          States
UN                        GDP and Infant Mortality
```

Fig. 5.5 Listing of datasets in the car package

If you only want a list of datasets in a particular package, you can use the
package argument. For example, if you have installed the car package [11]
(from John Fox's Companion to Applied Regression book) you can load it from
the library and see the datasets only it has (Fig. 5.5) with the following statements. Recall that R is case sensitive, so using a lower case "un" would not
work.

```
> library("car")
> data(package="car")
> head(UN)
                infant.mortality   gdp
Afghanistan                  154  2848
Albania                       32   863
Algeria                       44  1531
American.Samoa                11    NA
Andorra                       NA    NA
Angola                       124   355
```

To see all the datasets available in all installed packages, even those not
loaded from the library, enter the following function call.

```
data( package=.packages( all.available=TRUE ) )
```

Chapter 6
Running R

There are several ways you can run R:

- Interactively using its programming language. You can see the result of each command immediately after you submit it.
- Interactively using one of several graphical user interfaces (GUIs) that you can add on to R. Some of these use programming and some use menus like SPSS or SAS Enterprise Guide.
- Non-interactively in batch mode using its programming language. You enter your program into a file and run it all at once.
- From within SPSS.

You can ease your way into R by continuing to use SAS, SPSS, or your favorite spreadsheet program to enter and manage your data, and then use one of the methods below to import and analyze it. As you find errors in your data (and you know you will), you can go back to your other software, correct them, and then import it again. It is not an ideal way to work, but it does get you into R quickly.

6.1 Running R Interactively on Windows

You can run R programs interactively in several steps:

1. Start R by choosing *Start> All Programs> R> R 2.7.0* (the version number at the time this was written). The main *R Console* window will appear looking like the left window in Fig. 6.1. Then enter your program choosing one of the methods described in steps 2 and 3 below.
2. Enter R functions into the R console. You can enter commands into the console one line at a time at the ">" prompt. R will execute each line when you press the *Enter* key. If you enter them into the console, you can retrieve them with the up arrow key and edit them to run again. I find it much easier to use the program editor described in the next step.
   If you type the beginning of an R function, such as "me" and press Tab, R will show you all the R functions that begin with those letters, such as mean or median. If you enter the name of a function and an open parenthesis,

R.A. Muenchen, *R for SAS and SPSS Users*, DOI: 10.1007/978-0-387-09418-2_6,   21
© Springer Science+Business Media, LLC 2009

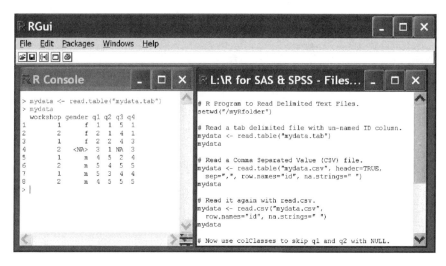

Fig. 6.1 The R graphical user interface on Microsoft Windows

such as "mean(", R will show you the arguments that you can use with that function.

3. Enter R functions into the R Editor. Open the *R Editor* by choosing *File> New Script.* You can see on the right side of Fig. 6.1. You can enter programs as you would in the *SAS Program Editor* or the *SPSS Syntax Editor.*

4. Submit your program from the R Editor. To submit just the current line, you can hold the *Ctrl* key down and press "r", or choose *Edit> Run line or selection.* To run a block of lines, select them first, and then submit them the same way. To run the whole program, choose *Edit> Run All.*

5. As you submit program statements, they will appear in the *R Console* along with results and/or error messages. Make any changes you need and submit the program again until finished. You can clear the console results by choosing *Edit> Clear console* or by holding the *Ctrl* key down and pressing "l". See *Help> Console* for more keyboard shortcuts.

6. Save your program and output. Click on either the console window or the R Editor window to make it active and choose *File> Save to file.* The console output will contain the commands and their output blended together like an SPSS output file rather than the separate log and listing files of SAS.

7. Save your data and any functions you may have written. The data and/or function(s) you created are stored in an area called your *workspace.* You can save that with the command *File> Save Workspace. . . .* In a later R session you can retrieve it with *File> Load Workspace. . . .* You can also save your workspace using the save.image function:

```
save.image(file="myWorkspace.RData")
```

Later, you can read the workspace back in with the command:

```
load("myWorkspace.RData")
```

See Chap. 19 for more details.

8. Optionally save your history. R has a history file that saves all of the functions you submit in a given session. This is just like the SPSS journal file. SAS has no equivalent. Unlike SPSS, the history file is not cumulative on Windows computers. You can save the session history to a file using *File> Save History…* and you can load it into future session with *File> Load History….* You can also use R functions to do these tasks.

```
savehistory(file="myHistory.Rhistory")
loadhistory(file="myHistory.Rhistory")
```

Note that the filename can be anything you like but the extension should be ".Rhistory". The entire default filename, if you do not provide one, is just ".Rhistory". I prefer to always save a cumulative history file automatically. For details, see Appendix C.

9. To quit R choose *File> Exit*, or submit the function quit() or just q(). R offers to save your workspace automatically upon exit. If you are using the save.image and load functions to tell R where to save/retrieve your workspace, you can answer, *No.*

   If you answer *Yes*, it will save your work in the file ".RData" in your default working directory. Next time you start R, it will load the contents of the .RData file *automatically.* Creating a .RData file this way is a convenient way to work. However, I prefer naming each project myself.

6.2 Running R Interactively on Macintosh

You can run R programs interactively on the Macintosh in several steps:

1. Start R by choosing *R* in the *Applications folder.* The R console window will appear (see left window in Fig. 6.2). Then enter your program choosing one of the methods described in steps 2 and 3 below.
2. Enter R functions in the console window. You can enter commands into the console one line at a time at the ">" prompt. R will execute each line when you press the *Enter* key. If you enter them into the console, you can retrieve them with the up arrow key and edit them to run again. I find it much easier to use the program editor described in the next step.

   If you type the beginning of an R function name like "me" at the command prompt and press Tab or hold the Command key down and press ".", R will show you all the R functions that begin with those letters, such as mean or median. When you type a whole function name, the functions arguments will appear below it in the console window.

Fig. 6.2 The R graphical user interface on Macintosh

3. Enter R functions into the R Editor. Open the *R Editor* by choosing *File>
   New Document*. The *R Editor* will start with an empty window. You can see it
   in the center of Fig. 6.2. You can enter R programs as you would in the *SAS
   Program Editor* or the *SPSS Syntax Editor*

4. Submit your program from the R Editor. To submit one or more lines,
   highlight them, then hold the Command key and press Return, or choose
   or choose *Edit> Execute*. To run the whole program, select it by holding
   down the Command key and pressing "a", then choose *Edit> Execute*.

5. As you submit program statements, they will appear in the *R Console* along
   with results and/or error messages. Make any changes you need and submit
   the program again until finished.

6. Save your program and output. Click on a window to make it the active
   window and choose *File> Save to file*. The commands and their output are
   blended together like an SPSS output file rather than the separate log and
   listing files of SAS.

7. Save your data and any functions you may have written. The data and/or
   function(s) you created are stored in an area called your *workspace*. You can
   save your workspace with *Workspace> Save Workspace File....* In a later R
   session you can retrieve it with *Workspace> Load Workspace File....* You
   can also perform these functions using the R functions save.image and
   load.

```
save.image(file="myWorkspace.RData")
load("myWorkspace.RData")
```

See Chap. 19 for more details.

8. Optionally save your history. R has a history file that saves all of the functions you submit in a given session (and not the output). This is just like the SPSS journal file. SAS has no equivalent. Unlike SPSS, the history file is not cumulative on Macintosh computers. You can view your history by clicking on the *Show/Hide R command history* icon in the console window (to the right of the lock icon). You can see the command history window on the right side of Fig. 6.2. Notice that it has alternating stripes, matching its icon. Clicking the icon once makes the history window slide out to the right of the console. Clicking it again causes it to slide back and disappear. You can see the various buttons at the bottom of the history, such as *Save History* or *Load History*. You can use them to save your history or load it from a previous session. You can also use R functions to do these tasks.

```
savehistory(file="myHistory.Rhistory")
loadhistory(file="myHistory.Rhistory")
```

The filename can be anything you like but the extension should be ".Rhistory". The entire default filename, if you do not provide one, is just ".Rhistory". I prefer to always save a cumulative history file automatically. For details, see Appendix C.

9. Exit R by choosing *R> Quit R*. Users of any operating system can quit by submitting the function quit() or just q(). R will offer to save your workspace automatically upon exit. If you are using the save.image and load functions to tell R where to save/retrieve your workspace in step 4 above, you can answer, *No*.

   If you answer *Yes*, it will save your work in the file ".RData" in your default working directory. The next time you start R, it will load the contents of the .RData file *automatically*. Creating a .RData file this way is a convenient way to work. However, I prefer naming each project myself.

6.3 Running R Interactively on Linux or UNIX

You can run R programs interactively in several steps:

1. Start R by entering the command "R", which will bring up the ">" prompt where you enter commands. For a wide range of options, see An Introduction to R [11], *Appendix B* Invoking R, available at http://www.r-project.org/ under Manuals, or in your R help files [2]. You can enter R functions using either of the methods described in steps 2 and 3 below.

2. Enter R functions into the console one line at a time at the ">" prompt. R will execute each line when you press the *Enter* key. You can retrieve a command with the up arrow key and edit it, and press Enter to run again. You can include whole R programs from files with the source function. For details, see Sect. 6.4 below.

If you type the beginning of an R function name like "me" and press the Tab key, R will show you all the R functions that begin with those letters, such as mean or median. If you type the function name and an open parenthesis like "mean(" and press Tab, R will show you the arguments you can use to control that function.

3. Enter R functions into a text editor. Although R for Linux or UNIX does not come with its own GUI or program editor, a popular alternative is to use the Emacs editor in ESS mode. It color-codes your commands to help find syntax errors. You can submit your programs directly from Emacs to R. See the R FAQ at http://www.r-project.org/ under *R for Emacs* for details.

4. Save your program and output. Linux or UNIX users can route input and output to a file with the sink function. You must specify it in advance of any output you wish to save.

```
sink("myTranscript.txt", split=TRUE)
```

The argument split = TRUE tells R to display the text on the screen as well as route it to the file. The file will contain a *transcript*, of your work. The commands and their output are blended together like an SPSS output file rather than the separate log and listing files of SAS.

5. Save your data and any functions you may have written. The data and/or function(s) you created are stored in an area called your *workspace*. Users of any operating system can save it by calling the save.image function

```
save.image(file="myWorkspace.RData")
```

Later, you can read the workspace back in with the command:

```
load("myWorkspace.RData")
```

See Chap. 19 for more details.

6. R has a history file that saves all of the functions you submit in a given session. This is just like the SPSS journal file. SAS has no equivalent. The Linux/UNIX version of R saves a cumulative set of commands across sessions. You can also save or load your history at any time with the savehistory and loadhistory functions.

```
savehistory(file="myHistory.Rhistory")
loadhistory(file="myHistory.Rhistory")
```

Note that the filename can be anything you like but the extension should be ".Rhistory". The entire default filename, if you do not provide one, is just ".Rhistory".

7. Quit R by submitting the function quit() or just q(). R offers to save your workspace automatically upon exit. If you are using the save.image and load functions to tell R where to save/retrieve your workspace in step 4 above, you can answer, *No*.

If you answer *Yes*, it will save your work in the file ".RData" in your default working directory. Next time you start R, it will load the contents of the .RData file *automatically*. Creating a .RData file this way is a convenient way to work. However, I prefer naming each project myself.

6.4 Running Programs that Include Other Programs

When you find yourself using the same block of code repeatedly in different programs, it makes sense to save it to a file and include it into the other programs where it is needed. SAS does this with form %INCLUDE 'myprog.sas'; and SPSS does it almost identically with INCLUDE 'myprog.sps'.

To include a program in R, use the source function.

```
source("myprog.R")
```

One catch to keep in mind is that by default R will not display any results that sourced files may have created. Of course, any objects they create – data, functions, etc. – will be available to the program code that follows. If the program you source creates output that you want to see, you can source the program in the following manner:

```
source("myprog.R", echo=TRUE)
```

This will show you all output created by the program. If you prefer to see only some results, you can use the print function around only those that you do want displayed. For example, if you sourced the following R program, it would display the standard deviation, but not the mean.

```
x <- c (1, 2, 3, 4, 5)
mean (x)          # This result will not display.
print (sd(x))   # This one will.
```

An alternative to using the source function to include bits of programs you are reusing is to create your own R package. However, that is beyond the scope of this book.

6.5 Running R in Batch Mode

You can write a program to a file and run it all at once, routing its results to another file (or files). This is called batch processing. If you had a program named myprog.sas, you would run it with the following command.

```
SAS myprog
```

SAS would run the program, and place the log messages in myprog.log and the listing of the output in myprog.lis. Similarly, SPSS runs batch programs with the spssb batch command.

```
spssb -f myprog.sps -out myprog.txt
```

If the SPSS program uses the SPSS-R Integration Package, you must add the "-i" parameter. In its GUI versions, SPSS also offers batch control through its Production Facility, SPSS.

In R, you can find the details of running batch on your operating system by starting R and entering the following command, in which the letters of BATCH must be all upper case.

```
help (BATCH)
```

The following operating system command is an example of running an R batch job on Microsoft Windows. You will need to change the path of Rterm. exe to reflect its location on your computer.

```
"C:\Program Files\R\R-2.7.0\bin\Rterm.exe" --no-restore
  --no-save < myprog.r > myprog.out
```

The command wraps to two lines in this book, but enter it as a single line. It is too long to fit in a standard cmd.exe window, so you will need to change its default width from 80 to something wider, like 132. You will need to open a new cmd.exe window for that change to take effect. R will execute myprog.r and write the results to myprog.out.

It is easier to write a small batch file like myR.bat:

```
"C:\Program Files\R\R-2.7.0\bin\Rterm.exe" --no-restore
  --no-save < %1 > %1.Rout 2>& 1
```

Once you have saved that in the file myR.bat, you can then submit batch programs with the following command. It will route your results to myprog.Rout.

```
myR myprog.R
```

UNIX users can run a batch program with the following command. It too will write your output to myprog.Rout.

```
  R CMD BATCH myprog.R
```

There are of course many options to give you more control over how your batch programs run. See the help file for details.

6.6 Running R from SPSS

SPSS has a very useful interface to R which allows you to transfer data back and forth, run R programs and get R results back into nicely formatted SPSS pivot tables. You can even add R programs onto the SPSS menus so that people can use R without knowing how to program. Since SPSS does not need to keep its data in the computer's main memory as R does, you can read vast amounts of data into SPSS, select the subset of variables and/or cases you need and then pass them to R for analysis. This approach also lets you make the most of your SPSS know-how, calling on R only after the data is cleaned up and ready to analyze. Even if that is what you plan to do, the remainder of this book will still provide valuable information. The way R deals with many aspects of programming and analysis – handling missing values and selecting variables to name two – is so different between the two that you will need to know a fair amount about R to take full advantage of this feature.

This interface is called the SPSS Statistics-R Integration Package and it is documented fully in a manual of the same name. The package plug-in and its manual are available from http://www.spss.com/devcentral/. Full installation instructions are also at that site, but it is quite easy as long as you follow the steps in order. First install SPSS. You will need version 16 or later to run R and 17 or later to add R procedures to the SPSS menus. Then you install the latest version of R that SPSS supports and finally the plug-in. The version of R that SPSS supports at the time you download it may be a version or two behind R's production release. Older versions of R are available at http://cran.r-project.org/, so that should not be a problem.

To understand how the SPSS Statistics-R Integration Package works requires discussing topics that we have not yet covered. If this is your first time reading this book, you might want to skip this section for now and return to it when you have finished the book.

To see how to run an R program within an SPSS program, let us step through an example. First, you must do something to get a dataset into SPSS. We will use our practice data set mydata.sav, but any valid SPSS data set will do. Creating the data could be as simple as choosing *File> Open> Data* from the menus, or some SPSS programming code like the commands below. If you use the commands, adjust your path specification to match your computer.

```
CD 'C:\myRfolder'.
GET FILE='mydata.sav'.
```

Now that you have data in SPSS, you can do any type of modifications you like, perhaps creating new variables or selecting subsets of observations before passing the data to R.

For the next step, you must have an SPSS syntax window open. Therefore, if you used menus to open the file, you must now choose *File> New> Syntax* to open a program editor window. Enter the program statement below.

```
BEGIN PROGRAM R.
```

From this command on, we will enter R programming statements. To get the whole current SPSS data set and name it mydata in R, we can use the spssdata. GetDataFromSPSS function.

```
mydata <- spssdata.GetDataFromSPSS()
```

However, it is often helpful to select variables by adding two arguments to this R function. The variables argument lets you list variables similar to the way

SPSS does except that it encloses the list within the c function. We will discuss that function in detail later. You can use the form c ("workshop", "gender", "q1 to q4") or simply c ("workshop to q4"). I used the longer form to demonstrate that you must enclose in quotes each variable name or set of contiguous names connected by the keyword TO. You can also use syntax that is common to R, such as c(1:6). That uses the fact that workshop is the 1st variable and q4 is 6th variable in the data set.

```
mydata <- spssdata.GetDataFromSPSS(
    variables=c("workshop", "gender", "q1 to q4"),
    row.label="id" )
```

You can include the optional row.label argument to specify an ID variable that R will use automatically in procedures that may identify individual cases. If the data set had SPLIT FILE turned on, this step would have been retrieved only data from the first split group. See the manual for details about the aptly named function, GetSplitDataFromSPSS.

Now that we have transferred the data to R, we can write any R statements we like. Below we print all the data. Notice that missing values for numeric variables are converted to *NaN* , Not a Number, a missing value. As we will discuss later, R also uses *NA* to represent missing data . If we had set a blank to represent missing data for gender in SPSS, we would have seen a <NA> where that blank is for subject 4. Getting the data from SPSS to R is a snap but getting an R data set to SPSS is more complicated. See the manual for details.

```
print ( mydata )

  workshop gender q1 q2  q3 q4
1        1      f  1  1   5  1
2        2      f  2  1   4  1
3        1      f  2  2   4  3
4        2         3  1 NaN  3
5        1      m  3  5   2  4
6        2      m  5  4   5  5
7        1      m  5  3   4  4
8        2      m  4  5   5  5
```

Notice that the variable ID is not labeled. Its values are used in the far left column to label the rows. If we had not specified the row.label="id" argument, we would have seen the ID variable listed before workshop and labeled "id". However, the row labels would still appear the same because R *always* labels them and if you do not provide a variable that contains labels to use, it defaults to simply sequential numbers, 1, 2, 3. ...

Now let us do some descriptive statistics on variables q1 to q4. There are a number of different ways to select variables in R. One way is to use mydata [3:6] since q1 is the 3rd variable in the data set and q4 is the 6th. Keep in mind that if we had not listed ID on the row.label argument, then ID would

have appeared as the 1st variable and our example would become `mydata`
[4:7] instead. R can select variables by name, but we will save that topic for
later. The `summary` function in R gets descriptive statistics. Therefore, if we
were running this interactively in R rather than in SPSS, we would submit this
command:

```
summary( mydata[ 3:6 ] )
```

And we would immediately see the results. However, when submitting this from
within SPSS, you will not see the results unless each function call is enclosed
within a call to the `print` function. Therefore, to see the summary results, you
would submit this.

```
print( summary( mydata[ 3:6 ] ) )

q1                    q2                    q3                    q4
Min.    :1.000        Min.    :1.00         Min.    :2.000        Min.    :1.00
1st Qu.:2.000         1st Qu.:1.00          1st Qu.:4.000         1st Qu.:2.50
Median :3.000         Median :2.50          Median :4.000         Median :3.50
Mean    :3.125        Mean    :2.75         Mean    :4.143        Mean    :3.25
3rd Qu.:4.250         3rd Qu.:4.25          3rd Qu.:5.000         3rd Qu.:4.25
Max.    :5.000        Max.    :5.00         Max.    :5.000        Max.    :5.00
                                            NA' s   :1.000
```

If you read the section *Running Programs that Include Other Programs* in this
chapter, you will see that the `source` function has the same requirement. R is
essentially sourcing the program from SPSS, hence the similarity. Advanced
programmers often want to run programs and do not see the results right away.
Instead they use the unseen results as input to other programs. They will of
course eventually print some final results.

Finally, we will do a linear regression model using standard R commands
that we will discuss in detail much later. Our goal here is just to see what the
`spsspivottable.Display` function does.

```
myModel <- lm(q4 ~ q1+q2+q3, data=mydata)
myAnova <- anova(myModel)

spsspivottable.Display(myAnova,
  title="My Anova Table",
  format=formatSpec.GeneralStat)
```

Table 6.1 An example pivot table created by R and transferred to SPSS

	My Anova Table				
	Df	Sum Sq	Mean Sq	F value	Pr(>F)
q1	1.000	12.659	12.659	34.890	.010
q2	1.000	3.468	3.468	9.557	.054
q3	1.000	.213	.213	.587	.499
Residuals	3.000	1.089	.363		

Table 6.2 Example program showing how to run R from within SPSS

| SPSS programming Statements | ```
* Example Program for Running R from within SPSS.
* SPSSrunsR.sps

CD 'C:\myRfolder'.
GET FILE='mydata.sav'.

BEGIN PROGRAM R.

mydata <- spssdata.GetDataFromSPSS(
 variables=c("workshop", "gender", "q1 to q4"),
 row.label="id" )

print( mydata )
print( summary( mydata[ 3:6] )

myModel <- lm(q4 ~ q1+q2+q3, data=mydata)
myAnova <- anova(myModel)

spsspivottable.Display(myAnova,
title="My Anova Table",
format=formatSpec.GeneralStat)

END PROGRAM.
``` |

The output of the function call created above is shown in Table 6.1. I routinely tell SPSS to put my output in CompactAcademicTimesRoman style. That style draws only horizontal lines in tables, as most scientific journals prefer. If you copy this table and paste it into a word processor, it should maintain its nice formatting and be a fully editable table.

When I ran the program, this table appeared first in the SPSS output window even though it was the last analysis run. SPSS puts its pivot tables first. Finally, I ended the program using the following statement and exited SPSS in the usual way.

```
END PROGRAM.
```

If your program contains some R code, then some SPSS code, then more R code, any data sets or variables you created in the earlier R session(s) will still exist. If the program you submit from SPSS to R uses R's quit function, it will cause both R and SPSS to terminate.

To learn how to add R functions to the SPSS graphical user interface, see the SPSS help file topic, Custom Dialog Builder in SPSS Statistics 17 or later.

6.7 Graphical User Interfaces

The main R installation provides an interface to help you enter programs. It does not include a point-and-click *GUI* for running analyses. There are however several GUIs written by R users. You can learn about several at
```

the main R website, http://www.r-project.org/ under *Related Projects* and then *R GUIs*.

6.7.1 R Commander

My favorite GUI is John Fox's *R Commander user interface*, [13] which looks similar to the SPSS GUI. It provides menus for many analytic and graphical methods and shows you the R commands that it enters, making it easy to learn the commands as you use it. Since it does not come with the main R installation, you have to install it one time with the `install.packages` function.

```
install.packages ("Rcmdr", dependencies=TRUE)
```

Let us review the steps of a basic R Commander session. Below are the steps I followed to create the screen image you see in Fig. 6.5.

1. I started R (for details see the section, *Running R Interactively*, above.)
2. Then, from within R itself I started R Commander by loading its package from the library. That brought up the window that looks something like Fig. 6.3.

```
library ("Rcmdr")
```

3. I then chose *Data> Load a data set*, and browsed to myRfolder where our practice datasets are stored. I had to tell it to look for *All Files* because by default it looks for .RDA file types and ours are .RData. I then chose the file, mydata.RData.

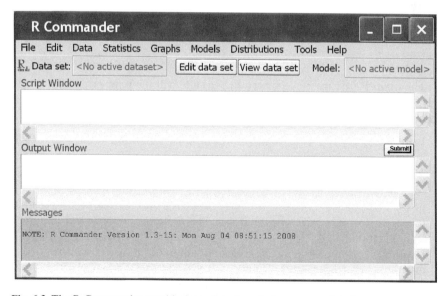

Fig. 6.3 The R Commander graphical user interface, before any work is done

4. Unlike the SPSS GUI, the data did not appear. So I clicked on the *View data set* button. The data appeared in Fig. 6.4.
5. I then chose *Statistics> Summaries> Active Data Set.* The output you see on the bottom of the screen in Fig. 6.5.
6. Finally, I chose Statistics> Means>... and you see the menu is still open showing that I can choose various *t*-tests and analysis of variance (ANOVA) procedures.

You can learn more about R Commander from http://socserv.mcmaster.ca/jfox/Misc/Rcmdr/.

6.7.2 Rattle for Data Mining

If you do data mining, you may be interested in the *Rattle* user interface from http://rattle.togaware.com/ [14]. Its name stands for the *R a*nalytical *t*ool *to learn e*asily. It is a point-and-click interface that writes and executes R programs for you. Before you install the `rattle` package, you must install some other tools. See the website for directions. Once it is installed, you load it from your library in the usual way.

```
> library("rattle")

Rattle, Graphical interface for data mining
   using R, Version 2.2.64.
Copyright (C) 2007 Graham.Williams@togaware.com, GPL
Type " rattle ()" to shake, rattle, and roll your data.
```

As the instructions tell you, simply enter the call to the `rattle` function to begin.

```
> rattle ()
```

The main Rattle interface shown in Fig. 6.6, will then appear. It shows the steps it uses to do an analysis on the tabs at the top of its window. You move from left to right, clicking on each tab to do the following steps:

1. Data – you choose your data from a comma separated value (CSV) file, attribute-relation file format (ARFF), open database connectivity (ODBC), .RData file, R data object already loaded or created before starting RATTLE or even manual data entry. Then choose your variables and the roles they play in the analysis. In Figure 6.6, I have chosen gender as the target variable (dependent variable) and the other variables as inputs (independent variables or predictors).
2. Explore – examine the variables using summary statistics, distributions, interactive visualization via GGobi, correlation, hierarchical cluster analysis of variables, and principal components. A very interesting feature in distribution analysis is the application of Benford's law, an examination of the

Fig. 6.4 The Data Viewer window that appears when you click the *View data set* button

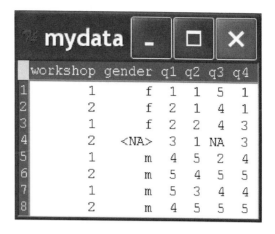

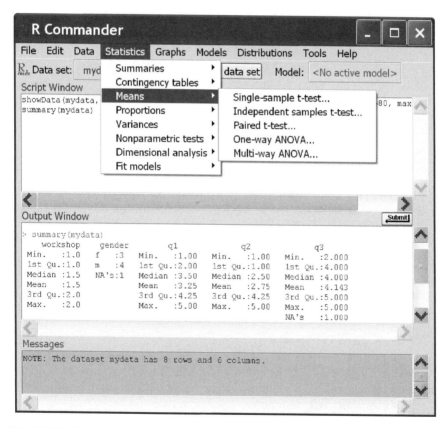

Fig. 6.5 The R Commander graphical user interface

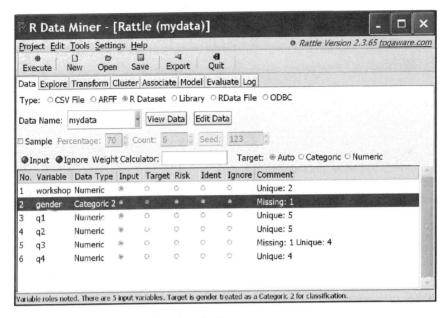

Fig. 6.6 RATTLE data mining interface for R

initial digits of data values that people use to detect fraudulent data (faked expense account values, etc.).

3. Transform – replace missing values with reasonable estimates (imputation), convert variables to factors or look for outliers.
4. Cluster – finds groups of similar cases.
5. Associate – finds association patterns.
6. Model – apply models from tree, boost, forest, SVM, regression, or all.
7. Evaluate – see how good your model is using confusion tables, lift charts, ROC curves, etc.
8. Log – see the R program that RATTLE wrote for you to do all the steps.

Figure 6.7 shows an R program that RATTLE wrote when asked for boxplots of mydata (boxplots not shown).

6.7.3 JGR Java GUI for R

The Java GUI for R, JGR (pronounced "jaguar") is very similar to R's own simple interface, making it very easy to learn [15]. It adds some helpful tools, like syntax checking in its program editor. It also provides the help files in a way that lets you execute any part of an example you select. That is very helpful when trying to understand a complicated example.

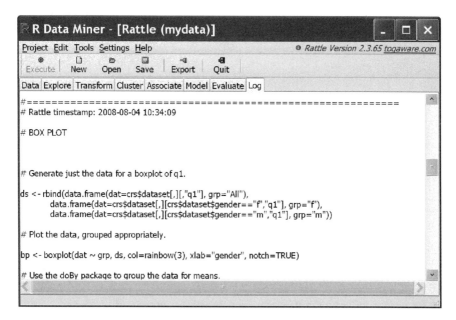

Fig. 6.7 An R program written by RATTLE based upon a menu selection to do a boxplot

To install it on Microsoft Windows or Apple Macintosh, you must download and run its installer from its website http://rosuda.org/JGR/ [16]. Linux users have some additional minor steps that are described at the site.

In Fig. 6.8, the JGR program editor has automatically color-coded my comments, function names, and arguments, making it much easier to spot errors.

In this next example (Fig. 6.9), when I typed `"cor("` it offered a box showing the various arguments that control the `cor` function for doing correlations. That is very helpful when you are learning!

JGR's package manager makes it easier to control which packages you are using (Fig. 6.10). Simply checking the boxes under "loaded" will load those packages from the library. If you also check it under "default" JGR will load them every time you start JGR. Without JGR's help, that feature would require editing your .Rprofile.

JGR's object browser makes it easy to manage your workspace (Fig. 6.11). Selecting different tabs across the top enable you to see the different types of objects in your workspace. Below I right clicked on gender, this brought up the box listing the number of males, females, and missing values (NAs). If you have a list of models, you can sort them easily by various measures, like their R-squared value.

Double-clicking on a data frame in the object browser starts a data editor (Fig. 6.12), which is much nicer than the one built into R. It lets you rename variables, search for values, sort by clicking on variable names, cut and paste values and add or delete rows or columns.

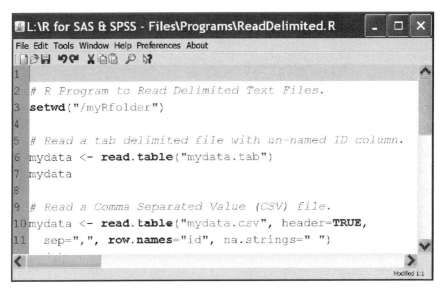

Fig. 6.8 Color-coded editor in JGR helps prevent typing errors

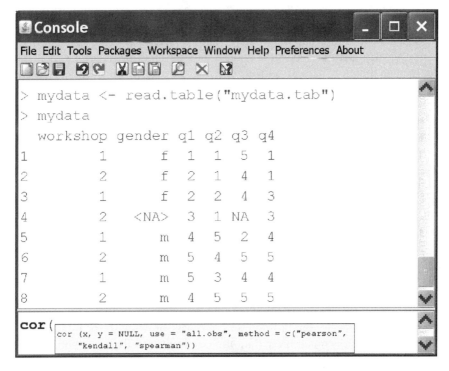

Fig. 6.9 JGR showing arguments that you might choose for the cor function

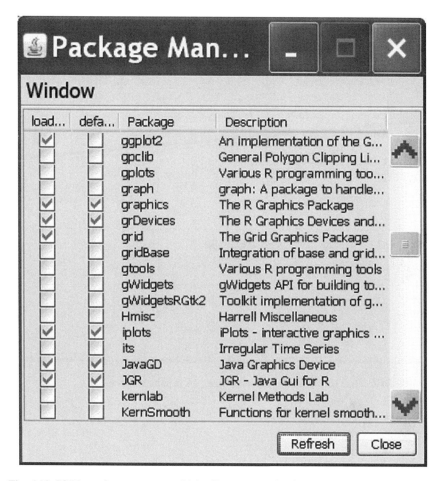

Fig. 6.10 JGR's package manager which allows you to load packages from the library on demand and/or at startup

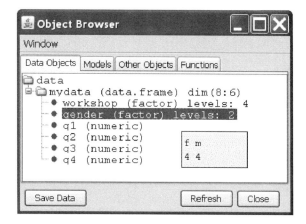

Fig. 6.11 JGR's object browser shows information about each object in your workspace

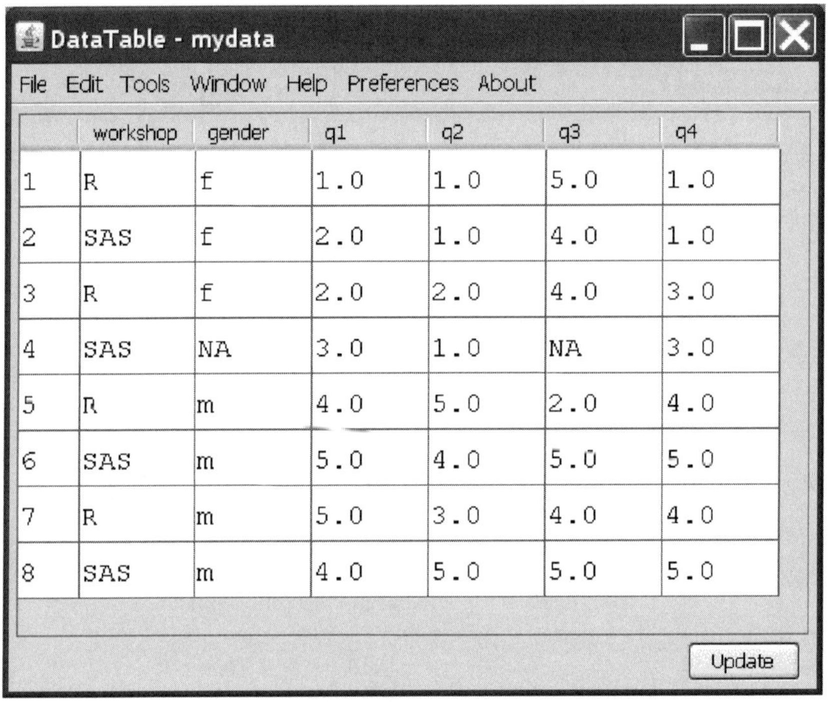

Fig. 6.12 JGR's data editor, an improvement over R's primitive one

Chapter 7
Help and Documentation

R has an extensive array of help files and documentation. However, they can be somewhat intimidating at first, since many of them assume you already know a lot about R. By the time you finish this book, the help files and other documentation should make much more sense.

7.1 Help Files

The help files in R are an important source of valuable information. However, they are written for intermediate to advanced users.

For example, the help file for the `print` function, used to print your data (among other things) says, that its use is to "Print Values", which is clear enough. However, it then goes on to say, "print prints its argument and returns it invisibly (via <u>invisible</u>(x)). It is a generic function which means that new printing methods can be easily added for new <u>classes</u>."

That requires a much higher level of knowledge that does the SPSS description of its similar command, "LIST displays case values for variables in the active dataset." However, when you are done with this book, you should have no problem understanding most help files.

7.2 Starting Help

You can start the help system by choosing *Help> HTML Help* on Windows or *Help> R Help* on Macintosh. On any operating system you can submit the `help.start` function in the R console. That is the way Linux/UNIX users start it since they lack menus.

```
> help.start ()
```

Regardless of how you start it, you will get a help window that looks something like Fig. 7.1.

To get help for a certain function such as `summary`, use the form:

```
help (summary)
```

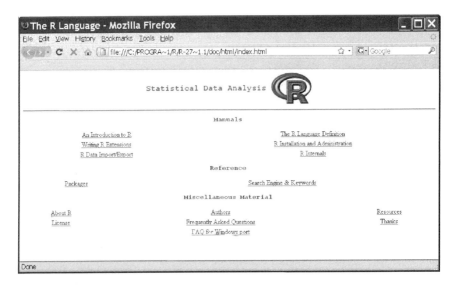

Fig. 7.1 R's help window

or prefix the topic with a question mark:

```
? summary
```

To get help on an operator, enclose it in quotes. For example, to get help on the assignment operator (equivalent to the equal sign in SAS or SPSS), enter:

```
(help "<- ")
```

If you do not know the name of a command or operator, use the `help.search` function to search the help files.

```
help.search ("your search string")
```

A particularly useful help file is the one on extracting and replacing parts of an object. That help file is opened with the following. The capital letter in Extract is necessary.

```
help (Extract)
```

It is best to read that one *after* you have read the chapters on *Selecting Variables* and *Selecting Observations*.

7.3 Help Examples

Most help files include examples that will execute. You can cut and paste them into a script window to submit in easily digestible chunks. You can also have R execute all of the examples at once with the `example` function. Here are the examples for the `mean` function, but do not try to understand them now. We will cover the `mean` function later.

```
> example (mean)

mean> x <- c (0:10, 50)

mean> xm <- mean (x)

mean> c (xm, mean (x, trim = 0.10))
[1] 8.75 5.50

mean> mean (USArrests, trim = 0.2)
  Murder   Assault  UrbanPop      Rape
    7.42    167.60     66.20     20.16
```

R changes its prefix of each example command from ">" to "mean >" to let you know that it is still submitting examples from the mean help files. Note that when an example is labeled, "*Not run*" it means that while it is good to study, it will not run unless you adapt it to your needs.

A very nice feature of the JGR GUI is that you can execute most help file example programs by selecting them, right-clicking on the selection, and then choosing "run line or selection".

R has add-on packages that you must load from its library before getting help. For example, Frank Harrell's Hmisc package [6] has many useful functions that add SAS-like capabilities to R. One of these is the contents function. Let us try to get help on it before loading the Hmisc package.

```
> help (contents)

No documentation for 'contents' in specified packages
    and libraries:

you could try 'help.search ("contents")'
```

The help system does not find it and it even reminds you how you might search the help files.

```
> help.search ("contents")
```

The window in Figure 7.2 pops up showing you the packages that might contain the contents function. This might remind us that we have not yet loaded the Hmisc package. We can do that with the following command:

```
> library ("Hmisc")
```

R responds with a warning about masked objects discussed in Section 5.2, Loading an Add-on package. It does not cause a problem for our purposes.

```
Attaching package: 'Hmisc'
    The following object(s) are masked from package:base :
        format.pval,
        round.POSIXt,
        trunc.POSIXt,
        units
```

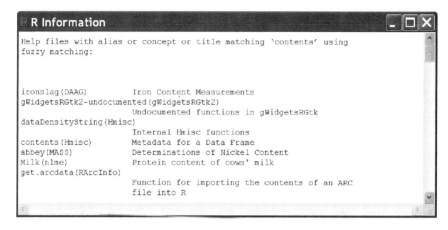

Fig. 7.2 Search results from the function call, help.search("contents")

Now that the Hmisc package is loaded, we can get help on the contents function with the command help(contents). We do not need to look at the actual help file at the moment. We will cover that function much later.

7.4 Help for Functions that Call Other Functions

R has functions, called *generic functions*, that call other functions. In many cases the help file for the generic function will refer you to those other functions, providing all the help you need. However, in some cases you need to dig for such help in other ways. We will discuss this topic in Chap. 8, in Sect. 8.5.3. We will also examine an example of this in Chap. 21, in Sect. 21.9.9.

7.5 Help for Packages

You can get help on packages by using the help argument on the library function. For example, to get help on the foreign package [17], use

```
> library(help=foreign)
```

The window in Fig. 7.3 is only a partial view of the information R provides. To get help on a package, you must first install it and load it. However, not all packages provide help for the whole package. Most do provide help on the functions that the package contains.

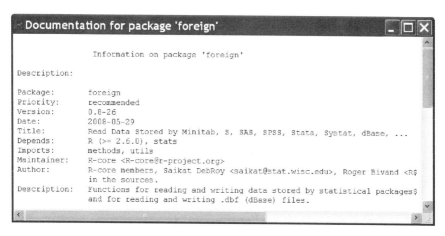

Fig. 7.3 Help window for the `foreign` package

7.6 Help for Datasets

If a dataset has a help file associated with it, you can see it with the `help` function. For example,

```
help(esoph)
```

will tell you that this dataset is, "Data from a case-control study of (o)esopha-geal cancer in Ile-et-Vilaine, France."

7.7 Books and Manuals

Other books on R are available free at http://cran.r-project.org/ under documenta-tion. We will use a number of functions from the `Hmisc` package. Its manual is *An Introduction to S and the Hmisc and Design Libraries*, by Alzola and Harrell at http:// biostat.mc.vanderbilt.edu/twiki/pub/Main/RS/sintro.pdf [6]. The most widely recommended advanced statistics book on R is *Modern Applied Statistics with S*, (abbreviated MASS) by Venables and Ripley [18]. Note that R is almost identical to the S language and books on S usually point out what the differences are.

An excellent book on managing data in R is Phil Spector's *Data Manipula-tion in R* [19].

We will discuss books on graphics in the chapters on that topic.

7.8 E-mail Lists

There are several different e-mail discussion lists regarding R that you can read about and sign up for at http://www.r-project.org/ under *mailing lists*. I recommend signing up for the r-help listserv. There you can learn a lot by reading answers to the myriad of questions people post. If you post your own questions on the list, you are likely to get an answer in an hour or two. However, please read the posting guide, http://www.R-project.org/posting-guide.html, before sending your first question. Taking the time to write a clear and concise question and providing a clear subject line will encourage others to take the time to respond. Sending a small example that demonstrates your problem clearly is particularly helpful. See Chap. 17 for ways to make up a small dataset for that purpose. Also include the version of R you are using and your operating system. You can generate all the relevant details with the following command:

```
> R.version
```

```
platform       i386-pc-mingw32
arch           i386
os             mingw32
system         i386, mingw32
status
major          2
minor          6.1
year           2007
month          11
day            26
svn rev        43537
language       R
version.string R version 2.6.1 (2007-11-26)
```

7.9 Searching the Web

Searching the web for information on R using generic search engines such as Google can be frustrating, since the letter *R* refers to many different things. However, if you add the letter *R* to other keywords, it is surprisingly effective. Adding the word *package* to your search will also narrow it down.

An excellent site that searches just for R topics is Jonathon Barron's R Site Search [20] at http://finzi.psych.upenn.edu/search.html.

You can search just the R site while in R itself by entering the RSiteSearch function,

```
RSiteSearch ("your search string")
```

or go to http://www.r-project.org/ and click *search*.

If you use the Firefox web browser, there is a free plug-in called *Rsitesearch* [21] you can use. Download it from http://addictedtor.free.fr/rsitesearch/.

7.10 Vignettes

Another kind of help is a vignette, a short description. People who write packages can put anything into its vignette. The command

```
vignette (all=TRUE)
```

will show you vignettes for all the packages you have installed. To see the vignette for a particular package, enter it in the function with its name in quotes:

```
vignette ("mypackage")
```

Unfortunately, many packages do not have vignettes.

Chapter 8
Programming Language Basics

R is an object-oriented language. Everything that exists in it – variables, datasets, functions (procedures) – are all objects.

Object names in R can be any length consisting of letters, numbers, underscores "_" or the period "." and should begin with a letter. However, if you always put quotes around a variable or dataset (actually any object) name, it can contain any characters, including spaces. Unlike SAS, the period has no meaning in the name of a dataset. However, given that my readers will often be SAS users, I avoid using the period. Case matters, so you can have two variables, one named myvar and another named MyVar in the same dataset, although that is not a good idea! Some add-on packages tweak names like the capitalized "Save" to represent a compatible, but enhanced, version of a built-in function like the lowercased "save". As in any statistics package, it is best to avoid names that match function names like "mean" or that match logical conditions like "TRUE".

Commands can begin and end anywhere on a line and R will ignore any additional spaces. R will try to execute a command when it reaches the end of a line. Therefore, to continue a command on a new line, you must ensure that the fragment you leave behind is not already a complete command by itself. Continuing a command on a new line after a comma is usually a safe bet. As you will see, R commands frequently use commas, making them a convenient stopping point. The R console will tell you that it is continuing a line when it changes the prompt from ">" to "+". If you see "+" unexpectedly, you may have simply forgotten to add the final close parenthesis, ")". Submitting only that character will then finish your command.

If you are getting the "+" and cannot figure out why, you can cancel the pending command with the Escape key on Windows, or CTRL-C on Macintosh or Linux/UNIX. For CTRL-C, hold the CTRL key down (Linux/UNIX) or the command key (Macintosh) while pressing the letter C.

You may end any R command with a semicolon just like SAS. That is not required though, except when entering multiple commands on a single line.

R.A. Muenchen, *R for SAS and SPSS Users*, DOI: 10.1007/978-0-387-09418-2_8,

8.1 Simple Calculations

Although few people would bother to use R just as a simple calculator, you can
do so with commands like

```
> 2+3

  [1] 5
```

The [1] tells you the resulting value is the first result. It is only useful when
your results run across several lines. We can tell R to generate some data for us
to see how the numbering depends upon the width of the output. The form 1:50
will generate the integers from 1 to 50.

```
> 1:50

  [1] 1   2  3  4  5  6  7  8  9 10 11 12 13 14 15 16 17 18
  [19]19 20 21 22 23 24 25 26 27 28 29 30 31 32 33 34 35 36
  [37]37 38 39 40 41 42 43 44 45 46 47 48 49 50
```

Now it is obvious that the numbers in square brackets are counting or
indexing the values. I have set my line width to 64 characters to help things fit
in this book. We can use the options function to change the width to 40 and
see how the bracketed numbers change.

```
> options(width=40)

> 1:50

  [1] 1   2  3  4  5  6  7  8  9 10 11 12
  [13] 13 14 15 16 17 18 19 20 21 22 23 24
  [25] 25 26 27 28 29 30 31 32 33 34 35 36
  [37] 37 38 39 40 41 42 43 44 45 46 47 48
  [49] 49 50

> options(width=64) #Set it wider again.
```

In SAS, that setting is done with OPTIONS LIZESIZE = 64. SPSS uses SET
WIDTH 64.

You can assign the values to symbolic variables like x and y using
the assignment operator, a two character sequence "<–". You can use
the equal sign as SAS and SPSS do, but more people seem to prefer
"<–".

```
> x <- 2

> y <- 3

> x+y

  [1] 5
```

```
> x* y
[1] 6
```

I have added extra spaces in the above commands and extra lines in the output for legibility. Additional spaces do not affect the commands.

8.2 Data Structures

SAS and SPSS both use one main data structure, the dataset. Instead, R has several different data structures including vectors, factors, data frames, matrices, arrays, and lists. The data frame is most like a dataset in SAS or SPSS. R also has data structures specifically for time series, but those are beyond the scope of this book.

8.2.1 Vectors

A vector is an object that contains a set of values called its *elements*. You can think of it as a SAS/SPSS variable, but that would imply that it is a column in a dataset. It is not. It exists by itself and is neither a column nor a row. In practice, it is usually one of two things: a variable or a set of parameter settings you can use to control a function.

All our examples will use the same dataset, a pretend survey about how people liked various workshops on statistics packages. Let us enter the responses to the first question, "Which workshop did you take?"

```
workshop <- c(1,2,1,2,1,2,1,2)
```

All the values of workshop are numeric, so the vector's *mode* is *numeric*. SAS and SPSS both refer to that as a variable's *type*. As in SAS and SPSS, even if one value were alphabetic (character or string) then the mode would be *coerced*, or forced, to be *character*.

R does all its work with *functions*, which are similar to SAS statements and procedures, or SPSS commands and procedures. Functions have a name followed by its parameters (or keywords in SPSS jargon), called *arguments*, in parentheses. The c function's job is to combine multiple values into a single vector. Its arguments are just the values to combine, in this case 1,2,1,2....

To print our vector, we can use the print function. However, this function is used so often, it is the default when you type the name of any object. So when working interactively, these two commands do exactly the same thing:

```
> print(workshop)
[1] 1 2 1 2 1 2 1 2

> workshop
[1] 1 2 1 2 1 2 1 2
```

We will run all the examples in this book interactively. That is, we submit commands and see the results immediately.

Although typing out the `print` function for most of our examples is not necessary, we will do it occasionally when showing how the R code looks in a typical analysis. Writing out the print function is required when running R from SPSS, see section 6.6 and when running a program included from a separate file, see section 6.4.

Let us create a character variable. Using R jargon, we would say we are going to create a character vector, or a vector whose *mode* is character. These are the genders of our hypothetical students:

```
> gender <- c("f","f","f",NA,"m","m","m","m")

> gender

 [1] "f" "f" "f" NA"m" "m" "m" "m"
```

NA stands for Not Available, which R uses to represent missing values. Even when entering string values for gender, *never enclose the NA in quotes*. Now let us enter the rest of our data:

```
q1 <- c(1,2,2,3,4,5,5,4)
q2 <- c(1,1,2,1,5,4,3,5)
q3 <- c(5,4,4,NA,2,5,4,5)
q4 <- c(1,1,3,3,4,5,4)
```

To get a simple table of frequencies we can use the `table` function.

```
> table(workshop)

workshop
1 2
4 4

> table(gender)

gender
f m
3 4
```

The first thing you will notice about the output is its sparseness. There are no percents and no lines drawn to form a table. This highlights a major difference in perspective between R and SAS or SPSS. The output is in a form that other functions can use immediately. Other functions exist that provide more output, like percents. Still others format output into publication quality form.

Let us get the mean of the responses to question 3:

```
> mean(q3)

[1] NA
```

The result is NA or Not Available. Many R functions handle missing values in an opposite manner from SAS or SPSS. R will usually provide output that is NA when performing an operation on data that contains any missing values. It will typically provide an answer only when you tell it to override that perspective. There are several ways to do this in R. For the mean function, you set the NA remove argument, na.rm, equal to TRUE.

```
> mean(q3,na.rm=TRUE)

[1] 3
```

8.2.2 Factors

Two of the variables we entered above, workshop and gender, are clearly categorical. R has a special data structure called a *factor* for such variables. Regardless of whether the original data is numeric or character, when it becomes a factor, its *mode* is *numeric*. Let us enter workshop again (just to see its values) and convert it to a factor:

```
workshop <- c(1,2,1,2,1,2,1,2)

workshop <- factor(
    workshop,
    levels=c(1,2,3,4),
    labels=c("R","SAS","SPSS","STATA")
)
```

The factor function has three arguments:

1. The name of a vector to convert to a factor.
2. The levels or values that the data can have. This allows you to specify values that are not yet in the data. In our case, workshop is limited to the values 1 and 2 but we can include the values 3 and 4 for future expansion.
3. Optionally, the labels for the levels. The factor function will match the labels to the levels in the order they both appear here; the order of the values in the dataset is irrelevant. If you do not provide the labels argument, R will use the values themselves as the labels.

Now when we print the data, it shows us that the people in our practice dataset have only taken workshops in R and SAS. It also lists the levels so you can see what labels are possible:

```
> workshop

 [1] R    SAS R    SAS R    SAS R    SAS

Levels: R SAS SPSS STATA
```

The `table` function now displays the workshop labels and how many people took each.

```
> table(workshop)

workshop
    R    SAS  SPSS STATA
    4     4     0     0
```

You can convert character vectors to factors in a similar manner. Let us review entering our practice data for gender and printing it back out:

```
> gender<- c("f","f","f",NA,"m","m","m","m

> gender

 [1] "f" "f" "f" NA "m" "m" "m" "m" "m"
```

Notice that the missing value, NA, does not have quotes around it. R drops them to let you know it is not a valid character string that might stand for something like North America.

If we are happy with those labels, we can convert gender to a factor by using the simplest form of the factor function:

```
> gender <- factor(gender)

> gender

 [1] f f f NA m m m m
Levels: f m
```

If instead we want nicer labels, we can use the longer form. It works the same way as for workshop, but the values on the levels command need to be in quotes:

```
> gender <- factor(
+     gender,
+     levels=c("m","f"),
+     labels=c("Male","Female")
+ )

> gender

 [1] Female Female Female NA Male   Male   Male   Male
Levels: Male Female
```

We will examine factors and compare them to SPSS value labels and SAS formats in Chap. 15.

8.2.3 Data Frames

The data structure in R that is most like a SAS or SPSS dataset is the *data frame*. SAS and SPSS datasets are always rectangular, with *variables* in the columns and *records* in the rows. SAS calls these records *observations* and SPSS calls them *cases*. A data frame is also rectangular. In R terminology, the columns are called vectors, variables, or just columns. R calls the rows observations, cases, or just rows.

A data frame is a generalized *matrix*, one that can contain both character and numeric columns. A data frame is also a special type of *list*, one which requires each element to have the same length. We will discuss matrices and lists in the next two sections below.

We have already seen that R can store variables in vectors and factors. Why does it need another data structure? R can generate almost any type of analysis or graph from data stored in vectors or factors. For example, getting a scatterplot of the responses to q1 versus q4 is easy. R will pair the first number from each vector as the first (x,y) pair to plot and so on down the line. However, it is up to you to make sure that this pairing makes sense. If you sort one vector independent of the others, or remove the missing values from vectors independently, the critical information of how the pairs should form is lost. The data frame helps maintain this critical information.

The most common way to create a data frame is to read it from another source such as a text file, spreadsheet, or database. You can usually do that with a single command. We will do that later. For the moment, we will create one from our original vectors. Here is our program so far:

```
workshop <- c(1,2,1,2,1,2,1,2)

workshop <- factor(workshop,
    levels = c(1,2,3,4),
    labels = c("R","SAS","SPSS","STATA") )

gender <- c("f","f","f",NA,"m","m","m","m")

gender <- factor(gender)

q1 <- c(1,2,2,3,4,5,5,4)
q2 <- c(1,1,2,1,5,4,3,5)
q3 <- c(5,4,4,NA,2,5,4,5)
q4 <- c(1,1,3,3,4,5,4,5)
```

Now we will use the `data.frame` function to combine our variables (vectors and factors) into our data frame. Its arguments are simply the names of the objects we wish to combine.

```
> mydata <- data.frame(workshop,gender,q1,q2,q3,q4)

> mydata

  workshop gender q1 q2 q3 q4
1        R      f  1  1  5  1
2      SAS      f  2  1  4  1
3        R      f  2  2  4  3
4      SAS   <NA>  3  1 NA  3
5        R      m  4  5  2  4
6      SAS      m  5  4  5  5
7        R      m  5  3  4  4
8      SAS      m  4  5  5  5
```

Notice that the missing value for gender is now shown as <NA>. When R prints data frames, it drops the quotes around character values and so must differentiate missing value NAs from valid character strings that happen to be the letters "NA."

If we wanted to rename the vectors as we created the data frame, we could do so with the following form. Here the vector "gender" will be stored in mydata with the name "sex" and the others will keep their original names. Of course, we could have renamed every variable using this form.

```
mydata <- data.frame(workshop, sex=gender, q1, q2,
   q3, q4)
```

Although we had made gender into a factor, the data.frame function will force, or *coerce*, all character variables to become factors when the data frame is created. You do not always want that to happen, for example when you have vectors that store people's names and addresses. To block that from happening, you can add the stringsAsFactors=FALSE argument to the call to the data frame function.

R data frames have a formal place for an ID variable it calls *row names*. These can be informative text labels like subject names, but by default, they are sequential numbers stored as character values. The row.names function will display them:

```
> row.names(mydata)

[1] "1" "2" "3" "4" "5" "6" "7" "8"
```

SAS and SPSS users typically have an ID variable containing an observation/case number or perhaps a subject's name. However, this variable is like any other unless you manually supply it to a procedure that identifies observations. In R, procedures that identify observations will do so automatically using row labels. If you set an ID variable to be row labels while reading a text file, the variable's original name (id, subject, SSN...) vanishes. Since functions that do things like identify outliers will use the information automatically, you usually do not need the name. We will discuss row names further when we read text files and in Sect. 14.6.

8.2.4 Matrices

A matrix is a two-dimensional data object that looks like a SAS or SPSS dataset, but it is actually one long vector wrapped into rows and columns. Because of this, its values must be of the same mode, that is, all numeric, all character, or all logical (more on logical vectors later). This constraint makes matrices more efficient than data frames for some types of analyses. The cbind function binds columns together into a matrix:

```
> mymatrix <- cbind(q1, q2, q3, q4)

> mymatrix

      q1 q2 q3 q4
[1,]   1  1  5  1
[2,]   2  1  4  1
[3,]   2  2  4  3
[4,]   3  1 NA  3
[5,]   4  5  2  4
[6,]   5  4  5  5
[7,]   5  3  4  4
[8,]   4  5  5  5
```

As you can see, a matrix is a two-dimensional array of values. The numbers to the left side in brackets are the row numbers. The form [1,] means that it is row number one and the blank following the comma means that R has displayed all the columns. We can get the dimensions of the matrix with the dim function.

```
> dim(mymatrix)

[1] 8  4
```

The first dimension is the number of rows, 8, and the second is the number of columns, 4.

To create a matrix, you do not need to start with vectors as we did. You can create one directly with the matrix function. It has four arguments. The data to place into a matrix, which you must enter enclosed in the c function, and the number of rows and columns. The byrow = TRUE argument tells it to read the matrix row by row instead of by columns, which is the default.

```
> mymatrix <- matrix(
+    c( 1, 1, 5, 1,
+       2, 1, 4, 1,
+       2, 2, 4, 3,
+       3, 1,NA, 3,
+       4, 5, 2, 4,
+       5, 4, 5, 5,
```

```
+           5, 3, 4, 4,
+           4, 5, 5, 5),
+      nrow=8, ncol=4, byrow=TRUE)
> mymatrix
      [,1] [,2] [,3] [,4]
[1,]    1    1    5    1
[2,]    2    1    4    1
[3,]    2    2    4    3
[4,]    3    1   NA    3
[5,]    4    5    2    4
[6,]    5    4    5    5
[7,]    5    3    4    4
[8,]    4    5    5    5
```

Now let us see what the `table`, `mean` and `cor` functions do with matrices.

```
> table(mymatrix)

mymatrix
1 2 3 4 5
6 4 4 8 8

> mean( mymatrix, na.rm=TRUE )

 [1] 3.266667

> cor( mymatrix, use="pairwise")

            [,1]         [,2]         [,3]         [,4]
[1,]   1.0000000   0.7395179  -0.12500000   0.88040627
[2,]   0.7395179   1.0000000  -0.27003086   0.85063978
[3,]  -0.1250000  -0.2700309   1.00000000  -0.2613542
[4,]   0.8804063   0.8506398  -0.02613542   1.00000000
```

The `table` function counts the responses across all survey questions and the `mean` function gets the mean response of them all. Neither of those is of much interest, but you might find cases where it would be of value. The `cor` function correlates each item with the others, which is a very common thing to do.

If you put a matrix into a data frame, its columns will become individual vectors. For example, now that we have mymatrix, we can create our practice data frame in two ways that have an identical result:

```
mydata <- data.frame(workshop, gender, q1, q2, q3, q4)
```

or

```
mydata <- data.frame(workshop, gender, mymatrix)
```

The latter can be much easier to do under some circumstances.

8.2.5 *Arrays*

An array is a multi-dimensional extension of the matrix structure. You can think of it as a set of matrices. The use of arrays is beyond our scope.

8.2.6 *Lists*

A list is a very flexible data structure. You can use it to store combinations of any other objects, even other lists. The objects stored in a list are called its *components*. That is a broader term than variables, reflecting the wider range of objects possible. Although you can use a list to store any R object, they are most often used to store data in various forms (as we do below), results from functions such as regression equation parameters (see Chap. 23 for examples), and sets of arguments to control functions (we will do that when aggregating data by workshop and gender). Here we will focus on storing data in a list.

We can combine our variables into a list using the `list` function.

```
> mylist <- list(workshop, gender, q1, q2, q3, q4,
    mymatrix)
```

Now let us print it.

```
> mylist
[[1]]
[1] R     SAS R     SAS R     SAS R     SAS
Levels: R SAS SPSS STATA

[[2]]
[1] f     f     f     <NA> m     m     m     m
Levels: f m

[[3]]
[1] 1 2 2 3 4 5 5 4

[[4]]
[1] 1 1 2 1 5 4 3 5

[[5]]
[1]  5  4  4 NA  2  5  4  5

[[6]]
[1] 1 1 3 3 4 5 4 5

[[7]]
```

```
       q1 q2 q3 q4
[1,]   1   1   5   1
[2,]   2   1   4   1
[3,]   2   2   4   3
[4,]   3   1  NA   3
[5,]   4   5   2   4
[6,]   5   4   5   5
[7,]   5   3   4   4
[8,]   4   5   5   5
```

Notice how the vector components of the list print sideways now. That allows each variable to have a different length, or even to have a totally different structure, like a matrix. Also notice that it counts the components of the list with an additional index value in double brackets [[1]], [[2]]... Then each component has its usual index in single brackets. We will use these later to select data from lists.

If we wanted to rename the objects as we created the list, we could do so with the following form. Here the vector "gender" will be stored in mylist with the name "sex":

```
mylist <- list(workshop, sex=gender, q1, q2, q3, q4,
  mymatrix)
```

Many R functions store their output in lists, especially those that fit models. For example, the results from a regression analysis might store the equation parameters in one component and the sums of squares in another. Since they are from the same model, it makes sense for them to have a data structure that can store these items with different lengths.

8.3 Saving Your Work So Far

When learning any new computer program, always do a small amount of work, save it and get completely out of the software. Then go back in and verify that you really did know how to save your work.

This is a good point to stop, clean things up, and save your workspace. You never know when a power outage might erase it. If you have done the examples above, you can use the ls function to see all the data objects you have created. If you put no arguments between the ls function's parentheses, you will get a list of all your objects. Another, more descriptive name for this function is objects.

```
> ls()

 [1] "gender"   "mydata"   "mylist"   "mymatrix"  "q1"
 [6] "q2"       "q3"       "q4"       "workshop"  "x"
[11] "y"
```

Once we have combined our vectors into a data frame, we do not need the individual vectors any more. In fact, having them in two places can become quiet confusing as we will see later. We can remove the ones we do not want by listing them as arguments separated by commas on the rm function.

```
> rm(x,y,workshop,gender,q1,q2,q3,q4)

> ls()

[1] "mydata" "mylist" "mymatrix"
```

We would like to save the remaining objects. These three reside in R's workspace in the computer's random access memory. While SAS and SPSS users typically only save one dataset to a file (SAS has exceptions that are used by experts), R users often save multiple objects to a single file.

The directory or folder that R will store files in is called its *working directory*. Unless you tell it otherwise, R will put your program, output, and data in that directory. The default working directory on Windows is My Documents, on Macintosh is /Users/yourname. The setwd function tells R where you would like your files to go, and the getwd function extracts it for you to see.

```
> getwd()

[1]"C:/Documents and Settings/yourname/My Documents"

> setwd("/myRfolder")

> getwd()

[1] "/myRfolder"
```

Notice that we use a forward slash in /myRfolder. R can use forward slashes in filenames even on computers running Windows. We will discuss why later.

The save.Image function will save all objects in your workspace to the file you specify:

```
save.image(file="myWorkspace.RData")
```

When you exit R, it will ask if you want to save your workspace. Since we are saving it to a file we have named, you can tell it no. The next time you start R, you can load your work with the load function.

```
> load("myWorkspace.RData")
```

If you want to see what you have loaded, use the ls function:

```
> ls()

[1] "mydata" "mylist" "mymatrix"
```

For more details, see Chap. 19.

8.4 Comments to Document Your Programs

No matter how simple you may think a program is when you write it, it is good to sprinkle in comments liberally to remind yourself later. R uses the # operator to begin a comment, which then continues until the end of the line. You can put comments in the middle of statements, so long as they are at the end of the line:

```
> #This comment is on its own line, between functions.

> workshop <- c(1,2,1,2,  #A comment in arguments.
+                1,2,1,2) #And this is at the end.
```

Unlike the SAS and SPSS /*...*/ style comments, there is no way to comment out a whole block of code that you want to ignore. R users get around that by pretending to define a function whose only goal is to have R ignore the code. For example,

```
BigComment <- function(x)
  {
    # Here is code I do not want to run,
    # but I might need it later.
mean(x, na.rm=TRUE)
sd(x, na.rm=TRUE)
}
```

While SAS and SPSS would allow errors to exist in the code that has been commented out, this approach in R does not. R is actually creating the function, so the code within it must be correct.

Another way to comment out a large block of text is to use a text editor that can easily add (and later remove) the "#" character to the front of each line in a selected block of code.

8.5 Controlling Functions (Procedures)

SAS and SPSS both control their procedures though statements like GLM and related sub-statements such as CLASS to specify which variables are factors (categorical). Those statements have options that control exactly what appears in the output. Modeling statements have a formula syntax. R has analogs to these methods, plus a few unique ones.

8.5.1 Controlling Functions with Arguments

SAS and SPSS use options to control what procedures do. R does too, using slightly different terminology. R uses arguments to control functions. Let us look at the help file for the mean function. The following command will call up its help file.

```
>help(mean)
```

In Fig. 8.1, in the section labeled *Usage*, the help file tells us that the overall form of the function is mean(x,...). That means you have to provide an R object represented by x, followed by arguments represented by "...."

The *##Default S3 Method* section tells us the arguments used by the mean function itself as well as their initial, or default, settings. So if you do not tell it otherwise, it will not trim any data (trim=0) and will not remove missing values (na.rm=FALSE). That means the presence of any missing values will result in the mean being missing or NA too. The "..." means that more arguments are possible, but the mean will pass those along to other functions that it calls. More on that later.

The *Arguments* section gets into the details. It tells you that x can be a numeric data frame, numeric vector, or date vector. It also has a comment about complex vectors, which are beyond our scope. The trim argument tells R the percent of the extreme values to exclude before calculating the mean. The zero indicates the default value, so it will trim none of your data unless you change that.

The na.rm argument appears in many R functions. R uses NA to represent Not Available, or missing values. The default value of "FALSE" tells us that R will not remove them unless you tell it otherwise. Therefore, as we have seen, the result will be NA if any missing values are present. Many functions in R are set

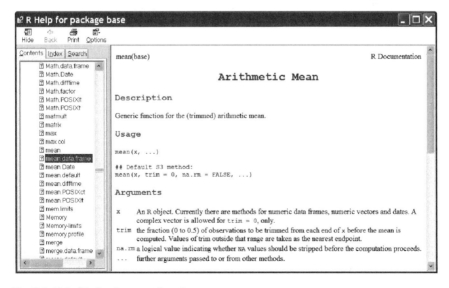

Fig. 8.1 Help file for the mean function

this way by default. That is quite the opposite from SAS and SPSS, which assume you want to use all the data you have unless you tell it otherwise. The "..." is finally defined as representing arguments that you can list here that will be passed on to other functions called by the mean function.

We can run this on a variable called myVar1 by naming each argument.

```
mean(x=myVar1, trim=.25, na.rm=TRUE)
```

If you name all the arguments, you can use them in any order:

```
mean(na.rm=TRUE, x=myVar1, trim=.25)
```

We can also run it by listing *every* argument in their proper positions but without the argument names:

```
mean(myVar1, .25, TRUE)
```

However, people usually run R functions by listing the object to analyze first without its name, followed by the names and values of only those arguments they want to change:

```
mean(myVar1, na.rm=TRUE)
```

You can also abbreviate some argument names, but it is a bit tricky. The abbreviation you choose must have enough letters to be unique. However, some functions pass arguments they do not recognize to other functions they control. This is indicated by "..." as the function's last argument in the help file. See the help file for the mean function used above for an example. Once you have started passing arguments to other functions, R will pass them all unless it sees the full name of an argument it uses.

People sometimes abbreviate the values TRUE or FALSE as T or F. This is a bad idea, as you can define T or F to be anything you like, leading to undesired results. You may avoid that trap yourself, but if you write a function that others will use, they may use those variable names.

So the function below will also run, but I do not recommend running it this way.

```
mean(myVar1, na=T)
```

8.5.2 Controlling Functions with Formulas

An important type of argument is the formula. It is the first parameter in functions that do modeling. For example, we can do linear regression, predicting q4 from the others with the following call to the lm function for *l*inear *m*odels.

```
lm(q4 ~ q1+q2+q3)
```

Some modeling functions accept arguments in the form of both formulas and vectors. For example, both of these function calls will compare the genders on the mean of variable q1.

```
t.test(q1 ~ gender, data=mydata)

t.test( q1[gender=="f"],
        q1[gender=="m"], data=mydata) #Data ignored!
```

However, there is one *very important difference*. When using a formula, the data argument can supply the name of a data frame that R will search before looking elsewhere for variables. When not using a formula, as in the second example above, the data argument is ignored! So that example would not work without first telling R which data frame to use. We will discuss that in detail later.

The symbols that R uses for formulas are different from SAS and SPSS. Table 8.1 shows some common examples using a, b, and c as categorical factors and y, x1, and x2 as continuous numeric variables.

8.5.3 Controlling Functions with an Object's Class

Since SAS and SPSS have only one main data structure, the idea of controlling procedures based on data structure is quite alien. The *class* of an object reflects what type of data structure it has. In R, each data structure stores its class as an

Table 8.1 Example formulas in SAS, SPSS, and R

Model	SAS	SPSS	R
Simple regression	MODEL y = x;	/DESIGN x.	y ~ x
Multiple regression with interaction	MODEL y = x1 x2 x1 *x2;	/DESIGN x1 x2 x1* x2.	y ~ x1 + x2 + x1:x2
	MODEL y = x1|x2;		y ~ x1*x2
Regression without intercept	MODEL y = x1 /noint;	/DESIGN x. /INTERCEPT=EXCLUDE.	y ~ -1 + x
One-way analysis of variance	MODEL y = a;	/DESIGN a.	y ~ a
Two-way analysis of variance with interaction	MODEL y = a b a*b;	/DESIGN a b a*b.	y ~ a + b + a:b y ~ a*b
	MODEL y = a|b;		
Analysis of covariance	Model y = a x;	/DESIGN a x.	Y ~ a x
Analysis of variance with b nested within a	MODEL y = b(a);	/DESIGN A B(A). /DESIGN A B WITHIN A.	y ~ b %in% A y ~ a/b

attribute that functions use to determine how to process the object. For objects whose mode is numeric, character, or logical an object's class is its mode. However, for matrices, arrays, factors, lists, or data frames, other values are possible (see Table 8.2).

You can display an object's class with the class function:

```
> workshop <- c(1, 2, 1, 2, 1, 2, 1, 2)

> class(workshop)

 [1] "numeric"

> summary(workshop)

    Min. 1st Qu.  Median    Mean 3rd Qu.    Max.
     1.0     1.0     1.5     1.5     2.0     2.0
```

The class "numeric" indicates that this version of workshop is a numeric vector, not yet a factor. The summary function provided us with inappropriate information because we failed to tell it that workshop is a factor. Note that when we convert workshop into a factor, we are changing its class to factor and summary gives us the more appropriate counts instead:

```
> workshop <- factor(workshop,
+     levels=c(1,2,3,4),
+     labels=c("R","SAS","SPSS","STATA") )

> class(workshop)

 [1] "factor"

> summary(workshop)

    R   SAS  SPSS STATA
    4     4     0     0
```

When we first created gender, it was a character vector so its class was character. Later we made its class factor. Numeric vectors like q1 have a class of numeric. The names of some other classes are obvious: factor, data.frame, matrix, list, and array. Objects created by functions have many other classes. For example, the linear model function lm, stores its output in objects with a class of lm.

R has some functions called *generic functions*. They accept multiple classes of objects, and change their processing accordingly. These functions are tiny. Their task is simply to determine the class of the object and then pass it off to another that will do the actual work. The methods function will tell you what other functions a generic function, will call. Let us look at the methods that the summary function uses.

```
> methods(summary)

 [1] summary.aov           summary.aovlist
 [3] summary.connection    summary.data.frame
```

```
 [5] summary.Date              summary.default
 [7] summary.ecdf*             summary.factor
 [9] summary.glm               summary.infl
[11] summary.lm                 summary.loess*
[13] summary.manova            summary.matrix
[15] summary.mlm               summary.nls*
[17] summary.packageStatus*  summary.POSIXct
[19] summary.POSIXlt           summary.ppr*
[21] summary.prcomp*           summary.princomp*
[23] summary.stepfun           summary.stl*
[25] summary.table            summary.tukeysmooth*
    Non-visible functions are asterisked
```

So when we enter summary(mydata), the summary function sees that mydata is a data frame and then passes it to the function summary.data. frame. The functions marked with asterisks above are "non-visible". Visible functions can be seen by typing their name (without any parentheses). That makes it easy to copy and change them, although only an advanced user would want to do that.

When we discussed the help files, we saw that the mean function ended with an argument of "…." That indicates that the function will pass arguments on to other functions. While it is very helpful that generic functions automatically do the "right thing" when you give it various objects to analyze, it complicates the process of using help files.

When written well, the help file for a generic function will refer you to other functions, providing a clear path to all you need to know. However, it does not always go so smoothly. We will see a good example of this in Chap. 21. The plot function is generic. When we call it with our data frame it will give us a scatterplot matrix. However, to find out all the arguments we might use to improve the plot, we have to use methods (plot) to find that plot.data. frame exists. We could then use help(plot.data.frame) to find that plot.data.frame calls the pairs function, then finally help (pairs) to find the arguments we seek. Luckily this is a worst case scenario, but it is important to realize that this situation does occasionally arise.

As you work with R, you may occasionally forget the mode or class of an object you created. This can result in unexpected output. You can always use the mode or class functions to remind yourself. Table 8.2 shows several R objects and their modes and classes.

8.5.4 Controlling Functions with Extractor Functions – ODS, OMS

Procedures in SAS and SPSS typically display all their output at once. SAS has some interactive procedures that let you request additional output

Table 8.2 Modes and classes of various R objects

Object	Mode	Class
Numeric vector	numeric	numeric
Character vector	character	character
Factor	numeric	factor
Data frame	list	data.frame
List	list	list
Numeric matrix	numeric	matrix
Character matrix	character	matrix
List created by `lm`	list	lm
Table created by `table`	numeric	table

The bottom two are examples of objects created by functions. Other functions can create objects of various modes and classes.

once you have seen the initial output, but SAS users rarely use it that way. R has simple functions, like the `mean` function, that show all their results at once. However, R functions that model relationships among variables tend to show you very little output initially. You save the output to a *model object* and then use *extractor functions* to get more information when you need it.

This section is poorly named from an R expert's perspective. Extractor functions do not actually control other functions the way parameters control SAS/SPSS output. Instead they show us what the other function has *already* done. In essence, most modeling in R is done through its equivalent to the SAS Output Delivery System (ODS) or the SPSS Output Management System (OMS).

Let us look at an example of predicting q4 from the other variables with linear regression using the `lm` function.

```
> lm( q4 ~ q1+q2+q3, data=mydata)

Call:
lm(formula= q4 ~ q1 + q2 + q3, data=mydata)

Coefficients:
  (Intercept)           q1           q2           q3
       -1.770        0.362        0.746        0.409
```

The output is extremely sparse, lacking the usual tests of significance. Now instead we will store the results in a model object called myModel and we will check its class with the `class` function.

```
> myModel <- lm(q4 ~ q1+q2+q3, data=mydata)

> class(myModel)

 [1] "lm"
```

The class function tells us that myModel has a class of "lm" for linear model. We have seen that R functions offer different results (called methods) for different types (called classes) of objects. So let us see what the summary function does with this class of object:

```
> summary(myModel)

Call:
lm(formula= q4 ~ q1+ q2+ q3, data= mydata)

Residuals:
        1        2        3        5        6        7
  -0.3822  -0.3349   0.9191  -0.2243  -0.0664   0.0885

Coefficients:
             Estimate Std. Error t value Pr(>|t|)
(Intercept)    -1.770      2.105   -0.84     0.49
q1              0.362      0.383    0.94     0.44
q2              0.746      0.476    1.57     0.26
q3              0.409      0.430    0.95     0.44

Residual standard error: 0.763 on 2 degrees of freedom
  (2 observations deleted due to missingness)
Multiple R-Squared: 0.917,    Adjusted R-squared: 0.792
F-statistic: 7.34 on 3 and 2 DF, p-value: 0.122
```

This is the type of output that SAS and SPSS print immediately. There are many other functions that we might use, including anova to extract an analysis of variance table, plot for diagnostic plots, predict to get predicted values, resid to get residuals, and so on. We will discuss those in Chapter 23.

Why use such a different approach? You get only what you need, in a form that is very easy to reuse and you use methods that are consistent across functions. Rather than learning different ways of saving residuals or predicted values in every procedure as SAS or SPSS does, you learn one approach that works on all.

8.5.5 How Much Output Is There?

In the previous section we discussed saving output and using extractor functions to get more results. But how do we learn what all is in an output object? Previously, the print function showed us what was in our objects, so let us give that a try. We can do that by simply typing its name, or by explicitly using the print function. In this case, let us actually type the print function.

```
> print(myModel)
```

```
Call:
lm(formula = q4 ~ q1+ q2+ q3, data = mydata)
```

```
Coefficients:
  (Intercept)            q1            q2            q3
      -1.770         0.362         0.746         0.409
```

Now let us check the mode and class of myModel.

```
> mode(myModel)

 [1] "list"

> class(myModel)

 [1] "lm"

> names(myModel)

 [1] "coefficients" "residuals" "effects" "rank"
 [5] "fitted.values" "assign" "qr" "df.residual"
 [9] "na.action" "xlevels" "call" "terms"
[13] "model"
```

So we see that myModel is a list, or collection of objects. More specifically, it is a list of class "lm." The names function shows us the names of all the objects in it. Why did the print function not show them to us? Because the print function has a predetermined method for displaying lm class objects. That method says basically, "If an object's class is lm, then print only the original formula that created the model and its coefficients." When we put our own variables together into a list back in section 8.2, that list did not have a class, so the print function printed it all. We can strip away the class of any object with the unclass function. If we do that, we can print everything.

```
> print( unclass(myModel) )

$coefficients
(Intercept)            q1            q2            q3
  -1.32426       0.42975       0.63101       0.31498

$residuals
        1             2             3             5             6
-0.311392    -0.426160   0.942827  -0.179747   0.076582
        7             8
 0.022574    -0.124684

$effects
(Intercept)            q1            q2            q3
  -8.69318       3.67333      -1.44758      0.78610       0.28015

   0.79299        -0.71722
```

```
$rank
[1] 4

$fitted.values
      1      2      3      5      6      7      8
 1.3114 1.4262 2.0572 4.1797 4.9234 3.9774 5.1247

$assign
[1] 0 1 2 3

$qr
$qr
    (Intercept)       q1       q2        q3
1    -2.64575 -8.69318 -7.93725 -10.96097
2     0.37796  3.92792  3.30964  -0.32733
3     0.37796  0.16771 -2.65449   0.72205
5     0.37796 -0.34146  0.43562   2.49573
6     0.37796 -0.59605 -0.33214  -0.10516
7     0.37796 -0.59605 -0.70886   0.44719
8     0.37796 -0.34146  0.43562  -0.41869
attr(,"assign")
[1] 0 1 2 3

$qraux
[1] 1.3780 1.1677 1.0875 1.7834

$pivot
[1] 1 2 3 4

$tol
[1] 1e-07

$rank
[1] 4

attr(,"class")
[1] "qr"

$df.residual
[1] 3

$na.action
4
4
attr(,"class")
[1] "omit"

$xlevels
list()
```

```
$call
lm(formula = q4 ~ q1 + q2 + q3, data = mydata)

$terms
q4 ~ q1 + q2 + q3

attr(,"variables")

list(q4, q1, q2, q3)

attr(,"factors")
   q1 q2 q3
q4  0  0  0
q1  1  0  0
q2  0  1  0
q3  0  0  1

attr(,"term.labels")
[1] "q1" "q2" "q3"

attr(,"order")
[1] 1 1 1

attr(,"intercept")
[1] 1

attr(,"response")
[1] 1

attr(,".Environment")
<environment: R_GlobalEnv>

attr(,"predvars")
list(q4, q1, q2, q3)

attr(,"dataClasses")
        q4          q1          q2          q3
"numeric"   "numeric"   "numeric"   "numeric"

$model
  q4 q1 q2 q3
1  1  1  1  5
2  1  2  1  4
3  3  2  2  4
5  4  4  5  2
6  5  5  4  5
7  4  5  3  4
8  5  4  5  5
```

It looks like the print function was doing us a big favor by not printing everything! When you explore the contents of any object, you can take this approach, or just given the names, explore things one at a time. For example, we saw that myModel contained the object named $coefficients. We can examine that just by appending the second name to the first.

```
> myModel$coefficients
```

```
(Intercept)              q1              q2              q3
  -1.32426         0.42975         0.63101         0.31498
```

That looks like a vector. Let us use the class function to check.

```
> class( myModel$coefficients)
```

```
[1] "numeric"
```

Yes, it is a numeric vector. So we can use it with anything that accepts such data. For example, we might get a barplot of the coefficients with the following (plot not shown). We will discuss barplots more in chapter 21.

```
> barplot( myModel$coefficients )
```

For many modeling functions, it is very informative to perform a similar exploration on the objects created by them.

8.5.6 *Writing Your Own Functions (Macros)*

In SAS or SPSS, if you wanted to use the same set of procedures repeatedly, you would write a macro, or in SPSS, a Python program. This entails using languages are separate from their main programming statements, and the result operates quite differently from the procedures that come with SAS or SPSS. In R, you write functions using the same language you use for anything else. The resulting function is used in exactly the same way as a function that came with R.

Let us write some variations of a simple function, one that calculates the mean and standard deviation at the same time. For this example, we will apply it to just the numbers 1, 2, 3, 4, and 5.

```
> myvar<- c(1,2,3,4,5)
```

We will begin the function called mystats and tell R that it is a function of x. What follows in curly brackets is the function itself. We will create this with an error to see what happens.

```
> # A bad function.
> mystats <- function(x)
+ {
+    mean(x, na.rm=TRUE)
```

```
+        sd(x, na.rm=TRUE)
+ }
```

Now let us apply it like any other function.

```
> mystats(myvar)

 [1] 1.5811
```

We got the standard deviation, but what happened to the mean? When I introduced the print function, I said you usually could just type an object's name rather than say print(myobject). Well, this is one of the cases where we need to explicitly tell R to print the result. Let us add that to the function.

```
> # A good function that just prints.
> mystats <- function(x)
+ {
+    print( mean(x, na.rm=TRUE) )
+    print(    sd(x, na.rm=TRUE) )
+ }
```

Now let us run it.

```
> mystats(myvar)
 [1] 3
 [1] 1.5811
```

That looks better. Let us create our function in a slightly different way, so that it will write our results to a vector for further use.

```
> mystats <- function(x)
+ {
+    mymean<- mean(x, na.rm=TRUE)
+    mysd<-      sd(x, na.rm=TRUE)
+    c(mean=mymean, sd=mysd)
+ }
```

Now when we run it, we get the results in vector form.

```
> mystats(myvar)

   mean        sd
3.0000 1.5811
```

As with any R function that creates a vector, we can assign the result to a variable to use in any way we like.

```
> myVector <- mystats(myvar)

> myVector

   mean        sd
3.0000 1.5811
```

Many R functions return their results in the form of a list. Recall that each member of a list can be any data structure. Let us use a list to save the original data, as well as the mean and standard deviation. We will use the `list` function as we did when we combined all our variables into one list.

```
> mystats <- function (x)
+ {
+    myinput <- x
+    mymean  <- mean(x, na.rm=TRUE)
+    mysd     <-   sd(x, na.rm=TRUE)
+    list(data=myinput, mean=mymean, sd=mysd)
+ }
```

Now let us run it to see how the results look.

```
mystats (myvar)

$data
 [1] 1 2 3 4 5

$mean
 [1] 3

$sd
 [1] 1.5811
```

We can save the result to mylist and then print just the data.

```
> myStatlist <- mystats(myvar)

> myStatlist$data

 [1] 1 2 3 4 5
```

If we want to see the function itself, simply type the name of the function without any parentheses following.

```
> mystats

function(x)
{
  myinput <- x
  mymean  <- mean(x, na.rm=TRUE)
  mysd     <-    sd(x, na.rm=TRUE)
  list(data=myinput,mean=mymean,sd=mysd)
}
```

You could easily copy this function into a script editor window and change it. You can see and change many R functions this way.

Chapter 9
Data Acquisition

This chapter covers data entry, import and export, emphasizing files in text, SAS and SPSS formats. For a more comprehensive discussion of data acquisition, see the *R Data Import/Export* manual.

9.1 The R Data Editor

R has a simple spreadsheet-style data editor. Unlike SAS and SPSS, *you cannot use it to create a new data frame.* You can only edit an existing one. However, it is easy to create an empty data frame, which you can then fill in using the editor. The following command will create the variables we will use in most other examples.

We can create a data frame named mydata that contains only our first variable, id, with the first observation's value set to zero.

```
mydata <- data.frame (id=0)
```

Since the data editor does not allow you to enter row names, you can create an id variable and make them the row names later. We can start the data editor with the fix function

```
fix (mydata)
```

The window in Fig. 9.1 will appear. Clicking on the variable name "var1" brings up a *Variable editor* window that allows you to change the name of the variable. We will change it to "id" and click the "numeric" button to make it a numeric variable. Then close the variable editor window by clicking the usual X in the upper right corner.

Follow the steps above until you have created this data frame in Fig. 9.2. Make sure to click "character" when defining a character variable. When you come to the NA values for observation 4, leave them blank. You could enter the two-character string, "NA" for numeric variables, but R will not recognize that as a missing value for character variables here. Exit the editor and save changes by choosing *File> Close* or by clicking the Windows X button. There is no *File> Save* option, which feels quite scary the first time you use it, but R does indeed save the data.

R.A. Muenchen, *R for SAS and SPSS Users*, DOI: 10.1007/978-0-387-09418-2_9,   77
© Springer Science+Business Media, LLC 2009

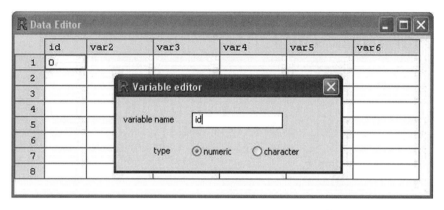

Fig. 9.1 Adding a new variable in the data editor

Before using this data, you would want to use the `factor` function to make workshop and gender into factors. We did that previously in Sect. 8.2.2. In a more realistic example, we would also want to move id into the row names attribute, so that functions that identify observations will do so automatically. See Sect. 14.6 for details.

You now have a data frame that you can analyze, save as a permanent R data file or export in text, SAS or SPSS format.

To save yourself from the trouble of setting the names and variable types, you can enter the command below instead:

```
mydata <- data.frame (id=0., workshop=0.,
gender=" ", q1=0., q2=0., q3=0., 4=0.)
```

	id	workshop	gender	q1	q2	q3	q4
1	1	1	f	1	1	5	1
2	2	2	f	2	1	4	1
3	3	1	f	2	2	4	3
4	4	2	<NA>	3	1	NA	3
5	5	1	m	4	5	2	4
6	6	2	m	5	4	5	5
7	7	1	m	5	3	4	4
8	8	2	m	4	5	5	5

Fig. 9.2 The data editor with our practice data entered

This is a major time saver when you have to create more than one copy of the data, or if you will create a similar dataset in the future.

Note that the `fix` function actually calls the more aptly named `edit` function and then writes the data back to your original data frame. What it is doing for you is:

```
mydata <- edit (mydata)
```

This tells R to edit mydata and when finished, write it back into mydata. I recommend not using the `edit` function on data frames as I find it all too easy to begin editing with just:

```
edit (mydata) #Do NOT do this!
```

It will look identical on the screen, but this does not tell edit where to save your work. When you exit, your work will appear to be lost. However, R stores the last value you gave it in the object .Last.value. So you can retrieve the data with this command.

```
mydata <- .Last.value  #Recovers data if you did "edit
  (mydata)"
```

We will use the `edit` function later when renaming variables.

9.2 Reading Delimited Text Files

Delimited text files use some delimiter such as spaces, tabs, or commas to separate each value. We will read this type of file using the `read.table` function. The default delimiter is one or more tabs, spaces, new lines, or carriage returns. Therefore, if you need to use a blank space as a missing value, you cannot use blanks or tabs as a delimiter. You must use some other delimiter such as commas. We will provide the variable names on the first row, but if you leave them out, R will name them V1, V2... similar to SAS and SPSS's VAR1, VAR2... approach.

Below is the first type of text file we will read (Table 9.1). Its layout is the easiest possible type of text file for R to read. There are three important things to notice about it.

1. ID numbers are in the left-most column, but that column does not have a name on top.
2. Values are separated by any number of tabs and/or spaces.
3. "NA" represents missing values.

Variable names are usually in the first row of a delimited text file and there is usually a variable name for every column. However, if you are lacking *just one* variable name, R assumes you have an ID variable in the first column. It will

Table 9.1 Tab/space delimited
text from practice data file
mydata.tab

	workshop	gender	q1	q2	q3	q4
1	1	f	1	1	5	1
2	2	f	2	1	4	1
3	1	f	2	2	4	3
4	2	NA	3	1	NA	3
5	1	m	4	5	2	4
6	2	m	5	4	5	5
7	1	m	5	3	4	4
8	2	m	4	5	5	5

then automatically assign the values it finds in the first column as the row names. You always want an ID number in your data to help you trace data entry errors back to their source.

When you read a tab-delimited file using SAS or SPSS, two tabs in a row indicate that there was a missing value. But in R, the default delimiter is *any number* of spaces and/or tabs. So trying to represent a missing value by entering two tabs in a row, or by putting a space between two tabs will generate the error message, "line 4 did not have 7 elements" when you try to read it. In the examples that follow, we will consider several ways to get around that problem. The dataset we are using now represents missing values with the string, "NA."

You can then read the file above and print it with the commands:

```
> setwd("/myRfolder")

> mydata <- read.table("mydata.tab")

> mydata
  workshop gender q1 q2 q3 q4
1        1      f  1  1  5  1
2        2      f  2  1  4  1
3        1      f  2  2  4  3
4        2   <NA>  3  1 NA  3
5        1      m  4  5  2  4
6        2      m  5  4  5  5
7        1      m  5  3  4  4
8        2      m  4  5  5  5
```

Given how easy that type of file is to read, it is too bad that few programs export data in this format! See that for observation 4 both gender and q3 have missing values (NA).

Our example program below will read the type of text file that almost any other program can create: a *c*omma *s*eparated *v*alue (CSV) file like the one below (Table 9.2). There are several important things to notice about it:

Table 9. 2 Comma delimited practice data
mydata.csv

```
id,workshop,gender,q1,q2,q3,q4
1,1,f,1,1,5,1
2,2,f,2,1,4,1
3,1,f,2,2,4,3
4,2, ,3,1, ,3
5,1,m,4,5,2,4
6,2,m,5,4,5,5
7,1,m,5,3,4,4
8,2,m,4,5,5,5
```

1. An ID variable name appears on first line.
2. Commas separate the values.
3. One or more blanks represent missing values.

You can read this file using the `read.table` function call below. If you have already set your working directory in your current R session, you do not need to set it again.

```
> setwd("/myRfolder")

> mydata <- read.table ("mydata.csv", header=TRUE,
+    sep=",", row.names="id", na.strings=" ")

> mydata
```

```
  workshop gender q1 q2 q3 q4
1        1      f  1  1  5  1
2        2      f  2  1  4  1
3        1      f  2  2  4  3
4        2   <NA>  3  1 NA  3
5        1      m  4  5  2  4
6        2      m  5  4  5  5
7        1      m  5  3  4  4
8        2      m  4  5  5  5
```

This function call uses five arguments:

1. The `"mydata.csv"` argument is the name of the file to read.
2. The `header = TRUE` argument tells the function that the file has the names in the first row. It is only necessary to use this argument when you do not have an un-named ID variable in the first column. Since we actually have ID named, we have to tell it to read the header.
3. The `sep = ","` argument tells it to use one comma between fields. You can any single character in this argument.
4. The `row.names = "id"` argument tells it that we are reading a variable named "id" and that we want its values to become the row names of the data frame. Using this form, the ID variable does not have to be the first column in the file.

5. The na.strings=" " argument tells it that any blanks it finds should be set to missing values. It will know that blanks in numeric variables must be missing, but blanks in character variables are valid characters unless you include this option.

A similar function, read.csv, assumes commas are delimiters and a header line contains variable names. So it could read the same file with the following statement. Is briefer but less flexible:

```
mydata <- read.csv("mydata.csv",
  row.names="id", na.strings=" ")
```

SAS and SPSS store their datasets on your computer's hard drive. If you read columns of data that you do not need, it is not a problem. You can always drop them later if disk space gets tight. However, R's read.table function must hold all the data in your computer's main memory. This makes skipping columns while reading them particularly important.

Here is the R command to read the data while skipping the fourth and fifth columns by adding the colClasses argument. If you have already set your working directory in your current R session, you do not need to set it again.

```
> setwd("/myRfolder")

> myCols <- read.table("mydata.csv", header=TRUE,
+   sep=",", row.names="id", na.strings=" ",
+   colClasses=c("integer", "integer", "character",
+   "NULL", "NULL", "integer", "integer") )

> myCols
```

	workshop	gender	q3	q4
1	1	f	5	1
2	2	f	4	1
3	1	f	4	3
4	2	<NA>	NA	3
5	1	m	2	4
6	2	m	5	5
7	1	m	4	4
8	2	m	5	5

I used the name myCols to avoid overwriting mydata. You use the col-Classes argument to specify the "class" of each column. The classes include logical (TRUE/FALSE), integer (whole numbers), numeric (can include decimals), character (alphanumeric string values), factor (categorical values like gender). See the help file for other classes like dates. The NULL class is the one we use to skip a column.

However, colClasses requires you to specify the classes of *all* columns, including any initial ID or row names variable. The classes must be included within quotes since they are character strings. The colClasses argument is also helpful for reading other variable types such as dates. For example programs that demonstrate these topics in SAS, SPSS and R, see Table 9.3.

Table 9.3 Example programs to read delimited text files

SAS programming statements	```* SAS Program to Read Delimited Text Files;```
	```* ReadDelimited.sas;```
	
	```LIBNAME myLib 'C:\myRfolder ';```
	
	```DATA myLib.mydata;```
	```INFILE '\myRfolder \mydata.csv' delimiter = ','```
	``` MISSOVER DSD lrecl=32767 firstobs=2 ;```
	```INPUT id workshop gender$q1 q2 q3 q4;```
	
	```PROC PRINT;```
	```RUN;```

SPSS programming statements	```* SPSS Program to Read Delimited Text Files.```
	```* ReadDelimited.sps.```
	
	```CD 'C: \myRfolder'.```
	```GET DATA /TYPE = TXT```
	``` /FILE = 'mydata.csv'```
	``` /DELCASE = LINE```
	``` /DELIMITERS = ","```
	``` /ARRANGEMENT = DELIMITED```
	``` /FIRSTCASE = 2```
	``` /IMPORTCASE = ALL```
	``` /VARIABLES = id F2.1 workshop F1.0 gender A1.0```
	``` q1 F1.0 q2 F1.0 q3 F1.0 q4 F1.0 .```
	```LIST.```
	```SAVE OUTFILE='C:\myRfolder\mydata.sav'.```

R programming statements	```# R Program to Read Delimited Text Files.```
	```# ReadDelimited.R```
	
	```setwd("/myRfolder")```
	
	```# Read a tab delimited file with un-named ID column.```
	```mydata <- read.table("mydata.tab")```
	```mydata```

Table 9.3 (continued)

```
# Read a Comma Separated Value (CSV) file.
mydata <- read.table("mydata.csv", header=TRUE,
  sep=",", row.names=" id", na.strings=" ")
mydata
# Read it again with read.csv.
mydata <- read.csv("mydata.csv",
  row.names=" id", na.strings=" ")
mydata

# Now use colClasses to skip q1 and q2 with NULL.
myCols <- read.table("mydata.csv", header=TRUE,
  sep=",", row.names="id", na.strings=" ",
  colClasses=c("integer", "integer", "character",
  "NULL", "NULL", "integer", "integer") )
myCols

# Clean up and save workspace.
rm(myCols)
save.image(file=" myWorkspace.RData")
```

9.3 Reading Text Data Within a Program (Datalines, Cards, Begin Data…)

Now that we have seen how to read a text file in the section above, we can more easily understand how to read data within a program. R works by putting the unprocessed text data into objects and then processing those objects with functions. In this case, we will put the data into a character vector, named "mystring."

```
mystring <-
"id,workshop,gender,q1,q2,q3,q4
 1,1,f,1,1,5,1
 2,2,f,2,1,4,1
 3,1,f,2,2,4,3
 4,2, ,3,1, ,3
 5,1,m,4,5,2,4
 6,2,m,5,4,5,5
 7,1,m,5,3,4,4
 8,2,m,4,5,5,5"
```

Mystring has only one very long value. Then we can read it with a call to the read.table function if we nest a call to the textConnection function within it.

```
mydata <- read.table( textConnection(mystring),
```

```
header=TRUE, sep=",",
row.names="id", na.strings=" ")
```

The textConnection function converts mystring into the equivalent of a file, which R then processes the same way the file was. The shorter read.csv program works on this data too.

```
mydata <- read.csv( textConnection(mystring),
    row.names="id", na.strings="")
```

For example programs that demonstrate these topics in SAS, SPSS and R, see Table 9.4.

Table 9. 4 Example programs to read text data within a program

SAS programming statements	```* SAS Program to Read Data Within a Program; * ReadWithin.sas ; LIBNAME myLib 'C:\myRfolder'; DATA myLib.mydata; INFILE DATALINES DELIMITER = ','; MISSOVER DSD firstobs=2 ; INPUT id workshop gender$q1 q2 q3 q4; DATALINES; id,workshop,gender,q1,q2,q3,q4 1,1,f,1,1,5,1 2,2,f,2,1,4,1 3,1,f,2,2,4,3 4,2, ,3,1, ,3 5,1,m,4,5,2,4 6,2,m,5,4,5,5 7,1,m,5,3,4,4 8,2,m,4,5,5,5 PROC PRINT; RUN;```
SPSS programming statements	```* SPSS Program to Read Data Within a Program. * ReadWithin.sps . DATA LIST / id 2 workshop 4 gender 6 (A) q1 8 q2 10 q3 12 q4 14. BEGIN DATA. 1,1,f,1,1,5,1 2,2,f,2,1,4,1 3,1,f,2,2,4,3 4,2, ,3,1, ,3 5,1,m,4,5,2,4 6,2,m,5,4,5,5 7,1,m,5,3,4,4 8,2,m,4,5,5,5 END DATA.```

Table 9. 4 (continued)

	LIST. SAVE OUTFILE=';C:\myRfolder\mydata.sav';.
R programming statements	

```
# R Program to Read Data Within a
Program.
#ReadWithin.R.
# This stores the data as one long text
string.
mystring <-
"id,workshop,gender,q1,q2,q3,q4
 1,1,f,1,1,5,1
 2,2,f,2,1,4,1
 3,1,f,2,2,4,3
 4,2, ,3,1, ,3
 5,1,m,4,5,2,4
 6,2,m,5,4,5,5
 7,1,m,5,3,4,4
 8,2,m,4,5,5,5"

# Read with more flexible read.table.
mydata <- read.table( textConnection(mystring),
    header=TRUE, sep=",",
    row.names="id", na.strings=" ")
mydata

# Read again with shorter read.csv.
mydata <- read.csv( textConnection(mystring),
    row.names="id", na.strings="")
mydata

# Set working directory & save workspace.
setwd ("/myRfolder")

save.image(file="myWorkspace.RData")
```

9.4 Reading Data from the Keyboard

If you want to enter data from the keyboard line-by-line using SAS or SPSS, you would do so as we did in the previous Sect. 9.3. They do not have a special data entry mode in their program editors (of course, you would probably use their data editors instead). You can put R's scan into a special data entry mode by providing it no arguments. It then prompts you for data one line at a time, but once you hit the Enter key you cannot go back and change it in that mode. I rarely use this approach, but it is quick for small variables. Here is an example.

```
> id <- scan()

1: 1 2 3 4 5 6 7 8
9:
Read 8 items
```

R prompts with "1:" indicating that you can type the first observation. When I entered the first line (just the digits 1 through 8), it prompted with "9:" indicating that I had already entered 8 values. When I entered a blank line, scan stopped reading and saved the vector named id.

To enter character data, we have to add the what argument. Since spaces are separating the values, to enter a value that includes a space you would enclose it in quotes like "R.A. Fisher".

```
> gender <- scan(what="character")

1: f f f NA m m m m
9:
Read 8 items
```

When finished with this approach, we could use the data.frame function to combine the vectors into a data frame:

```
mydata <- data.frame(id,workshop,gender,q1,q2,q3,q4)
```

9.5 Reading Fixed-Width Text Files, One Record per Case

Files that separate data values with delimiters such as spaces or commas are convenient for people to work with, but they make a file larger. So many text files dispense with such conveniences and instead keep variable values locked into the exact same column(s) of every record.

If you have a non-delimited text file with one record per case, you can read it using the following approach. R has nowhere near the flexibility in reading fixed-width text files that SAS and SPSS have. As you will soon see, making an error specifying the width of one variable will result in reading the wrong columns for all those that follow. While SAS and SPSS offer approaches that would do that too, I do not recommend their use. Other Open Source languages such as Perl or Python are extremely good at reading text files and converting them to a form that R can easily read.

Below is the same data that we used in other examples but now it is in fixed-width format (Table 9.5). The important things to notice about this file are:

1. No names appear on first line.
2. Nothing separates values.

Table 9.5 Fixed-width practice data file mydataFWF.txt

```
11f1151
22f2141
31f2243
42 31 3
51m4524
62m5455
71m5344
82m4555
```

3. The first value of each record appears in column 2 just to give us a 2-column value to read.
4. Blanks represent missing values, but we could use any other character that would fit into the fixed number of columns.
5. The last line of the file contains data. That is what SAS and SPSS expect, but R generates a warning that there is an "incomplete final line found". It works fine though. If the warning in R bothers you, simply edit the file and press Enter once at the end of the last line.

The R function that reads fixed-width files is read.fwf. Here is an example of it reading the data in Table 9.5.

```
> setwd("/myRfolder")
> mydata <- read.fwf(
+     file="mydataFWF.txt",
+     width=c(2,-1,1,1,1,1,1),
+     col.names=c("id", "gender", "q1", "q2", "q3",
      "q4"),
+     row.names="id",
+     na.strings="",
+     fill=TRUE,
+     strip.white=TRUE)

Warning message:
In readLines(file, n = thisblock) :
    incomplete final line found on 'mydataFWF.txt'

> mydata

    gender  q1 q2 q3 q4
1        f   1  1  5  1
2        f   2  1  4  1
3        f   2  2  4  3
4     <NA>   3  1 NA  3
5        m   3  5  2  4
6        m   5  4  5  5
```

```
7          m    5    3    4    4
8          m    4    5    5    5
```

The read.fwf function uses seven arguments:

1. The file argument lists the name of the file. It will read it from your current working directory. You can set the working directory with setwd ("path") or you can specify a path as part of the file specification.
2. The width argument provides the width, or number of columns, required by each variable *in order*. The widths we supplied as a numeric vector are created using the c function. The first number, 2 tells R to read ID from columns 1 and 2. The next number, −1, tells R to skip one column. In our next example, we will not need to read the workshop variable, so we will put in a −1 to skip it now. The remaining pattern of 1,1,1,1,1 tells R that each of the remaining five variables will require one column each. Be careful at this step. If you made an error and told R that ID was 1 column wide, then read.fwf would read all the other variables from the wrong columns. For quick ways to generate patterns of column widths, see Chap. 17.
3. The col.names argument provides the column or variable names. Those we provide in a character vector. We create it using the c function, c ("id","gender","q1","q2","q3","q4"). Since the names are character (string) data, we must enclose them in quotes.
4. The row.names argument tells R that we have a variable that stores a name or identifier for each row. It also tells it which of the variable names from the col.names argument that is, "id".
5. The na.strings argument tells R that a blank is a missing value. It already is for numeric data but, as in SAS or SPSS, a blank is a valid character value. Note that there is *no* blank between the quotes! That is because we set the strip.white option to strip out extra blanks from the end of strings (below). As you see, R displays missing data for character data within angle brackets as <NA>.
6. The fill argument tells R to fill in blank spaces if the file contains lines that are not of the full length (like the MISSOVER SAS option). Now is a good time to stop and enter help (read.fwf). Note that the help file offers no fill argument. It does however list its last argument as "...." This means that it will pass any additional arguments onto another function that read.fwf might call. In this case, it is the read.table function. Clicking the link to that function will reveal the fill argument and what it does.
7. The strip.white argument tells R to remove any additional blanks it finds in character data values. Therefore, if we were reading a long text string like "Bob   ", it will delete the additional spaces and store just "Bob". That saves space and makes logical comparisons easier.

The file was read just fine. The warning message about an "incomplete final line" is caused by an additional line feed character at the end of the last line of the file. Neither SAS nor SPSS would print a warning about such a condition.

9.5.1 Macro Substitution

The example above is a good one to use to learn what SAS or SPSS would call macro substitution. This approach makes your programs much easier to write and maintain. The most interesting aspect to "macro substitution" is that unlike SAS or SPSS, there is no separate macro language. R is powerful enough to do this using its standard features.

Since file paths often get quite long, we will store it in a character vector named myfile. This approach also lets you put all the file references you use at the top of your programs, so you can change them easily. We do this with the command:

```
myfile <-  "mydataFWF.txt"
```

Next, we will store our variable names in another character vector, myVariableNames. This makes it much easier to manage when you have a more realistic dataset that may contain hundreds of variables:

```
myVariableNames <- c("id", "gender", "q1", "q2", "q3",
  "q4")
```

Now we will do the same with our variable widths. This makes our next example, which reads multiple records per case, much easier.

```
myVariableWidths <- c(2, -1, 1, 1, 1, 1, 1)
```

Now we will put it all together in a call to the read.fwf function.

```
> mydata <- read.fwf(
+     file=myfile,
+     width=myVariableWidths,
+     col.names=myVariableNames,
+     row.names="id",
+     na.strings="",
+     fill=TRUE,
+     strip.white=TRUE)

Warning message:
In readLines(file, n = thisblock) :
   incomplete final line found on 'mydataFWF.txt'
```

For example programs that demonstrate these topics in SAS, SPSS and R, see Table 9.6.

Table 9.6 Example programs to read fixed-width text files with one record per case. These do not save the data as they skip workshop for demonstration purposes

SAS programming statements	

```
* SAS Program for Reading a Fixed-Width Text
File,
* 1 Record per Case;
* ReadFWF1.sas ;

LIBNAME myLib 'C:\myRfolder';
DATA myLib.mydata;
INFILE '\myRfolder\mydataFWF.txt' MISSOVER;
INPUT id 1-2 gender$4
  q1 5   q2 6   q3 7   q4 8;
RUN;
```

SPSS programming statements

```
* SPSS Program for Reading a Fixed-Width Text
File,
* 1 Record per Case.
* ReadFWF1.sps .

CD 'C:\myRfolder'.

DATA LIST FILE=';mydataFWF.txt' RECORDS=1
 /1 id 1-2 gender 4 (A)   q1 5   q2 6
q3 7   q4 8.
LIST.
```

R programming statements

```
# R Program for Reading a Fixed-Width Text
File,
# 1 Record per Case.
# ReadFWF1.R

setwd("/myRfolder")
mydata <- read.fwf(
   file="mydataFWF.txt",
   width=c(2,-1,1,1,1,1,1),

col.names=c("id","gender","q1","q2","q3","q4"),
   row.names="id"
na.strings="",
   fill=TRUE,
   strip.white=TRUE)
mydata

# Now we'll use "macro" substitution to do the
same thing.

myfile <- "mydataFWF.txt"
myVariableNames   <-
c("id","gender","q1","q2","q3","q4")
```

Table 9.6 (continued)

```
myVariableWidths <- c(2,-1,1,1,1,1,1)
mydata <- read.fwf(
file=myfile,
width=myVariableWidths,
col.names=myVariableNames,
row.names="id",
na.strings="",
fill=TRUE,
strip.white=TRUE)
mydata
```

9.6 Reading Fixed-Width Text Files, Two or More Records per Case

This section builds upon the example above. We will only use the macro substitution form in this example.

First, we will store the filename in the character vector named myfile:

```
myfile <- "/mydataFWF.txt"
```

Next, we will store the variable names in another character vector. We will pretend that our same file now has two records per case with q1 to q4 on the first record and q5 to q8 in the same columns on the second. Even though id, group, and gender appear on every line, we will not read them again from the second line. Here are our variable names:

```
myVariableNames <- c("id", "group", "gender",
  "q1", "q2", "q3", "q4 ",
  "q5", "q6", "q7", "q8" )
```

Now we need to specify the columns to read. We must store the column widths for each line of data (per case) in its own vector. Note that on record 2 we begin with $-2, -1, -1$ to skip the values for id, group, and gender.

```
myRecord1Widths <- c( 2, 1, 1, 1, 1, 1, 1)
myRecord2Widths <- c(-2,-1,-1, 1, 1, 1, 1)
```

Next, we need to store both of the above variables in a list. The list function below combines the two record width vectors into one list named myVariableWidths:

```
myVariableWidths <- list( myRecord1Widths, myRecord
  2Widths )
```

Let us look at the new list:

```
> myVariableWidths
```

```
[[1]]
[1] 2 1 1 1 1 1 1

[[2]]
[1] -2 -1 -1  1  1  1  1
```

You can see that the entry [[1]] is the first numeric vector and [[2]] is the second. The fact that there are two entries tells the width argument we have two records per case. Now we are ready to use the read.fwf function to read the data file:

```
> mydata <- read.fwf(
+     file=myfile,
+     width=myVariableWidths,
+     col.names=myVariableNames,
+     row.names="id",
+     na.strings="",
+     fill=TRUE,
+     strip.white=TRUE)

Warning message:
incomplete final line found by readLines on
 'mydataFWF.txt' in: readLines(file, n = thisblock)

> mydata
  group gender q1 q2 q3 q4 q5 q6 q7 q8
1     1      f  1  1  5  1  2  1  4  1
3     1      f  2  2  4  3  3  1 NA  3
5     1      m  3  5  2  4  5  4  5  5
7     1      m  5  3  4  4  4  5  5  5
```

You can see we now have only four records and eight q variables, so it has worked well. It is also finally obvious that the row names do not always come out as simple sequential numbers. It just so happened that is what we have had until now. The warning message is caused by the lack of an additional line feed character at the end of the last line of the file. That does not cause problems. For example programs that demonstrate these topics in SAS, SPSS and R, see Table 9.7.

Table 9.7 Example programs to read fixed-width text files with two records per case

| SAS programming statement | ```
* SAS Program for Reading Fixed Width
Text Files,                .
* 2 Records per Case;
* ReadFWF2.sas ;

DATA temp;
INFILE '\myRfolder\mydataFWF.txt'
MISSOVER;
``` |
|---|---|
```

Table 9.7 (continued)

```
INPUT
 #1 id 1-2 workshop 3 gender 4
    q1 5 q2 6  q3 7  q4 8
 #2 q5 5 q6 6  q7 7  q8 8;


PROC PRINT;
RUN;
```

SPSS programming
statement

```
* SPSS Program for Reading Fixed Width Text Files,
* 2 Records per Case.
* ReadFWF2.sps .

DATA LIST FILE='\myRfolder\mydataFWF.txt' RECORDS=?
  /1 id 1-2 workshop 3 gender 4 (A)
       q1 5 q2 6 q3 7 q4 8
  /2 q5 5 q6 6 q7 7 q8 8.
LIST.
```

R programming
statement

```
# R Program for Reading Fixed Width
Text Files,
# 2 Records per Case.
# ReadFWF2.R
setwd("/myRfolder")

# Set all the values to use.
myfile <- "mydataFWF.txt"
myVariableNames  <- c("id","workshop",
"gender",
  "q1","q2","q3","q4",
  "q5","q6","q7","q8")
myRecord1Widths  <- c( 2, 1, 1, 1, 1,
1, 1)
myRecord2Widths  <- c(-2,-1,-1, 1, 1,
1, 1)
myVariableWidths <-
list(myRecord1Widths,myRecord2Widths)

# Now plug them in and read the data:
mydata <- read.fwf(
   file=myfile,
   width=myVariableWidths,
   col.names=myVariableNames,
   row.names="id",
   na.strings="",
   fill=TRUE,
   strip.white=TRUE )
mydata
```

9.7 Importing Data from SAS

R can read a SAS dataset in xport format and, if you have SAS installed,
directly from a regular SAS dataset with the extension sas7bdat. Although
the foreign package is the most widely documented approach, it lacks important capabilities. Functions in the optional Hmisc package add the ability to
read formatted values, variable labels, and lengths.

SAS users rarely use the LENGTH statement, accepting the default storage
method of double precision. This wastes a bit of disk space but saves programmer time. However, since R saves all its data in memory, space limitations are far more important. If you use SAS' LENGTH statement to save space,
the sasxport.get function in Hmisc will take advantage of it. However,
unless you know a lot about how computers store data, it is probably best to
only shorten the length used to store integers. The Hmisc package does not
come with R but it is easy to install. For instructions, see Sect. 5.1.

The example below loads the two packages we need and then translates the
data.

```
library("foreign")
library("Hmisc")
mydata <- sasxport.get("/mydata.xpt")
```

Table 9.8 Example programs to export data from SAS and import it into R

SAS Export	

	```* SAS Program to Create Export Format File.```
	```* ExportToR.sas ;```
	
	```LIBNAME myLib 'C:\myRfolder';```
	```LIBNAME To_R xport '\myRfolder\mydata.xpt';```
	
	```DATA To_R.mydata;```
	``` SET myLib.mydata;```
	``` RUN;```

R Import	

	```# R Program to Read a SAS Export File```
	```# ImportFromSAS.R```
	
	```setwd("/myRfolder")```
	
	```library("foreign")```
	```library("Hmisc")```
	
	```mydata <- sasxport.get ("mydata.xpt")```
	```mydata```

Unlike most of our example programs, the SAS and R code here do opposite things rather
than the same thing.

The example in Table 9.8 assumes you have a SAS xport format file. For much more information on reading SAS files, see *An Introduction to S and the Hmisc and Design Libraries* at http://cran.r-project.org/doc/contrib/Alzo-latHarrell-Hmisc-Design-Intro.pdf.

Warning: In most other example programs, the SAS and R code do the same thing. However in Table 9.8, the SAS program creates a file for R to read.

9.8 Importing Data from SPSS

If you have SPSS 16 or later, the best way to import data into R from SPSS is by using the SPSS Statistics-R Integration Plug-in. That includes support of recent features such as long variable names. For an example, see Section 6.6.

If you do not have SPSS, the best way to import a data file from SPSS is with the Hmisc package. For instructions on installing Hmisc, see Sect. 5.1. You can read an SPSS portable file using the spss.get function.

```
library("Hmisc")
mydata <- spss.get("mydata.por",use.value.labels = TRUE)
```

Table 9.9 Example programs to export data from SPSS and import it into R

SPSS programming statements	Export
	```* SPSS Program to Create Export Format File.``` ```* ExportToR.sps .``` ```CD 'C:\myRfolder'.``` ```GET FILE='mydata.sav'.``` ```EXPORT OUTFILE='C:\myRfolder\mydata.por'.```
R programming statements	Import
	```# R Program to Import an SPSS Data File.``` ```# ImportFromSPSS.R``` ```library("Hmisc")``` ```setwd("/myRfolder")``` ```# Read & print the SPSS file.``` ```mydata<-``` ```spss.get("mydata.por",use.value.labels=TRUE)``` ```mydata``` ```# Save the workspace.``` ```save.image(file="myWorkspace.RData")```

Unlike most of our example programs, the SPSS and R code here do opposite things rather than the same thing.

The use.value.labels argument tells R to convert any SPSS variables that use value labels to R factors. That keeps the labels but essentially turns them into categorical variables. This is the default value so I list it here only to point out its importance. Setting it to FALSE will leave your variables as numeric, allowing you to calculate means and standard deviations more easily. SPSS users often have Likert scale 1 through 5 items stored as scale variables (numeric vectors in R) and have labels assigned to them. Read Chap. 15 for more details about factors.

Other useful arguments include lowernames = TRUE to convert all names to lower case and datevars to tell R about date variables to convert. After you have loaded the Hmisc package, you can use help(spss.get) for more information. For example programs that demonstrate these topics in SAS, SPSS and R, see Table 9.9.

9.9 Exporting Data

Now let us use our example text file above to export data. The first example uses the default write.table function to create a text file with row and column names, and with values separated by tabs. Tabs have an advantage over commas as some countries use commas as decimal points, and exported text strings often contain commas. Enter help(write.table) for output options.

```
write.table(mydata,
   file="mydata2.txt",
   quote=FALSE,
   sep="\t",
   na=" ",
   row.names=TRUE,
   col.names=TRUE)
```

This example uses seven arguments:

1. The name of the R data frame to export.
2. The file argument names the output text data file. R will write it to the working directory.
3. The quote = FALSE argument tells R not to write quotes around character data like "m" and "f." By default, it will write the quotes.
4. The sep="\t" tells that the separator (delimiter) to use between values is the tab.
5. The na=" " argument tells R to use blanks to represent missing values. By default, it will write out "NA" instead. That is what you want only if you plan to read the data back into R. Few other packages recognize that as a code for missing values. SAS and SPSS will convert it to missing but

they will generate a lot of irritating messages, so it is probably best to use blank.

6. The `row.names = TRUE` argument tells R to write row names in the first column of the file. In other words, it will write out an ID-type variable. This is the default value, so you do not actually need to list it here. If you do not want it to write row names, then you must use `row.names = FALSE`.

7. The `col.names = TRUE` argument tells R to write variable names in the first row of the file. This is the default value, so you do not actually need to list it here. If you do not want it to write row names, then you must use `col.names = FALSE`. Unlike most programs, R will not write out a name for an ID variable.

The second two examples use the `write.foreign` function to write out a comma delimited text file along with either a SAS or SPSS program file. To complete the importation into SAS or SPSS, you must edit the program file in SAS or SPSS and then execute it to read the text file and finally create a dataset.

To begin the process, you must load the `foreign` package.

```
library("foreign")
write.foreign(mydata,
    datafile="mydata2.txt",
    codefile="mydata.sas",
    package="SAS")
```

This uses four arguments:

1. The name of the R data frame you wish to export.
2. The datafile argument tells R the name of text data file. R will write it to the current working directory unless you specify the full path in the filename.
3. The codefile argument tells R the filename of a program that SAS or SPSS can use to read the text data file. You will have to use this file in SAS or SPSS to read the data file and create a SAS- or SPSS-formatted file. R will write it to the current working directory unless you specify the full path in the filename.
4. The package argument takes the values `"SAS"` or `"SPSS"` to determine which type of program R writes to the `codefile` location.

Note that these two examples write out the gender values as 1 and 2 for f and m, respectively. See Chap. 15 if you want to change those values.

9.9.1 Viewing an External Text File

All operating systems have ways to look at text files. Sometimes it is helpful to see one under program control. To look at the contents of any text file in R, you can use the `file.show` function.

```
>file.show("mydata2.txt")
```

Fig. 9.3 Output window from `file.show` function

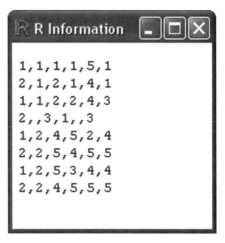

R Information

```
1,1,1,1,5,1
2,1,2,1,4,1
1,1,2,2,4,3
2,,3,1,,3
1,2,4,5,2,4
2,2,5,4,5,5
1,2,5,3,4,4
2,2,4,5,5,5
```

On Windows or Macintosh, it will open a read-only window like the one in Fig. 9.3. To end the command, close the window by clicking the X in the upper right corner.

Table 9.10 Example programs to write a text file from R for use in any program. Also, to write text files and matching SAS and SPSS programs to read them

R Export to Text	```# R Program to Write a Text File.```
	```# WriteToText.R```
	
	```setwd("/myRfolder")```
	
	```write.table(mydata,```
	``` file="mydata2.txt",```
	``` quote=FALSE,```
	``` sep="\t",```
	``` na="",```
	``` row.names=TRUE,```
	``` col.names=TRUE)```
	
	```# Look at the contents of our new file.```
	```file.show("mydata2.txt")```

R Export to SAS	```# R Program to Write a SAS Export File```
	```# and a program to read it into SAS.```
	```# ExportToSAS.R```
	
	```setwd("/myRfolder")```
	```library("foreign")```
	
	```write.foreign(mydata,```
	``` datafile="mydata2.txt",```
	``` codefile="mydata2.sas", package="SAS")```

Table 9.10 (continued)

```
# Look at the contents of our new files.
file.show("mydata2.txt")
file.show("mydata2.sas")
```

R Export
to SPSS

```
# R Program to Write an SPSS Export File
# and a program to read it into SPSS.
# ExportToSPSS.R
setwd("/myRfolder")
library("foreign ")

write.foreign (mydata,
 datafile="mydata2.txt ",
 codefile="mydata2.sps",
 package="SPSS")

# Look at the contents of our new files.
file.show("mydata2.txt")
file.show("mydata2.sps")
```

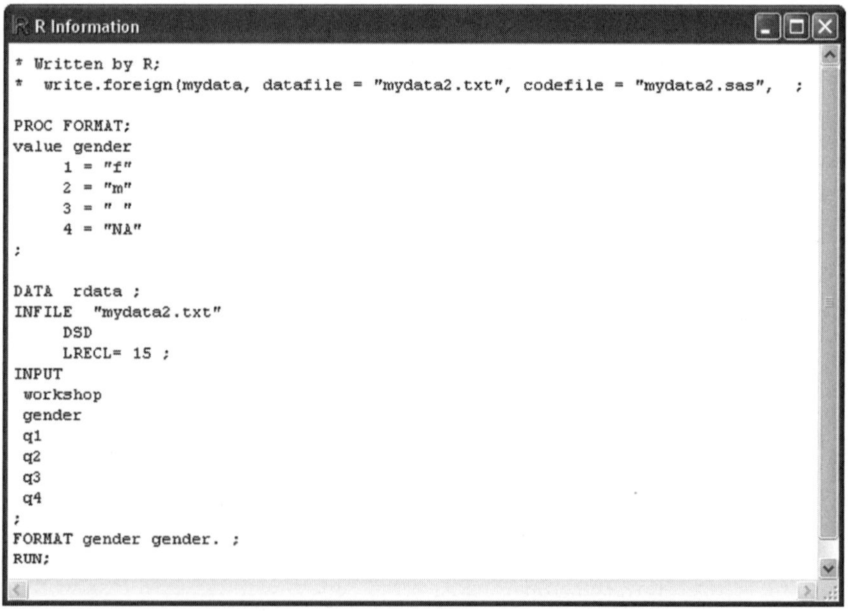

```
R Information                                                    _ □ ✕

* Written by R;
*  write.foreign(mydata, datafile = "mydata2.txt", codefile = "mydata2.sas",  ;

PROC FORMAT;
value gender
      1 = "f"
      2 = "m"
      3 = " "
      4 = "NA"
;

DATA  rdata ;
INFILE   "mydata2.txt"
      DSD
      LRECL= 15 ;
INPUT
 workshop
 gender
 q1
 q2
 q3
 q4
;
FORMAT gender gender. ;
RUN;
```

Fig. 9.4 Output window from file.show ("mydata2.sas")

On Linux/UNIX it will display the file in the R console.

Similarly, you can view the SAS program R created with the following command. The output is given in Fig. 9.4

```
>file.show("mydata2.sas"):
```

The example programs that demonstrate theses concepts, including the file.show function, are in Table 9.10

Chapter 10
Selecting Variables – Var, Variables =

In SAS and SPSS, selecting *variables* for an analysis is simple while selecting *observations* is often much more complicated. In R, these two processes can be almost identical. As a result, variable selection in R is both more flexible and quite a bit more complex. However, since you need to learn that complexity to select observations, it is not much added effort.

Selecting *observations* in SAS or SPSS requires the use of logical conditions with commands like IF, WHERE, SELECT IF, or FILTER. You do not usually use that logic to select *variables*. It is possible to do so, through the use of macros or, in the case of SPSS, Python, but it is not a standard approach.

If you have used SAS or SPSS for long, you probably know dozens of ways to select observations, but you did not see them all in the first introductory guide you read. With R, it is best to dive in and see all the methods of selecting variables because understanding them is the key to understanding other documentation, especially the help files and discussions on the R-help mailing list.

Even though you select variables and observations in R using almost identical methods, we will discuss them in two different sections, with different example programs. This chapter focuses only on selecting variables.

10.1 Selecting Variables in SAS and SPSS

Selecting *variables* in SAS or SPSS is quite simple. It is worth reviewing them now before discussing R's approach. Our example dataset contains the variables: `workshop`, `gender`, `q1`, `q2`, `q3`, `q4`. SAS lets you refer to them by individual name or in contiguous order separated by double dashes "--"as in

```
PROC MEANS DATA=myLib.mydata; VAR workshop--q4;
```

SAS also uses a single dash "–" to request variables that share a numeric suffix, even if they are not next to each other in the dataset:

```
PROC MEANS DATA=myLib.mydata; VAR q1-q4;
```

R.A. Muenchen, *R for SAS and SPSS Users*, DOI: 10.1007/978-0-387-09418-2_10,   103
© Springer Science+Business Media, LLC 2008

You can select any variable beginning with the letter "q" using the colon operator.

```
PROC MEANS DATA=myLib.mydata; VAR q: ;
```

Finally, if you do not tell it which variables to use, SAS uses them all.

SPSS allows you to list variable names individually or with contiguous variables separated by "TO," as in

```
DESCRIPTIVES VARIABLES=gender to q4.
```

If you want SPSS to analyze all variables in the dataset, you use the keyword ALL

```
DESCRIPTIVES VARIABLES=ALL.
```

Now let us turn our attention to how R selects variables. For our examples, we will use the summary function.

10.2 Selecting All Variables

In R, if you perform an analysis without selecting any variables, the function will use all the variables if it can. That is essentially how SAS works. For example, to get summary statistics on all variables (across all observations or rows), use

```
summary (mydata)
```

10.3 Selecting Variables by Index Number

Coming from SAS or SPSS, you would think a discussion of selecting variables in R would begin with various ways to select variables using their names. R can use variable names of course, but column index numbers are more fundamental to the way R works.

Our data frame has two dimensions, rows and columns. We reference the rows and columns using square brackets as

```
mydata [rows,columns]
```

If you leave out the row or column indexes, R will process all rows and all columns. Therefore, the following three statements have the same result:

```
summary( mydata )
summary( mydata[ ] )
summary( mydata[ , ] )
```

This section focuses on the second parameter, the columns (variables).

Our data frame has six variables or columns, which are automatically given index numbers, or indexes of 1, 2, 3, 4, 5, and 6. You can select variables by supplying *one index number or a vector of indexes (also called indices)*. For example:

```
summary( mydata[ ,3] )
```

selects all rows of the third variable or column, q1. If you leave out an index, it will assume you want them all. If you leave the comma out completely, R assumes you want a column, so

```
summary( mydata[3] )
```

is almost the same as

```
summary( mydata[ ,3] )
```

Both refer to our third variable, q1. While the summary function treats the presence or absence of the comma the same, some functions will have problems. That is because *with the comma*, the variable selection passes a numeric *vector* and *without the comma*, it passes a *data frame containing only one vector*. See Chap. 13 for details.

To select more than one variable using indexes, you combine the indexes into a vector using the c function. Therefore, this will analyze variables 3 through 6.

```
summary( mydata[ c(3,4,5,6) ] )
```

You will see the c function used in many ways in R. Whenever R requires one object and you need to supply it several, it combines the several into one. In this case, the several index numbers become a single numeric vector.

The colon operator ":" can generate a numeric vector directly, so

```
summary( mydata[3:6] )
```

will use the same variables. Unlike SAS'

```
workshop --q4
```

or SPSS'

```
workshop to q4
```

The colon operator is not just shorthand. We saw in an earlier chapter that entering 1:N at the R console will cause it to generate the sequence, 1,2,3, ...N.

If you use a negative sign on an index, you will *exclude* those columns. For example:

```
summary( mydata[ -c(3,4,5,6) ] )
```

will analyze all variables *except* for variables 3, 4, 5, and 6. Your index values must be either all positive or all negative. Otherwise, the result would be illogical. You cannot say, "include *only* these" and "include all *but* these" at the same time. Index values of zero are accepted, but ignored.

The colon operator can abbreviate sequences of numbers, but you need to be careful with negative numbers. If you want to exclude columns 3:6, the following approach will not work:

```
> -3:6
[1] -3 -2 -1 0 1 2 3 4 5 6
```

This would of course generate an error since you cannot exclude 3 and include 3 at the same time. Adding parentheses will clarify the situation, showing R that you want the minus sign to apply to the just the set of numbers from –3 through –6 rather than –3 through +6:

```
> -(3:6)
[1] -3 -4 -5 -6
```

Therefore, we can exclude variables 3 through 6 with:

```
summary(  mydata[ -(3:6) ] )
```

If you find yourself working with a set of variables repeatedly, you can easily save a vector of indexes so you will not have to keep looking up index numbers:

```
myQindexes <- c(3,4,5,6)
summary( mydata[myQindexes] )
```

You can list indexes individually or, for contiguous variables, use the colon operator. For a big dataset, you could use variables 1, 3, 5 through 20, 25, and 30 through 100 like this:

```
myindexes <- c(1,3,5:20,25,30:100)
```

This is an important advantage of this method of selecting variables. Most of the other variable selection methods do not easily allow you to select mixed sets of contiguous and non-contiguous variables as you are used to doing in either SAS or SPSS. For another way to do this, see Sect. 10.9.

If your variables follow patterns such as every other variable or every tenth, see Chap. 17 for ways to generate other sequences of index numbers.

A convenient way to have R list the index for each variable in a data frame is to extract the names using the names function and the use the data.frame function to list it along with the default row names of 1, 2, 3...

```
> data.frame( names(mydata) )

  names.mydata.
1      workshop
2        gender
3            q1
4            q2
```

```
5                    q3
6                    q4
```

It is easy to rearrange the variables to put the four q variables in the beginning of the data frame. That way, you will easily remember, for example, that q3 is index 3 and so on. Storing them in a separate data frame is another way to make indexes easy to remember for sequentially numbered variables like these. However, that approach runs into problems if you sort one data frame as the rows then no longer match up in a sensible way. Correlations between the two sets would be meaningless.

The ncol function will tell you the number of columns in a data frame. Therefore, another way to analyze all your variables is

```
summary( mydata[ 1:ncol(mydata)  ]  )
```

If you remember that q1 is the third variable and you want to analyze all the variables from there to the end, you can use

```
summary( mydata[ 3:ncol(mydata)  ]  )
```

10.4 Selecting Variables by Column Name

Variables in SAS and SPSS are required to have names and those names must be unique. In R, you do not need them since you can refer to variables by index number as described in the previous section. Amazingly enough, the names do not have to be unique, although having two variables with the same name would *not* be a good idea! R data frames usually include variable names, as does our example data: workshop, gender, q1, q2, q3, q4.

Both SAS and SPSS store their variable names within their datasets. However, you do not know exactly *where* they reside within the dataset. Their location is irrelevant. They are in there somewhere, and that is all you need to know. However, in R, they are stored within a data frame in place called the *names attribute*. The names function accesses that attribute, and you can display them by entering:

```
> names(mydata)

 [1] "workshop"  "gender"   "q1"    "q2"    "q3"    "q4"
```

To select a column by *name*, you put it in quotes, as in

```
summary( mydata["q1"] )
```

R is still using the form

```
mydata[row,column]
```

However, when you supply only one index value, it assumes it is the column. So

```
summary( mydata[ ,"q1"] )
```

works as well. Note the addition of the comma before the variable name is the only difference between the two examples of summary above. While the summary function treats the presence or absence of the comma the same, some functions will have problems. That is because *with the comma* the selection results in a numeric *vector* and *without the comma* the selection is a *data frame containing only that vector*. See Chap. 13 for details.

If you have more than one name, combine them into a single character vector using the c function. For example,

```
summary( mydata[ c("q1","q2","q3","q4") ] )
```

Unfortunately, the colon operator does not work directly with character prefixes as it does with indexes. So the form q1:q4 does not work in this context. However, you can paste the letter "q" onto the numbers you generate using the paste function.

```
myQnames <- paste( "q", 1:4, sep="")
summary( mydata[myQnames] )
```

The paste function call above has three arguments:

1. The string to paste, which for this example is just the letter "q."
2. The object to paste it to, which is the numeric vector 1, 2, 3, 4 generated by the colon operator 1:4.
3. The separator character to paste between the two. Since this is set to "", the function will put nothing between "q" and "1," then "q" and "2" and so on.

R will store the resulting names "q1", "q2", "q3", "q4" in the character vector, myQnames. You can use this approach to generate variable names to use in a variety of circumstances. Note that merely changing the 1:4 above to 1:400 would generate the sequence from q1 to q400.

See Sect. 10.9, for another way to select variables by name using the colon operator.

10.5 Selecting Variables Using Logic

You can select a column by using a *logical vector* of TRUE/FALSE values. For example:

```
summary( mydata[ c(FALSE,FALSE,TRUE,
    FALSE,FALSE,FALSE) ] )
```

will select the third column, q1 because the third value is TRUE and the third column is q1. In SAS or SPSS, the digits 1 and 0 can represent TRUE and

FALSE, respectively. They can do this in R, but they first require processing by the as.logical function. Therefore, we could also select the third variable with

```
summary( mydata[ as.logical( c(0,0,1,0,0,0) ) ] )
```

If we had not converted the 0/1 values to logical FALSE/TRUE, the above function call would have first asked for two variables with index values of zero. Zero is a valid value, which is ignored. It would have then asked for the variable in column 1, which is workshop. Finally, it would have asked for three more variables in column zero. The result would have been an analysis only for the first variable, workshop. It would have been a valid, if odd, request and so would have generated no error message!

Luckily, you do not have to actually enter logical vectors like the ones above. Instead, you will generate a vector by entering a logical statement such as:

```
names(mydata)=="q1"
```

That logical comparison will generate this logical vector for you:

```
FALSE, FALSE, TRUE, FALSE, FALSE, FALSE
```

Therefore, another way of analyzing q1 is

```
summary( mydata[ names(mydata)=="q1" ] )
```

While that example is good for educational purposes, in actual use you would prefer the shorter approach using variable names:

```
summary( mydata["q1"] )
```

Once you have mastered the various approaches of variable selection, you will find yourself alternating among the methods as each has its advantages in different circumstances.

The "==" operator compares every element of a vector to a value and returns a logical vector of TRUE/FALSE values. The vector length will match the number of variables, not the number of observations, so we cannot store it as a variable in our data frame. So if we have assigned it to an object name, it would just exist as a vector in our R workspace.

The "!" sign represents NOT, so you can also use that vector to get all the variables except for q1 using the form:

```
summary( mydata[ !names(mydata)=="q1" ] )
```

To use logic to select multiple variable names, we can use the OR operator, "|". For example, select q1 through q4 with the following approach. Complex selections like this are much easier when you do it in two steps. First, create the logical vector and store it, then use that vector to do your selection. In the name myQtf below, I am using the "tf" part to represent TRUE/FALSE to remind myself that this is a logical vector.

```
myQtf <- names(mydata)=="q1" |
         names(mydata)=="q2" |
         names(mydata)=="q3" |
         names(mydata)=="q4"
```

Then we can get summary statistics on those variables using:

```
summary( mydata[myQtf] )
```

Whenever you are making comparisons to many values, you can use the
%in% operator. This will generate exactly the same logical vector as the OR
example above.

```
myQtf <- names(mydata) %in% c("q1","q2","q3","q4")

summary( mydata[myQtf] )
```

You can easily convert a logical vector into index vector that will select the
same variables. For details, see Chap. 13.

10.6 Selecting Variables by String Search (varname: or varname1-varnameN)

You can select variables by searching their names for strings of text. This
approach uses the methods of selection by index number, name, and logic as
discussed above, so make sure you have mastered them before trying these.

SAS uses the form:

```
VAR q: ;
```

to select all the variables that begin with the letter q. SPSS does not off this type
of selection in its main language but can do it using a Python program. SAS also
lets you select variables in the form:

```
PROC MEANS; VAR q1-q4;
```

which gets only the variables q1, q2, q3, and q4 regardless of where they occur in
the dataset and how many variables may lie in between them. The searching
approach we will use in R handles both cases.

R searches variable names for patterns using the grep function to create
vectors containing variable selection criteria we need in the form of indexes,
names, or TRUE/FALSE logical values. The grep function and the rules that
it follows, called *regular expressions*, appear in many different software packages
and operating systems. SAS implements this type of search in the PRX function
(*P*erl *R*egular e*X*pressions), although it does not need it for this type of search.

Below we will use the grep function to find the index numbers for names for
those that begin with the letter q:

```
myQindexes <- grep("^q", names(mydata), value=FALSE)
```

The `grep` function call above uses three arguments:

1. The first is the command string, or regular expression, "^q" which means, "find strings that begin with lower case q." The symbol "^" represents "begins with." You can use any regular expression here, allowing you to search for a wide range of patterns in variable names. We will discuss using wildcard patterns later.
2. The second argument is the character vector that you wish to search, which in our case, is our variable names. Substituting `names (mydata)` here will extract those names.
3. The `value` argument tells it what to return when it finds a match. The goal of grep in any computer language or operating system is to find patterns. A value of TRUE here will tell it to return the variable names that match the pattern we seek. However, in R, indexes are more important than names so the default setting is FALSE to return indexes instead. We could leave it off in this particular case, but we will use it the other way in the next example so we will list it here for educational purposes.

The contents of `myQindexes` will be 3, 4, 5, 6. In all our examples that use that name, it will have those same values.

To analyze those variables, we can then use:

```
summary( mydata[myQindexes] )
```

Now let us do the same thing but have grep save the actual variable names. All we have to do is change to `value = TRUE`.

```
myQnames <- grep("^q", names(mydata), value=TRUE)
```

The character vector myQnames now contains the variable names "q1," "q2," "q3," "q4" and we can analyze those variables with:

```
summary(   mydata[myQnames]   )
```

This approach gets what we expected: variable names. Since it uses names, it makes much more sense to a SAS or SPSS user. So, why we did not do this first? Because in R, indexes are more flexible than variable names.

Finally, let us see how we would use this search method to select variables using logic. The %in% operator in R works just like the IN operator in SAS and the !IN keyword in SPSS. It finds things that occur in a list. We will use it to find when a member of all our variable names (stored in `mynames`) appears in the list of names beginning with "q" (stored in myQnames). The result will be a logical set of TRUE/FALSE values that indicate that the q variables are the last four:

```
FALSE, FALSE, TRUE, TRUE, TRUE, TRUE
```

We will store those values in the logical vector myQtf:

```
myQtf <- names(mydata) %in% myQnames
```

Now we can use the myQtf vector in any function like `summary`:

```
summary( mydata[myQtf] )
```

It is important to note that since we were searching for variables that begin with the letter "q," our program would have also found variables qA and qB if they had existed. We can narrow our search with a more complex search expression that says the letter "q" precedes at least one digit. This would give us the ability to simulate SAS' ability to refer to variables that have a numeric suffix, such as "var1-var100."

This is actually quite easy, although the regular expression is a bit cryptic. It requires changing the myQnames line in the example above to the following:

```
myQnames <- grep("^q[1-9]", names(mydata), value=TRUE)
```

This regular expression means "any string that begins (^) with "q," is followed by one or more numerical digits ([0–9]). Therefore, if they existed, this would select q1, q27, q1old but not qA or qB. You can use it in your programs by simply changing the letter "q" to the root of the variable name you are using.

You may be more familiar with the search patterns using wildcards in Microsoft Windows. That system uses "*" to represent any number of characters and "?" to represent any single character. So the wildcard version of any variable name beginning with the letter q is "q*." Computer programmers call this type of symbol a "*glob*," short for global. R lets you convert *globs* to *regular expressions* with the `glob2rx` function. Therefore, we could do our first grep again in the form:

```
myQindexes <- grep( glob2rx("q*"), names(mydata),
  value=FALSE)
```

Unfortunately, wildcards or globs are limited to simple searches and cannot do our example of q ending with any number of digits.

10.7 Selecting Variables Using $ Notation

You can select a column using $ notation, which combines the name of the data frame and the name of the variable within it, as in:

```
summary( mydata$q1 )
```

This is referred to several ways in R, including "$ prefixing," "prefixing by dataframe$," or "$ notation." When you use this method to select multiple variables, you need to combine them into a single object like a data frame, as in:

```
summary( data.frame( mydata$q1, mydata$q2 ) )
```

Having seen the `c` function, your natural inclination might be to use it for multiple variables as in:

```
summary( c( mydata$q1, mydata$q2 ) )  #Not good!
```

This would indeed make a single object, but certainly not the one a SAS or SPSS user expects. It would stack them both into a single variable with twice as many observations! Summary would then happily analyze the new variable.

10.8 Selecting Variables by Simple Name: `attach` and `with`

This section introduces using short names for variables stored in a data frame, like gender instead of mydata$gender. The technical details we will cover in Chap. 19.

In SAS and SPSS, you refer to variables by short names like gender, or q1. You might have many datasets that contain a variable named gender, but there is no confusion since you have to specify the dataset in advance.

In SAS, you can specify the dataset by adding `DATA=` option on every procedure. Alternatively, since SAS will automatically use the last dataset you created, or you can pretend you just created a dataset by using:

```
OPTIONS _LAST_=myLib.mydata;
```

Every variable selection thereafter would use that dataset.

In SPSS, you clarify which dataset you want to use by opening it with `GET FILE`. If you have multiple datasets open, you instead use `DATASET NAME`.

In R, the potential for confusing variable names is greater because it is much more flexible. For example, you can actually correlate X in one data frame with Y stored in another! All the variable selection methods discussed above made it perfectly clear which data frame to use, but they required extra typing. You can avoid this extra typing by using the `attach` function.

```
attach(mydata)
```

After running that function, you can refer to just q1 and R will know which one you mean. With this approach, getting summary statistics on multiple variables might look like:

```
summary(q1)
```

or

```
summary( data.frame(q1, q2, q3, q4) )
```

If you finish with that dataset and wish to use another, you can detach it with:

```
detach( mydata )
```

Objects will detach when you quit R, so using `detach` is not that important unless you need to use those variable names stored in a different data frame. In that case, detach one file before attaching the next.

The `attach` function works well when *selecting* existing variables but is best avoided when *creating* them. An attached data frame can be thought of as a temporary copy, so changes to *existing* variables will be lost. Therefore, when adding new variables to a data frame, you need to use any of the other above methods that make it absolutely clear where to store the variable. Afterwards you can detach the data and attach it again to gain access to the modified or new variables. We will look at the `attach` function more thoroughly in Chap. 19.

The `with` function is similar to the `attach` function, followed by any other single function and then followed by a `detach` function. Here is an example:

```
with( mydata, summary( data.frame(q1, q2, q3, q4) ) )
```

It lets you use simple names, and even lets you create variables safely. The downside is that you use it with every function, while you might need the `attach` function only once at the beginning of your program. The added set of parentheses also increases your odds of making a mistake. To help avoid errors, you can type this as:

```
with( mydata,
   summary( data.frame(q1, q2, q3, q4) )
)
```

10.9 Selecting Variables with the Subset Function (varname1-varnameN)

R has a `subset` function that you can use to select variables (and observations). It is the easiest way to select contiguous sets of variables by name such as in SAS

```
PROC MEANS; VAR q1--q4;
```

or in SPSS

```
DESCRIPTIVES VARIABLES=q1 to q4.
```

It follows the form:

```
subset(mydata, select=q1:q4)
```

For example, when used with the summary function, it would appear as:

```
summary( subset(mydata, select=q1:q4 ) )
summary( subset(mydata, select=c(workshop, q1:q4) ) )
```

The second example above contains three sets of parentheses. It is very easy to make mistakes with so many nested functions. A syntax-checking editor like

JGR's or Emacs will help. Another thing that helps is to split them across multiple lines:

```
summary(
  subset(mydata, select=c(workshop, q1:q4) )
)
```

10.10 Selecting Variables by List Index

Our data frame is also a list. The components of the list form the columns of the data frame. You can address these components of the list using two square brackets. For example, to select our third variable we can use:

```
summary( mydata[[3]] )
```

With this approach, the colon operator will not extract variables 3 through 6:

```
mydata[[3:6]]  # Will NOT get variables 3 through 6.
```

10.11 Generating Indexes A to Z from Two Variable Names

We have seen how the colon operator can help us analyze variables 3 through 6 using the form:

```
summary( mydata[3:6] )
```

With that method, you have to know the index numbers and digging through lists of variables can be tedious work. Now that we have learned about the names function and the which function, we can combine them to have R tell us the index values we need.

This call to the names function,

```
names(mydata) == "q1"
```

will generate the logical vector:

```
FALSE, FALSE, TRUE, FALSE, FALSE, FALSE, FALSE
```

because q1 is the third variable. The which function will tell us the index values of any TRUE values in a logical vector, so

```
which( names(mydata) == "q1" )
```

will yield a value of 3. Putting these ideas together, we can find the index number of the first variable we want, store it in myqA, then find the last variable, store it in myqZ and then use them with the colon operator to analyze our data from A to Z:

```
myqA <- which( names(mydata)=="q1" )
myqZ <- which( names(mydata)=="q4" )
summary( mydata[ ,myqA:myqZ ] )
```

10.12 Saving Selected Variables to a New Dataset

You can use any method of variable selection to create a new data frame that contains only those variables. If we wanted to create a new data frame that contained only the q variables, we could do so using any method described above. Here are a few variations:

```
myqs <- mydata[3:6]
myqs <- mydata[ c("q1","q2","q3","q4") ]
```

This next example will work, but R will name the variables "mydata.q1," "mydata.q2"... showing the data frame they came from.

```
myqs <- data.frame(mydata$q1, mydata$q2,
                    mydata$q3, mydata$q4)
```

You can add variable name indicators to give them any name you like. With this next one, we are manually specifying original names.

```
myqs <- data.frame(q1=mydata$q1, q2=mydata$q2,
                    q3=mydata$q3, q4=mydata$q4)
```

Using the attach function, the data.frame function leaves the variable names in their original form.

```
attach(mydata)
myqs <- data.frame(q1, q2, q3, q4)
detach(mydata)
```

Finally, we have the subset function with its unique and convenient use of the colon operator directly on variable names.

```
myqs <- subset(mydata, select=q1:q4)
```

10.13 Example Programs for Variable Selection

The examples below demonstrate many ways to select variables. In the examples above, we used the summary function to demonstrate how a complete analysis request would look. However, here we will use the print function to make it easier to see the result of each selection. Even though

```
mydata["q1"]
```

is equivalent to

```
print( mydata["q1"] )
```

because print is the default function, we will use the longer form because it is more representative of its look with most functions. As you learn R, you will quickly opt for the shorter approach.

The example programs for this chapter are in Table 10.1. For most of the programming examples in this book, the SAS and SPSS programs are shorter because the R programs demonstrate R's greater flexibility. However, in the case of variable selection, SAS and SPSS have a significant advantage in ease-of-use. These programs demonstrate roughly equivalent features.

Table 10.1 Example programs to select variables in various ways

SAS programming statements

```
* SAS Program for Selecting Variables;
* SelectingVars.sas ;

LIBNAME myLib 'C: \myRfolder ' ;
OPTIONS _LAST_=myLib.mydata;

PROC PRINT; RUN;
PROC PRINT; VAR workshop gender q1 q2 q3 q4; RUN;
PROC PRINT; VAR workshop--q4; RUN;
PROC PRINT; VAR workshop gender q1-q4; RUN;
PROC PRINT; VAR workshop gender q: ;

* Creating a data set from selected variables;
DATA myLib.myqs;
  SET myLib.mydata (KEEP=q1-q4);
  RUN;
```

SPSS programming statements

```
* SPSS Program for Selecting Variables.
* SelectingVars.sps .

CD ' C: \ myRfolder ' .
GET FILE='mydata.sav' .
LIST.
LIST VARIABLES=workshop,gender,q1,q2,q3,q4.
LIST VARIABLES=workshop TO q4.

* Creating a data set from selected variables.
SAVE OUTFILE=' C:\myRfolder\myqs.sav ' /KEEP=q1 TO q4.
```

R programming statements

```
# R Program for Selecting Variables.
# SelectingVars.R

# Uses many of the same methods as selecting observations.
setwd("/myRfolder")
load(file="myWorkspace.RData")
```

Table 10.1 (continued)

```
# This refers to no particular variables,
# so all are printed.
print(mydata)

#---SELECTING VARIABLES BY INDEX NUMBER---

# These also select all variables by default.
print( mydata[ ] )
print( mydata[ ,] )

# Select just the 3rd variable, q1.
print( mydata[ ,3] ) #Passes q3 as a vector.
print( mydata[ 3] # Passes q3 as a data frame.

# These all select the variables q1,q2,q3 and q4 by indexes.
print( mydata[ c(3, 4, 5, 6) ] )
print( mydata[ 3:6] )

# These exclude variables q1,q2,q3,q4 by indexes.
print( mydata[ -c(3, 4, 5, 6) ] )
print( mydata[ -(3:6) ] )

# Using indexes in a numeric vector.
myQindexes <- c(3, 4, 5, 6)
myQindexes
print( mydata[ myQindexes] )
print( mydata[ -myQindexes] )

# This displays the indexes for all variables.
print( data.frame( names(mydata) ) )

# Using ncol to find the last index.
print( mydata[ 1:ncol(mydata) ] )
print( mydata[ 3:ncol(mydata) ] )

#---SELECTING VARIABLES BY COLUMN NAME---

# Display all variable names.
names(mydata)

# Select one variable.
print( mydata[ "q1"] ) #Passes q1 as a data frame.
print( mydata[ ,"q1"] ) #Passes q1 as a vector.

# Selecting several.
print( mydata[ c("q1", "q2", "q3", "q4") ] )

# Save a list of variable names to use.
myQnames <- c("q1", "q2", "q3", "q4")
myQnames
print( mydata[ myQnames] )
```

Table 10.1 (continued)

```
# Generate a list of variable names.
myQnames<- paste("q", 1:4, sep="")
myQnames
print( mydata[ myQnames] )

#---SELECTING VARIABLES USING LOGIC---

# Select q1 by entering TRUE/FALSE values.
print( mydata[ c(FALSE,FALSE,TRUE,FALSE,FALSE,FALSE) ] )

# Manually create a vector to get just q1.
print( mydata[ as.logical( c(0,0,1,0,0,0) )] )

# Automatically create a logical vector to get just q1.
print( mydata[ names(mydata)=="q1" ] )

# Exclude q1 using NOT operator "!".
print( mydata[ !names(mydata)=="q1" ] )

# Use the OR operator, "" to select q1 through q4,
# and store the resulting logical vector in myqs.
myQtf <- names(mydata)=="q1"  |
         names(mydata)=="q2"  |
         names(mydata)=="q3"  |
         names(mydata)=="q4"
myQtf
print( mydata[ myQtf] )

# Use the %in% operator to select q1 through q4.
myQtf <- names(mydata) %in% c ("q1", "q2", "q3", "q4")
myQtf
print( mydata[ myQtf] )

#---SELECTING VARIABLES BY STRING SEARCH---.

# Use grep to save the q variable indexes.
myQindexes <- grep("^q", names(mydata), value=FALSE)
myQindexes
print( mydata[ myQindexes] )

# Use grep to save the q variable names (value=TRUE now).
myQnames <- grep("^q", names(mydata), value=TRUE)
myQnames
print( mydata[ myQnames] )

# Use %in% to create a logical vector
# to select q variables.
myQtf <- names(mydata) %in% myQnames
myQtf
print( mydata[ myQtf] )

# Repeat example above but searching for any
```

Table 10.1 (continued)

```
# variable name that begins with q, followed
# by one digit, followed by anything.
myQnames <- grep("^q[[:digit:]]\{1\}",
    names(mydata), value=TRUE)
myQnames
myQtf <- names(mydata) %in% myQnames
myQtf
print( mydata[ myQtf] )

# Example of how glob2rx converts q* to ^q.
glob2rx("q*")

#---SELECTING VARIABLES USING$NOTATION---

print( mydata$q1)
print( data.frame(mydata$q1, mydata$q2) )

# ---SELECTING VARIABLES BY SIMPLE NAME---

# Using the "attach" function.
attach(mydata)
print(q1)
print( data.frame(q1, q2, q3, q4) )
detach(mydata)

# Using the "with" function.
with( mydata,
  summary( data.frame(q1, q2, q3, q4) )
)

#---SELECTING VARIABLES WITH SUBSET FUNCTION---

print( subset(mydata, select=q1:q4) )
print( subset(mydata,
  select=c(workshop, q1:q4)
) )

#---SELECTING VARIABLES BY LIST INDEX---

print( mydata[[ 3]] )

#---GENERATING INDEXES A TO Z FROM TWO VARIABLES---

myqA <- which( names(mydata)=="q1")
myqA
myqZ<- which( names(mydata)=="q4")
myqZ
print( mydata[ myqA:myqZ] )

#---CREATING A NEW DATA FRAME OF SELECTED VARIABLES

# Equivalent ways to create a data frame
```

Table 10.1 (continued)

```
# of just the q vars.
myqs <- mydata[ 3:6]
myqs
myqs <- mydata[ c("q1","q2","q3","q4") ]
myqs
myqs <- data.frame(mydata$q1, mydata$q2,
                   mydata$q3, mydata$q4)
myqs
myqs <- data.frame(q1=mydata$q1, q2=mydata$q2,
                   q3=mydata$q3, q4=mydata$q4)

myqs

attach(mydata)
myqs <- data.frame(q1,q2,q3,q4)
myqs
detach(mydata)

myqs <- subset(mydata, select=q1:q4)
myqs
```

Chapter 11
Selecting Observations – Where, If, Select If, Filter

It bears repeating that the approaches that R uses to select observations are, for the most part, the same as those discussed previously for selecting variables. This chapter focuses only on selecting observations. The examples and descriptions parallel the previous chapter as closely as possible. While some may view this as dull repetition, I view it as a helpful review of topics that will initially seem quite odd to SAS and SPSS users. It also allows you read each topic independently for future reference.

11.1 Selecting Observations in SAS and SPSS

There are many ways to select observations in SAS and SPSS and it is outside our scope to discuss them all here. However, we will look at one method for comparison purposes. For both SAS and SPSS, if you do not select observations, it assumes you want to analyze all the data. So in SAS:

```
PROC MEANS;
VAR workshop--q4;
RUN;
```

will analyze all the observations. And in SPSS:

```
DESCRIPTIVES VARIABLES=workshop TO q4.
```

will also use all observations.

To select a subset of observations, for example the males, SAS uses the WHERE statement.

```
PROC MEANS;  VAR workshop--q4;
WHERE gender="m";
RUN;
```

It is also common to create a logical 0/1 value in the form:

```
female=gender='f';
```

R.A. Muenchen, *R for SAS and SPSS Users*, DOI: 10.1007/978-0-387-09418-2_11,   123
© Springer Science+Business Media, LLC 2009

Which you could then apply with,

```
PROC MEANS;  VAR workshop--q4;
WHERE female;
RUN;
```

SPSS does the same selection using both the TEMPORARY and the SELECT IF commands:

```
TEMPORARY.
SELECT IF(gender="m").
DESCRIPTIVES VARIABLES=workshop TO q4.
```

If we had not used the **TEMPORARY** command, the selection would have deleted the females from the dataset. We would have had to open the dataset again if we wanted to analyze both groups in a later step. R has no similar concept. Alternatively, we could create a variable that has a value of 1 for observations we want and zero otherwise. Using that variable on the FILTER command leaves a selection in place until a USE ALL brings the data back. As we will see, R uses a similar filtering approach.

```
COMPUTE male=(gender="m").
FILTER BY male.
DESCRIPTIVES VARIABLES=workshop TO q4.
* more stats could follow for males.
USE ALL.
```

11.2 Selecting All Observations

In R, if you perform an analysis without selecting any observations, the function will use all the observations it can. That is how both SAS and SPSS work. For example, to get summary statistics on all observations (and all variables) we could use:

```
summary(mydata)
```

The methods to select observations apply to all R functions. We will use the summary function so you will see the selection in the context of an analysis.

11.3 Selecting Observations by Index Number

Although it is easy to select observations by index number, you need to be careful doing it. This is because sorting a data frame is something you do often and sorting changes the index number of each row (if you save the sorted version of course). Variables rarely change order, so this approach is much more widely used to select them. That said, let us dive in and see how R does it.

Our data frame has two dimensions, rows and columns. R refers to these as

```
mydata[rows,columns]
```

If you leave out the row or column indexes, R will process all rows and all columns. Therefore, the following three statements have the same result:

```
summary( mydata )
summary( mydata [ ] )
summary( mydata [ , ] )
```

Since this section focuses on selecting observations, we will now discuss just the first index, the rows. Our data frame has eight observations or rows, which are automatically given index numbers, or indexes of 1, 2, 3, 4, 5, 6, 7, and 8. You can select observations by supplying *one index number* or a *vector of indexes*. For example:

```
summary( mydata [5 , ] )
```

selects all the variables for only row 5. There is not much worth analyzing with that selection! Note that the comma is very important, even though we request no columns in the example above. If only one number appears in the brackets (i.e., no comma either), R will assume that it is a column index. Since this selection goes across columns of a data frame, this selection method must return a one-row data frame. A data frame can contain variables that are numeric, character, or factor. A vector could not store such a mixture. That is the opposite of selecting the fifth *variable* with mydata[,5] because that would select a vector. In many cases this distinction would not matter, but it might.

To select more than one observation using indexes, you must combine them into a numeric vector using the c function. Therefore, this will select rows 5 through 8, which happen to be the males:

```
summary( mydata [ c (5,6,7,8) , ] )
```

You will see the c function used in many ways in R. Whenever R requires one object and you need to supply it several, it combines the several into one. In this case, the several index numbers become a single numeric vector. Again, take note of the comma that precedes the right square bracket. If we left that comma out, R would try to analyze *variables* 5 through 8 instead of *observations* 5 through 8! Since we have only six variables, that would generate an error message. I added extra spaces in this example to help you notice it. You do not need additional spaces in R, but you can have as many as you like to enhance readability.

The colon operator " : " can generate a numeric vector directly, so

```
summary( mydata[5:8, ] )
```

selects the same observations.

The colon operator is not just shorthand. Entering 1:N at the R console will cause it to generate the sequence, 1,2,3,...N.

If you use a negative sign on an index, you will *exclude* those observations. For example:

```
summary( mydata [ -c(1,2,3,4) , ] )
```

will exclude the first four records, three females, and one with a gender of NA. R will then analyze the males. Your index values must be either all positive or all negative. Otherwise, the result would be illogical. You cannot say, "include only these observations" and "include all but these observations" at the same time.

The colon operator can abbreviate sequences of numbers, but you need to be careful with negative numbers. If you want to exclude rows 1 through 4, the following sequence will not work:

```
> -1:4
[1] -1  0  1  2  3  4
```

This would of course generate an error because they must all have the same sign. Adding parentheses will clarify the situation, showing R that you want the minus sign to apply to the just the set of numbers from +1 through +4 rather than −1 through +4:

```
> -(1:4)
[1] -1 -2 -3 -4
> summary( mydata[ -(1:4) , ] )
```

If you find yourself working with a set of observations repeatedly, you can easily save a vector of indexes so you will not have to keep looking up index numbers. In this example, I am storing the indexes for the males in myMindexes (M for male).

```
myMindexes <- c(5,6,7,8)
```

From now on, I can use that variable to analyze the males:

```
summary( mydata[myMindexes, ] )
```

For a more realistic dataset, typing all the observation index numbers you need would be absurdly tedious and error prone. We will use logic to create that vector in Sect. 11.8.

You can list indexes individually or, for contiguous observations, use the colon operator. For a bigger dataset, you could use observations 1, 3, 5 through 20, 25, and 30 through 100 like this:

```
mySubset <- c(1,3,5:20,25,30:100)
```

If your observations follow patterns such as every other one or every tenth, see Chap. 17 for ways to generate other sequences of index numbers.

It is easy to have R list the index for each observation in a data frame. Simply create an index using the colon operator and append it to the front of the data frame.

```
> data.frame(myindex=1:8, mydata)
```

```
   myindex  workshop  gender  q1  q2  q3  q4
1        1         R       f   1   1   5   1
2        2       SAS       f   2   1   4   1
3        3         R       f   2   2   4   3
4        4       SAS   <NA>     3   1  NA   3
5        5         R       m   4   5   2   4
6        6       SAS       m   5   4   5   5
7        7         R       m   5   3   4   4
8        8       SAS       m   4   5   5   5
```

Note that the unlabeled column on the left contains the row names. In our case, the row names look like indexes. The row name could have been "Bob" so there is no relationship between row names and indexes.

You can use the nrow function to find the *n*umber of *rows* in a data frame. Therefore, another way to analyze all your observations is

```
summary( mydata[ 1:nrow(mydata) , ] )
```

If you remember that the first male is the fifth record, and you want to analyze all the observations from there to the end, you can use

```
summary( mydata[ 5:nrow( mydata ) , ] )
```

11.4 Selecting Observations by Row Name

SAS and SPSS datasets have variable names but not observation or case names. In R, data frames always name the observations and store those names in the *row names attribute*. When we read our dataset from a text file, we told it the first column would be our row names. The row.names function will display them:

```
row.names(mydata )
```

R will respond with

```
"1", "2", "3", "4", "5", "6", "7", "8"
```

The quotes around them show that R treats them as characters, not as numbers. If you do not provide an ID or name variable for R to use as row names, it will always create them in this form. Therefore, if we had not had an ID variable we would have ended up in exactly the same state. I included an ID variable because it emphasizes the need to be able to track your data back to its most original source when checking for data entry errors.

With such boring row names, there is little need to use them. Let us change the names so we will have an example that makes sense. We will create a new character vector of names:

```
mynames <- c("Ann","Cary","Sue","Carla",
             "Bob","Scott","Mike","Rich")
```

Now we will write those names into the row names attribute of our data frame:

```
row.names(mydata) <- mynames
```

This is a very interesting command! It shows that the `row.names` function does not just show you the names, it is showing you the vector itself. Assigning mynames to that vector renames all the row names! In Sect. 14.6 we will see this again with several variations.

Now that we have some interesting names to work with, let us see what we can do with them. If we want to look at the data for "Ann," we could use:

```
mydata ["Ann", ]
```

You might think that if we had several records per person, we could use row names to select the person. R, however, requires that row names be unique, which is a good idea. You could always use an id number that is unique for row names, then have the subjects' names on each record in their set and a counter like time 1, 2, 3, 4. We will look at just that structure in Sect. 14.16.

To select more than one row name, you must combine them into a single character vector using the c function. For example, we could analyze the females using

```
summary( mydata [ c("Ann","Cary","Sue","Carla"), ] )
```

With a more realistically size data frame, we would probably want to save the list of names to a character vector that we could use repeatedly. Here I use F to represent females and names to remind me of what is in the vector:

```
myFnames <- c("Ann","Cary","Sue","Carla")
```

Now we will analyze the females again using this vector:

```
summary( mydata[ myFnames, ] )
```

11.5 Selecting Observations Using Logic

You can select observations by using a *logical vector* of TRUE/FALSE values. For example, the following will print the first four rows of our dataset.

```
> myRows <- c(TRUE, TRUE, TRUE, TRUE,
+   FALSE, FALSE, FALSE, FALSE)
```

```
> print( mydata[myRows, ] )
> print( mydata[myRows, ] )
    workshop    gender    q1    q2    q3    q4
1          R         f     1     1     5     1
2        SAS         f     2     1     4     1
3          R         f     2     2     4     3
4        SAS     <NA>      3     1    NA     3
```

In SAS or SPSS, the digits 1 and 0 can represent TRUE and FALSE. Let us see what happens when we try this in R.

```
> myBinary <- c(1, 1, 1, 1, 0, 0, 0, 0)
> print( mydata[myBinary, ] )
        workshop    gender    q1    q2    q3    q4
1              R         f     1     1     5     1
1.1            R         f     1     1     5     1
1.2            R         f     1     1     5     1
1.3            R         f     1     1     5     1
```

What happened? Remember that putting a 1 in for the row index asks for row 1. So our request asked for row 1 four times in a row and then asked for row 0 four times. Index values of zero are ignored. We can get around this problem by using the as.logical function.

```
> myRows <- as.logical(myBinary)
```

Now myRows contains the same TRUE/FALSE values it had in the previous example and would work fine.

While the above examples make it clear how R selects observations using logic, they are not very realistic. Hundreds of records would require an absurd amount of typing. Rather than typing such logical vectors, you can generate them with logical statement such as:

```
> mydata$gender=="f"
```

```
[1]   TRUE    TRUE    TRUE      NA  FALSE  FALSE  FALSE  FALSE
```

The "==" operator compares every value of a vector (like gender) to a value (like "f") and returns a logical vector of TRUE/FALSE values. The length of the resulting logical vector will match the number of observations in our data frame. Therefore, we could store it our data frame as a new variable.

Unfortunately, we see that the fourth logical value is NA. That is because the fourth observation has a missing value for gender. Up until this point, we have been mirroring the previous chapter, Chap. 10. Logical comparisons of variable names did not have a problem with missing values. Now, however, we must take a different approach. First, let us look at what would happen if we continued down this track.

```
> print ( mydata[ mydata$gender=="f", ] )
```

	workshop	gender	q1	q2	q3	q4
1	R	f	1	1	5	1
2	SAS	f	2	1	4	1
3	R	f	2	2	4	3
NA	<NA>	<NA>	NA	NA	NA	NA

What happened to the fourth observation? It had missing values only for gender and q3. Now *all* the values for that observation are missing. R has noticed that we were selecting rows based upon only gender. Not knowing what we would do with the selection, it had to make all the other values missing too. Why? Because we might have been wanting to correlate q1 and q4. Those two had no missing values in the original data frame. If we want to correlate them only for the females, even their values must be set to missing.

We could select observations using this logic and then count on R's other functions to remove the bad observations as they would any others with missing values. However, there is little point in storing them. Their presence could also affect future counts of missing values for other analyses, perhaps when females are recombined with males.

Luckily, there is an easy way around this problem. The where function gets the index values for the TRUE values of a logical vector. Let us see what it does.

```
> which ( mydata$gender=="f" )

 [1] 1 2 3
```

It has ignored both the NA value and the FALSE values to show us that only the first three values of our logical statement were TRUE. We can save these index values in myFemales.

```
> myFemales <- which ( mydata$gender=="f" )
> myFemales
 [1] 1 2 3
```

We can then analyze just the females with the following:

```
summary ( mydata[ myFemales , ] )
```

Negative index values exclude those rows, so we could analyze the non-females (males and missing) with the following. We could of course get males and exclude missing the same way we got the females.

```
summary ( mydata[-myFemales , ] )
```

We can select observations using logic that is more complicated. For example, we can use the AND operator, "&" to analyze subjects who are both male and "strongly agree" that the workshop they took was useful. Compound selections like this are much easier when you do it in two steps. First, create the logical vector and store it, then use that vector to do your selection.

```
> HappyMales <- which(mydata$gender=="m"
+    & mydata$q4==5)
> HappyMales
  [1] 6 8
```

So we could analyze these observations with

```
summary( mydata[HappyMales , ] )
```

Whenever you are making comparisons to many values, you can use the %in% operator. Let us select observations who have taken the R or SPSS workshop. With just two target workshops could use a simple, workshop="R" | workshop="SPSS" but the longer the target list, the happier you will be to save all the repetitive typing.

```
> myRspss <-
+    which( mydata$workshop %in% c("R","SPSS") )
> myRspss

  [1] 1 3 5 7
```

Then we can get summary statistics on those observations using:

```
summary( mydata[myRspss, ] )
```

The various methods we described in Chap. 10 make a big difference in how complicated the logical commands to select observations appear. Here are several different ways to analyze just the females:

```
myFemales <- which( mydata$gender=="f")
myFemales <- which( mydata[2] == "f")
myFemales <- which( mydata["gender"] == "f")
with(mydata,
   myFemales <- which(gender=="f")
)
attach(mydata)
   myFemales <- which(gender=="f")
detach(mydata)
```

Any of which you could then analyze the data using

```
summary( mydata[ myFemales, ] )
```

You can easily convert a logical vector into index vector that will select the same observations. For details, see Chap. 13.

11.6 Selecting Observations by String Search

If you have string variables, or useful row names, you can select observations by searching their values for strings of text. This approach uses the methods of selection by indexes, row names, and logic discussed above, so make sure you have mastered them before trying these.

R searches strings for patterns using the grep function to create vectors containing variable selection criteria we need in the form of indexes, names, or TRUE/FALSE logical values. The grep function and the rules that it follows, called *regular expressions*, appear in many different software packages and operating systems. In fact, SAS implements this type of search in the PRX function (*P*erl *R*egular e*X*pressions).

In the last section, we replaced our original row names ("1," "2,"...) with more interesting ones ("Ann," "Cary,"...). We will use the grep function to find row names that begin with the letter "C":

```
myCindexes <- grep("^C", row.names(mydata),
  value=FALSE)
```

This grep function call uses three arguments:

1. The first is the command string, or regular expression, "^C" which means, "find strings that begin with a capital letter "C." The symbol "^" represents "begins with." You can use any regular expression here, allowing you to search for a wide range of patterns in variable names. We will discuss using wildcard patterns later.
2. The second argument is the character vector that you wish to search. In our case, we want to search the row names of mydata, so we use the row.names function here.
3. The value argument tells it what to store when if finds a match. The goal of grep in any computer language or operating system is to find patterns. A value of TRUE here will tell it to save the row names that match the pattern we seek. However, in R, indexes are more fundamental than names, which are optional, so the default setting is FALSE to save indexes instead. We could leave it off in this particular case, but we will use it the other way in the next example so we will list it here for educational purposes.

The contents of myCindexes will be 2 and 4 because Cary and Carla are the second and fourth observations. If we wanted to save this variable, it does not match the eight values of our other variables, so we cannot store it in our data frame. We would instead just store it in the workspace as a vector outside our data frame.

To analyze those observations, we can then use:

```
summary( mydata[myCindexes , ] )
```

Now let us do the same thing but have grep save the actual variable names. All we have to do is change to `value = TRUE`:

```
myCnames <- grep("^C", row.names(mydata),value=TRUE)
```

The character vector myCnames now contains the row names "Cary" and "Carla," and we can analyze those observations with:

```
summary(  mydata[myCnames ,  ]  )
```

Finally, let us see how we would use this search method to select observations using logic. The `%in%` function works just like the IN operator in SAS and the `! IN` keyword in SPSS. It finds things that occur in a set of values. We will use it to find whenever one of our row names appears in the set of names beginning with "C" (stored in myCnames). The result will be a logical set of TRUE/FALSE values that indicate that the names beginning with "C" are in the second and fourth position:

```
FALSE, TRUE, FALSE, TRUE, FALSE, FALSE, FALSE, FALSE
```

We will store those values in the logical vector myCtf:

```
myCtf <- row.names(mydata) %in% myCnames
```

Now we can use the myCtf vector in any analysis like summary:

```
summary( mydata[myCtf, ] )
```

You may be more familiar with the search patterns using wildcards in Microsoft Windows. They use "*" to represent any number of characters and? to represent any single character. So the wildcard version of any variable name beginning with the letter "C" is "C*". Computer programmers call this type of symbol a *glob*, short for global. R lets you convert globs to *r*egular e*x*pressions with the `glob2rx` function. Therefore, we could do our first grep again in the form:

```
myCindexes <- grep( glob2rx("C*")  , row.names(mydata),
    value=FALSE)
```

11.7 Selecting Observations with the Subset Function

You can select observations using the `subset` function. You simply list your logical condition under the `subset` argument as in:

```
subset(mydata, subset=gender=="f")
```

Note that an equal sign follows the subset argument because that is what R uses to set values. The `gender=="f"` comparison is still done using `"=="` because that is the symbol R uses for logical comparisons.

You can use subset to analyze your selection using the form:

```
summary(
   subset(mydata, subset=gender=="f")
)
```

In Sect. 10.9, we discussed how to select variables. By adding the `select` argument, we can analyze just variables q1 through q4 for the females with:

```
summary( subset( mydata, select=q1:q4,
   subset=gender=="f" )
)
```

Since the first argument to the `summary` function is the data frame to use, you do not have to write out the longer forms of names like mydata$q1 or mydata$gender. That is a very helpful function!

11.8 Generating Indexes from A to Z from Two Row Names

We have discussed various observation selection techniques. Now we are ready to examine combination methods that use a blend of row names, logic, and index numbers. If you have not mastered the previous examples, now would be a good time to review them!

We have seen how the colon operator can help us analyze the males who are observations 5 through 8 using the form:

```
summary( mydata[5:8, ] )
```

But you had to know the index numbers, and digging through lists of observation numbers can be tedious work. However, we can use the `row.names` function and the `which` function to get R to find the index values we need. The command

```
row.names(mydata) == "Bob"
```

will generate the logical vector:

```
FALSE, FALSE, FALSE, FALSE, TRUE, FALSE, FALSE, FALSE
```

because Bob is the fifth observation. The `which` function will tell us the index values of any TRUE values in a logical vector, so

```
which(FALSE, FALSE, FALSE, FALSE,
   TRUE, FALSE, FALSE, FALSE)
```

will yield a value of 5. Putting these ideas together, we can find the index number of the first observation we want, store it in myMaleA, then find the last variable, store it in myMaleZ and then use them with the colon operator to analyze our data from A to Z:

```
myqMaleA <- which( names(mydata)=="Bob" )
```

```
myqMaleZ <- which( names(mydata)=="Rich" )
summary( mydata[ myMaleA:myMaleZ, ] )
```

It is important to keep in mind that this example assumes the data is sorted by gender.

11.9 Variable Selection Methods with No Counterpart for Selecting Observations

As we have seen, the methods that R uses to select variables and observations are almost identical. However, there are several techniques for selecting variables that have no equivalent in selecting observations. They are: $ prefix form (e.g., mydata$gender), the attached form of variable names (e.g., just gender), and the list form (e.g., mydata [[2]]). When selecting observations, these three have no equivalents.

11.10 Saving Selected Observations to a New Data Frame

You can create a new data frame that is a subset of your original one by using any of the methods for selecting observations. You simply assign the data to a new data frame. The examples below all select the males and assign them to the myMales data frame:

```
myMales <- mydata[5:8, ]
myMales <- mydata[ which(mydata$gender=="m") , ]
myMales <- subset( mydata, subset=gender=="m" )
```

11.11 Example Programs for Selecting Observations

The examples in Table 11.1 demonstrate many ways to select observations. In the examples above, we used the summary function to demonstrate how a complete analysis request would look. Here we will instead use the print function to make it easier to see the result of each selection. Even though:

```
mydata[5:8, ]
```

is equivalent to:

```
print( mydata[5:8, ] )
```

because print is the default function, we will use the longer form because it is more representative of its look with most functions. As you learn R, you will quickly opt for the shorter approach.

Table 11.1 Example programs to select observations in various ways

SAS programming statements

```
* SAS Program to Select Observations;
* SelectingObs.sas ;

LIBNAME myLib 'C:\myRfolder';

PROC PRINT data=myLib.mydata;
  WHERE gender="m";
  RUN;

PROC PRINT data=myLib.mydata;;
  WHERE gender="m" & q4=5;

DATA myLib.males;
  SET myLib.mydata;
  WHERE gender="m";
  RUN;

DATA myLib.females;
  SET myLib.mydata;
  WHERE gender="f";
  RUN;
```

SPSS programming statements

```
* SPSS Program to Select Observations.
* SelectingObs.sps .

CD 'c:\myRfolder'.
GET FILE='mydata.sav'.

COMPUTE   male=(gender="m").
COMPUTE female=(gender="f").

FILTER BY male.
LIST.
* analyses of males could follow here.

FILTER BY female.
LIST.
* analyses of females could follow here.

USE ALL.

DO IF male.
XSAVE OUTFILE='males.sav'.
ELSE IF female.
XSAVE OUTFILE='females.sav'.
END IF.
EXECUTE.
```

Table 11.1 (continued)

R programming statements

```
# R Program to Select Observations.
# SelectingObs.R .

setwd("/myRfolder")
load(file="myWorkspace.RData")
print(mydata)

#---SELECTING OBSERVATIONS BY INDEX
# Print all rows.
print( mydata[ ] )
print( mydata[ , ] )
print( mydata[1:8, ] )

# Just observation 5.
print( mydata[5 , ] )

# Just the males:
print( mydata[ c(5,6,7,8) , ] )
print( mydata[ 5:8         , ] )

# Excluding the females with minus sign.
print( mydata[ -c(1,2,3,4), ] )
print( mydata[ -(1:4)      , ] )

# Saving the Male (M) indexes for reuse.
myMindexes <- c(5,6,7,8)
summary( mydata[myMindexes, ] )

# Print a list of index numbers for each observation.
data.frame(myindex=1:8,mydata)

# Select data using length as the end.
print( mydata[ 1:nrow(mydata),  ]  )
print( mydata[ 5:nrow(mydata),  ]  )

#---SELECTING OBSERVATIONS BY ROW NAME

# Display row names.
row.names(mydata)

# Select rows by their row name.
print( mydata[ c("1","2","3","4"), ] )

# Assign more interesting names.
mynames <- c("Ann","Cary","Sue","Carla",
             "Bob","Scott","Mike","Rich")
mynames
```

Table 11.1 (continued)

```
# Store the new names in mydata.
row.names(mydata) <- mynames
mydata

# Print Ann's data.
print( mydata[ "Ann" , ] )
mydata[ "Ann" , ]

# Select the females by row name.
print( mydata[ c("Ann","Cary","Sue","Carla"), ] )

# Save names of females to a character vector.
myFnames <- c("Ann","Cary","Sue","Carla")
myFnames

# Use character vector to select females.
print( mydata[ myFnames, ] )

# - - -SELECTING OBSERVATIONS USING LOGIC- - -

#Selecting first four rows using TRUE/FALSE.
myRows <- c(TRUE, TRUE, TRUE, TRUE,
 FALSE, FALSE, FALSE, FALSE)
print( mydata[myRows, ]  )

# Selecting first four rows using 1s and 0s.
myBinary <- c(1, 1, 1, 1, 0, 0, 0, 0)
print( mydata[myBinary, ] )
myRows <- as.logical(myBinary)
print( mydata[ myRows, ] )

# Use a logical comparison to select the females.
mydata$gender=="f"
print( mydata[ mydata$gender=="f", ] )
which( mydata$gender=="f" )
print( mydata[ which(mydata$gender=="f") , ] )

# Select females again, this time using a saved vector.
myFemales <- which( mydata$gender=="f" )
myFemales
print( mydata[ myFemales , ] )

# Excluding the females using the "!" NOT symbol.
print( mydata[-myFemales , ] )

 # Select the happy males.
HappyMales <- which(mydata$gender=="m"
 & mydata$q4==5)
HappyMales
print( mydata[ HappyMales , ] )
```

Table 11.1 (continued)

```
# Selecting observations using %in%.
myRspss <-
 which( mydata$workshop %in% c("R","SPSS") )
myRspss
print( mydata[myRspss , ] )

# Equivalent selections using different
# ways to refer to the variables.

print( subset(mydata, gender=='f') )

attach(mydata)
print( mydata[ gender=="f" , ] )
detach(mydata)

with(mydata,
print ( mydata[ gender=="f" )
)
print( mydata[ mydata["gender"] =="f" , ] )

print( mydata[ mydata$gender=="f" , ] )

# - - -SELECTING OBSERVATIONS BY STRING SEARCH- - -

# Search for row names that begin with "C".
myCindexes <- grep("^C", row.names(mydata), value=FALSE)
print( mydata[ myCindexes , ] )

# Again, using wildcards.
myCindexes <- grep( glob2rx("C*") ,
 row.names(mydata), value=FALSE)
print( mydata[myCindexes , ] )

# - - -SELECTING OBSERVATIONS BY subset Function- - -

subset(mydata,subset=gender=="f")

# - - -GENERATING INDEXES A TO Z FROM TWO ROW NAMES- - -
myMaleA <- which( row.names(mydata)=="Bob" )
myMaleA
myMaleZ <- which( row.names(mydata)=="Rich" )
myMaleZ
print( mydata[myMaleA:myMaleZ , ] )

#- - -CREATING A NEW DATA FRAME OF SELECTED OBSERVATIONS

# Creating a new data frame of only males (all equivalent).
myMales <- mydata[ 5:8, ]
myMales
myMales <- mydata[ which( mydata$gender=="m" ) , ]
myMales
```

Table 11.1 (continued)

```
myMales <- subset( mydata, subset=gender=="m" )
myMales

# Creating a new data frame of only females (all equivalent).
myFemales <- mydata[1:3, ]
myFemales
myFemales <- mydata[ which( mydata$gender=="f" ) , ]
myFemales
myFemales <- subset( mydata, subset=gender=="f" )
myFemales
```

Chapter 12
Selecting Both Variables and Observations

While the sections above have focused on selecting variables and observations separately, you can combine methods in most cases. For example, analyzing variables gender through q4 for just the females could be done with any of these approaches.

Here we select the females by using index numbers 1:4 and then the q variables by telling it that they are column indexes 2:6.

```
summary( mydata [ 1:4, 2:6 ] )
```

Here we do the same selection, but we select the females with a logical condition on rows and then select the q variables by supplying a character vector of their names. The only reason we need to attach the data on this one is that it refers to gender rather than mydata$gender.

```
attach (mydata)
summary( mydata [
  which(gender=="f"),
  c("gender", "q1", "q2", "q3", "q4")
] )
detach(mydata)
```

Here we use the `subset` function, with both its `subset` and `select` arguments. It is often the easiest to use.

```
summary (
  subset(mydata, subset=gender=="f",
    select=gender:q4)
)
```

R.A. Muenchen, *R for SAS and SPSS Users*, DOI: 10.1007/978-0-387-09418-2_12,   141
© Springer Science+Business Media, LLC 2009

Chapter 13
Converting Data Structures

In SAS and SPSS, there is only one data structure, the dataset. Within that, there is only one structure, the variable. It seems absurdly obvious, but we need to say it: all SAS and SPSS procedures accept variables as input. Of course, you have to learn that putting a character variable in where a numeric one is expected causes an error. Occasionally, a character variable contains numbers and we must convert them. Putting a continuous variable in where a categorical variable belongs may not yield an error message, but perhaps it should. Both packages have added variable classification methods (nominal, ordinal, or scale) to help you choose the correct analyses and graphs.

As we have seen, R has several data structures including vectors, factors, data frames, matrices, and lists. For many functions (what SAS/SPSS call procedures), R can automatically provide output optimized for the data structure you give it. Said more formally, *generic functions* apply different *methods* to different *classes* of objects. So to control a function, you have to know several things:

1. The classes of objects the function is able to accept.
2. What it will do with each. That is, what its method is for each. Although many important functions in R offer multiple methods (they are generic), not all are.
3. The data structure you are supplying to the function, that is, the class of your object. As we have seen, the way you select data determines its data structure or class.
4. How to convert from the data structure you have to one you may need.

In chapter 10 on *Selecting Variables*, we learned that both of these commands select our variable q1 and pass the data in the form of a data frame:

```
mydata [3]
mydata ["q1"]
```

while these, with their additional comma, also select variable q1 but instead pass the data in the form of a vector:

```
mydata [,3]
mydata [,"q1"]
```

Many functions would work just fine on either result. But some procedures are fussier than others and require very specific data structures. If you are having a problem figuring out which form of data you have, there are functions that will tell you. For example:

```
class (mydata)
```

will tell you its class is "data frame." Knowing a data frame is a type of list, you will know that functions that require either of those structures will accept it. As with the `print` function, it may produce different output, but it will accept it. There are also functions that test the status of an object, and they all begin with "is." For example:

```
is.data.frame ( mydata [3] )
```

Table 13.1 Data conversion functions

Conversion to perform	Example
Index vector to logical vector	`myQindexes <- c (3, 4, 5, 6)` `myQtf <- 1:6 %in% myQindexes`
Vectors to columns of a data frame	`data.frame (x,y,z)`
Vectors to rows of a data frame	`data.frame ( rbind(x,y,z) )`
Vectors to columns of a matrix	`cbind(x,y,z)`
Vectors to rows of a matrix	`rbind(x,y,z)`
Vectors combined into one long one	`c(x,y,z)`
Data frame to matrix (all variables must be same mode, e.g., all numeric or all character)	`as.matrix(mydataframe)`
Matrix to data frame	`as.data.frame(mymatrix)`
A vector to an r by c matrix	`matrix(myvector,nrow=r,ncol=c)` (note this is *not* as.matrix!)
Matrix to one very long vector	`as.vector(mymatrix)`
List to one long vector	`unlist (mylist, recursive = TRUE)`
List to separate vectors	`unlist(mylist)`
Lists (or data frames) combined into lists	`c(list1,list2)`
List of vectors or matrices to rows of matrix	`myMatrix <-( do.call(rbind,` `myList) )`
List of vectors or matrices to columns of matrix	`myMatrix <-( do.call(cbind,` `myList) )`
Logical vector to index vector	`myQtf <- c (FALSE, FALSE,TRUE,` `TRUE, TRUE, TRUE)` `myQindexes <- which(myQtf)`
Remove the class completely from any object that has one.	`unclass(myobject)`

will display TRUE but

```
is.vector ( mydata [3] )
```

will display FALSE.

Some of the functions you can use to convert from one structure to another are listed in Table 13.1. Let us apply one to our data. First, we will remind ourselves how the `print` function prints our data frame in a vertical format.

```
> print (mydata)

  workshop gender q1 q2 q3 q4
1        R      f  1  1  5  1
2      SAS      f  2  1  4  1
3        R      f  2  2  4  3
4      SAS   <NA>  3  1 NA  3
5        R      m  4  5  2  4
6      SAS      m  5  4  5  5
7        R      m  5  3  4  4
8      SAS      m  4  5  5  9
```

Now let us print it in list form by adding the `as.list` function. Essentially, the data frame becomes a list at the moment of printing, giving us the horizontal orientation that we have seen before when printing a true list.

```
> print ( as.list (mydata) )

$workshop
[1] R    SAS R    SAS R    SAS R    SAS
Levels: R SAS SPSS STATA

$gender
[1] f    f    f    <NA> m    m    m    m
Levels: f m

$q1
[1] 1 2 2 3 4 5 5 4

$q2
[1] 1 1 2 1 5 4 3 5

$q3
[1] 5  4  4 NA  2  5  4  5

$q4
[1] 1 1 3 3 4 5 4 9
```

13.1 Converting from Logical to Index and Back

We have looked at various ways to select variables and observations using both logic and indexes. Now let us look at how to convert from one to the other.

The which function will examine a logical vector and tell you which of them are true. We can create a logical vector that chooses our q variables many ways. Let us use the vector that selects the last four variables in our practice data frame. See Chaps. 10 or 11 for different ways to create a vector like this. We will just enter it manually.

```
myQtf <- c(FALSE, FALSE, TRUE, TRUE, TRUE, TRUE)
```

We can convert that to an index vector that gives us the index numbers for each occurrence of the value TRUE using the which function.

```
myQindexes <- which(myQtf)
```

Now myQindexes contains 3, 4, 5, 6 and we can analyze just the q variables using:

```
Summary ( mydata[myQindexes] )
```

To go in the reverse direction, we would want to know all the variable indexes: 1, 2, 3, 4, 5, 6, which of them were, in a logical sense, in our list of 3, 4, 5, 6.

```
myQindexes <- c (3, 4, 5, 6)
```

Now we will use the %in %function operator to create it using logical comparisons:

```
myQtf <- 1:6 %in % myQindexes
```

Now myQtf has the values FALSE, FALSE, TRUE, TRUE, TRUE, TRUE and we can analyze just the Q variables with:

```
Summary ( mydata[myQtf] )
```

Why are there two methods? The logical method is the most direct, since the which function is often an additional step. However, if you have 20,000 variables as researchers in genetics often have, the logical vector will have 20,000 values. The index vector will have only as many values as there are variables in your subset.

The which function also has a critical advantage. Since it looks only to see which values are TRUE, the NA missing values do not affect it. Using a logical vector, R will look at all values, TRUE, FALSE and – here is the problem – missing values of NA. However, the selection mydata [NA ,] is undefined, causing problems. See Chap. 11 for details.

Chapter 14
Data Management

14.1 Transforming Variables

Unlike SAS, R has no separation of phases for data modification (data step) and analysis (proc step). It is more like SPSS where as long as you have data read in, you can modify it. Anything that you have read into or created in your R workspace you can modify at any time.

R performs transformations such as adding or subtracting variables on the whole variable at once, as do SAS and SPSS. It calls that *vector arithmetic*. R has loops, but you do not need them for this type of manipulation.

R can nest one function call within any other. This applies to transformations as well. For example, taking the log of our q4 variable and then getting summary statistics on it, you have a choice of a two-step process like:

```
mydata$q4Log  <- log(mydata$q4)
summary( mydata$q4Log )
```

Or you could simply nest the `log` function within the `summary` function:

```
summary( log( mydata$q4 ) )
```

If you planned to do several things with the transformed variable, saving it under a new name would lead to less typing and quicker execution.

Table 14.1 shows basic transformations in R, SAS and SPSS.

In chapter 10, *Selecting Variables*, we chose variables using various methods: by index, by column name, by logical vector, using the style `mydata$myvar`, by using simply the variable name after you have attached a data frame and using the `subset` or `with` functions. Here are several examples that perform the same transformation using different variable selection approaches. The `within` function is a variation of the `with` function that has some advantages for variable creation that are beyond our scope. We have seen that R has a `mean` function but we will calculate the mean the long way just for demonstration purposes.

R.A. Muenchen, *R for SAS and SPSS Users*, DOI: 10.1007/978-0-387-09418-2_14,   147
© Springer Science+Business Media, LLC 2009

Table 14.1 Mathematical operators

	R	SAS	SPSS
Addition	x + y	x + y	x + y
Antilog, base 10	10^x	10**x	10**x
Antilog, natural	exp (x)	exp(x)	exp(x)
Division	x/y	x/y	x/y
Exponentiation	x^2	x**2	x**2
Logarithm, base 10	log10 **(x)**	log10(x)	lg10(x)
Logarithm, natural	log (x)	log(x)	ln(x)
Multiplication	x*y	x*y	x*y
Round off	round (x)	round(x)	rnd(x)
Square root	sqrt (x)	sqrt(x)	sqrt(x)
Subtraction	x−y	x−y	x−y

```
mydata$meanQ <-   (mydata$q1 + mydata$q2
                + mydata$q3 + mydata$q4)/4

mydata[,"meanQ"] <-   (mydata[ , "q1"] + mydata[ , "q2"]
                   + mydata[ , "q3"] + mydata[ , "q4"])/4

within( mydata,
  meanQ <-   (q1 + q2 + q3 + q4 )/4
)
```

Another way to use the shorter names is with the transform function. It is similar to attaching a data frame, performing as many transformations as you like using short variable names, and then detaching the data (we do that example next). It looks like:

```
mydata <- transform(mydata, meanQ=(q1+q2+q3+q4)/4 )
```

It may seem strange to use the " = " now in an equation instead of "<-", but in this form meanQ is a named argument, and those are always specified using " = ".

If you have many transformations, it is easier to read them on separate lines:

```
mydata <- transform( mydata,
  score1=(q1+q2)/2,
  score2=(q3+q4)/2
)
```

The transform function reads the data before it begins, so if you want to continue to transform variables you just created, you must do it in a second call to that function. For example, to get the means of score1 and score2, you cannot do the following:

```
mydata <- transform( mydata,
  score1=(q1+q2)/2,
  score2=(q3+q4)/2,
  meanscore=(score1+score2)/2   #does not work!
)
```

It will not know what score1 and score2 are for the creation of meanscore. You can do that in two steps.

```
mydata <- transform( mydata,
   score1=(q1+q2)/2,
   score2=(q3+q4)/2
)
mydata <- transform( mydata,
   meanscore=(score1+score2)/2   #this works.
)
```

You can create a new variable using the index method too, but it requires a bit of extra work. Let us load the dataset again since we already have a variable named meanQ in the current one.

```
> load(file="myWorkspace.RData")
```

Now we will add a variable at index position 7 (we currently have six variables). Using the index approach, it is easier to initialize a new variable by binding a new variable to mydata. Otherwise, R will automatically give it a column name of V7 that we would want to rename later. We used the column bind function, cbind, to create mymatrix before. Here we will use it to name the new variable, meanQ, initialize it to zero, and then bind it to mydata.

```
> mydata <- data.frame( cbind( mydata, meanQ=0.) )
```

Finally, we can add the values to column 7.

```
> mydata[7] <- (mydata$q1 + mydata$q2 +
+                mydata$q3 + mydata$q4)/4
```

Finally, let us examine what happens when you create variables using the attach function.

WARNING! *The attach function is hazardous to variable creation.* You can think of the attach function as creating a *temporary copy* of the data frame, so changing that is worthless. You can use the attach method to simplify naming variables only on the right side of the equation. This is a safe example, because the variable being created uses the long dataframe$varname style:

```
attach(mydata)
mydata$meanQ <- (q1+q2+q3+q4)/4
detach(mydata)
```

If you were to modify a variable in your data frame, you would have to re-attach it before you would see it. In the following example, we attach mydata and look at q1:

```
> attach(mydata)
> q1
[1] 1 2 2 3 4 5 5 4
```

So we see what q1 looks like. Next, we will see what it looks like squared and then write it to mydata$q1 (choosing a new name would be wiser but would not make this point clear). By specifying the full name mydata$q1, we know R will write it to the original data frame, not the temporary working copy:

```
> mydata$q1^2

[1]  1   4   4   9 16 25 25 16

> mydata$q1 <- q1^2
```

But what does the short name of q1 show us? The unmodified temporary version!

```
> q1

[1] 1 2 2 3 4 5 5 4
```

If we attach the file again, it will essentially make a new temporary copy and q1 finally shows that we did indeed square it:

```
> attach(mydata)

        The following object(s) are masked from mydata
           ( position 3 ):

        gender q1 q2 q3 q4 workshop
> q1

[1]  1   4   4   9 16 25 25 16
```

Just like SAS or SPSS, R does all its calculations in the computer's main memory. You can use them immediately, but they will exist only in your current session unless you use the save.image function to write out a new file:

```
setwd("/myRfolder")
save.image("myWorkspace.RData")
```

See chapter 19 for more ways to save new variables. For example programs that demonstrate these topics in SAS, SPSS and R, see Table 14.2.

Table 14.2 Example programs for transforming variables

SAS programming statements

```
                * SAS Program for Transforming Variables;
                * Transform.sas ;
                LIBNAME myLib  'C: \myRfolder' ;
                DATA myLib.mydata;
                SET  myLib.mydata;
                totalq=(q1+q2+q3+q4);
                logtot=log10(totalq);
                mean1=(q1+q2+q3+q4)/4;
```

Table 14.2 (continued)

```
mean2=mean(of q1-q4);
PROC PRINT; RUN;
```

SPSS programming statements

```
* SPSS Program for Transforming Variables.
* Transform.sps .
CD 'C: \myRfolder ' .
GET FILE=' mydata.sav ' .
COMPUTE Totalq=q1+q2+q3+q4.
COMPUTE Logtot=lg10(totalq).
COMPUTE Mean1=(q1+q2+q3+q4)/4.
COMPUTE Mean2=MEAN (q1 TO q4).
SAVE OUTFILE=' C: \myRfolder\mydata.sav ' .
LIST.
```

R programming
   statements

```
# R Program for Transforming Variables.
# Transform.R
setwd("/myRfolder")
load(file="myWorkspace.RData")
mydata
#Transformationin the middle of another function.
summary(log (mydata$q4)  )
#Creating meanQ with dollar notation.
mydata$meanQ  <-  (mydata$q1 + mydata$q2
                     + mydata$q3 + mydata$q4)/4
mydata
# Creating meanQ using attach.
attach(mydata )
mydata$meanQ <- (q1+q2+q3+q4)/4
detach(mydata)
mydata
# Creating meanQ using transform.
mydata  <- transform (mydata,
  meanQ=(q1 +q2+q3+q4)/4 )
mydata
# Creating two variables using transform.
mydata  <- transform( mydata,
  score1=(q1+q2)/2,
  score2=(q3+q4)/2  )
mydata
# Creating meanQ using index notation on the left.
load(file ="myWorkspace.RData")
mydata <- data.frame ( cbind( mydata, meanQ=0.) )
mydata[7]  <-  (mydata$q1 + mydata$q2+
                  mydata$q3 + mydata$q4)/4
mydata
save.image(file ="myWorkspace.RData")
```

14.2 Procedures or Functions? The Apply Function Decides

The last section described data transformations, but said little about statistical functions. SAS and SPSS each have two independent ways to calculate statistics: functions and procedures. Statistical functions work within each observation to calculate a statistic like the mean of our q variables for each observation. Statistical procedures work within a variable to calculate statistics like the mean of our q4 variable across all observations.

R on the other hand, has only one way to calculate: functions. What determines if the function is working on variables or observations is *how you apply it!*

In the previous section, we created a mean variable using:

```
mydata$meanQ <- (mydata$q1 + mydataq2
                 mydata$q3 + mydataq4) / 4
```

This approach gets tedious with long lists of variables. It also has a problem with missing values. The meanQ variable will be missing if any of the variables has a missing value. The mean function solves that problem.

14.2.1 Applying the mean Function

We have previously seen that R has both a mean function and a summary function. For numeric objects, the R mean function returns a single value while the summary function returns the minimum, first quartile, median, mean, third quartile, and maximum. We could use either of these functions to create a meanQ variable. However, the mean function returns only the value we need, so it is better for this purpose.

Let us start examining our options by first putting our q variables into a matrix. Simply selecting the variables with the command below will not work, because even though variables 3 through 6 are all numeric, it will maintain its form as a data frame.

```
myQmatrix <- mydata[ ,3:6]   #Not a matrix!
```

The proper way to convert the data is with the as.matrix function:

```
> myQmatrix <- as.matrix( mydata[ ,3:6] )
> myQmatrix
   q1 q2 q3 q4
1   1  1  5  1
2   2  1  4  1
3   2  2  4  3
4   3  1 NA  3
5   4  5  2  4
6   5  4  5  5
7   5  3  4  4
8   4  5  5  5
```

Let us review what happens if we use the mean function on myQmatrix:

```
> mean ( myQmatrix )
[1] NA
```

The result is NA, or missing, because the matrix contains an NA value. The default method of dealing with missing values in R is to set the resulting value to missing. To remove missing values, most basic statistical functions have the na.rm argument. The value is FALSE by default, so we will set it to TRUE to get our grand mean:

```
> mean ( myQmatrix, na.rm=TRUE )
[1] 3.451613
```

This is an interesting ability, but it is not that useful in our case. What is of much more interest is the means of the variables as a SAS/SPSS *procedure* would do, or the means of the observations to create a new variable, as a SAS/SPSS *function* would do. We can do either by using the apply function. Let us start by getting the means of the variables:

```
> apply(myQmatrix, 2, mean, na.rm=TRUE)
      q1       q2       q3       q4
3.250000 2.750000 4.142857 3.750000
```

The apply function has three arguments and passes a fourth to the mean function.

1. The name of the matrix (or array) you wish to analyze.
2. The margin you want to apply the function over, with 1 representing rows and 2 for columns. This is easy to remember since R uses the index order of [rows,columns], so the margins values are [1,2] respectively.
3. The function you want to apply to each row or column. In our case, this is the mean function.
4. The apply function passes any other arguments onto the function you are applying. In our case, na.rm=TRUE is an argument for the mean function, not the apply function. If you look at the help file for the apply function, you will see its form is apply (X, MARGIN, FUN, ...). That means it only has three arguments, but will pass other arguments, indicated by the ellipses "...", to the function "FUN" (mean in our case). So na.rm=TRUE means nothing to the apply function other than something to pass to the function it is applying.

Applying the mean function to rows is as easy as changing the value 2, representing columns, to 1, representing rows:

```
> apply(myQmatrix, 1, mean, na.rm=TRUE)
       1        2        3        4        5
2.000000 2.000000 2.750000 2.333333 3.750000
       6        7        8
4.750000 4.000000 5.750000
```

Since means and sums are such popular calculations, there are specialized functions to get them: `rowMeans`, `colMeans`, `rowSums`, and `colSums`. For example, to get the row means of myQmatrix, we can do:

```
> rowMeans(myQmatrix, na.rm=TRUE)
       1        2        3        4        5
2.000000 2.000000 2.750000 2.333333 3.750000
       6        7        8
4.750000 4.000000 5.750000
```

To add a new variable to our data frame that is the mean of the q variables, we could any *one* of these forms:

```
> mydata$meanQ <- apply(myQmatrix, 1, mean, a.rm=TRUE)
> mydata$meanQ <   rowMeans(myQmatrix, na.rm=TRUE)
> mydata <- transform(mydata,
+    meanQ=rowMeans(myQmatrix, na.rm=TRUE)
+ )
> mydata

  workshop gender q1 q2 q3 q4     meanQ
1        R      f  1  1  5  1 2.000000
2      SAS      f  2  1  4  1 2.000000
3        R      f  2  2  4  3 2.750000
4      SAS   <NA>  3  1 NA  3 2.333333
5        R      m  4  5  2  4 3.750000
6      SAS      m  5  4  5  5 4.750000
7        R      m  5  3  4  4 4.000000
8      SAS      m  4  5  5  5 5.750000
```

Finally, we can apply a function to each vector in a data frame by using the `lapply` function. A data frame is a type of list, and the letter "l" in `lapply` stands for *list*. The function applies other functions to lists, and it returns its results in a list. Since it is clear we want to apply the function to each component in the list, there is no need for a row/column margin argument.

```
> lapply(mydata[ ,3:6], mean, na.rm=TRUE)

$q1
[1] 3.25

$q2
[1] 2.75

$q3
[1] 4.1429

$q4
[1] 3.25
```

Since the output is in the form of a list, it takes up more space when printed than the vector output from the `apply` function. You can also use the `sapply` function on a data frame. Its simplified vector output would be much more compact. The "s" in `sapply` means it simplifies its output whenever possible to vector or matrix form.

```
> sapply(mydata[ ,3:6], mean, na.rm=TRUE)

    q1      q2      q3      q4
1.2500  2.7500  4.1429  3.2500
```

Since the result is a vector, it is very easy to get the mean of the means:

```
> mean(
+     sapply(mydata[ ,3:6], mean, na.rm=TRUE)
+   )

[1] 3.3482
```

Other statistical functions that work very similarly are `sd` for standard deviation, `var` for variance, and `median`. The `length` function is similar to the SAS n function, but different enough to deserve its own section below.

14.2.2 Finding N or NVALID

In SAS saying, `N (q1,q2,q3,q4)`, or in SPSS saying, `NVALID (Q1 TO Q4)`, would count the valid responses on those variables for each observation. Running descriptive statistical procedures would give you the number of valid observations for each variable. R has several variations on this theme. First, let us look at the `length` function.

```
> length(mydata[ ,"q3"] )
[1] 8
```

The variable q3 has seven valid values and one missing value. The `length` function is telling us the number of total responses. Another approach is to ask for values that are not missing. The "!" sign means not, so let us try this:

```
> !is.na( mydata[ ,"q3"] )

[1]  TRUE  TRUE  TRUE FALSE  TRUE  TRUE  TRUE  TRUE
```

That identified them logically. Since TRUE has a value of 1 and FALSE a value of zero, summing them will give us the number of valid values.

```
> sum( !is.na( mydata[ ,"q3"] ) )

[1] 7
```

Jim Lemon and Philippe Grosjean's `prettyR` package has a function that does that very calculation. Let us load that library and apply the function to our data frame using `sapply`.

```
> library("prettyR")
> sapply(mydata, valid.n)
workshop   gender        q1       q2       q3       q4
       8        7         8        8        7        8
```

That is the kind of output we would get from descriptive statistics procedures in SAS or SPSS. In the chapter on statistics, we will see functions that provide that information and much more, like means and standard deviations.

What about applying it across rows? Let us create a myQn variable that contains the number of valid responses on q1 through q4. First, we will pull those variables out into a matrix. That will let us use the apply function to the rows.

```
> myMatrix <- as.matrix( mydata[ ,3:6] )
> myMatrix

  q1 q2 q3 q4
1  1  1  5  1
2  2  1  4  1
3  2  2  4  3
4  3  1 NA  3
5  4  5  2  4
6  5  4  5  5
7  5  3  4  4
8  4  5  5  5
```

Now we use the `apply` function with the 1 argument which asks it to go across rows.

```
> apply(myMatrix, 1, valid.n)
1 2 3 4 5 6 7 8
4 4 4 3 4 4 4 4

> mydata$myQn <- apply(myMatrix, 1, valid.n)
> mydata
  workshop gender q1 q2 q3 q4 myQn
1        1      f  1  1  5  1    4
2        2      f  2  1  4  1    4
3        1      f  2  2  4  3    4
4        2   <NA>  3  1 NA  3    3
5        1      m  4  5  2  4    4
6        2      m  5  4  5  5    4
7        1      m  5  3  4  4    4
8        2      m  4  5  5  5    4
```

Another form of the `apply` function is `tapply`. It exists to apply the function repeatedly to groups in the data and then place its results into two- or multi-way *t*ables (of class matrix or array). For details, see the section, *Aggregating or Summarizing Data.*

Finally, there is the `mapply` function, which is a *m*ultivariate version of `sapply`. See `help(mapply)` for details.

The functions we have examined in this section are very basic. They are more like what SAS and SPSS call functions. For functions that provide much more information, see Chap. 23. There we will examine R functions that operate more like SAS or SPSS procedures. Still, R does not differentiate one type of function from another as SAS and SPSS do for their functions and procedures. For example programs that demonstrate these topics in SAS, SPSS and R, see Table 14.3.

Table 14.3 Example programs for applying the `mean` and `length` functions

SAS programming statements	<pre>* SAS Program for calculating N and Mean
	* across both variables and observations;
	* FuncProc.sas ;
	LIBNAME myLib 'C: \myRfolder ';
	DATA myLib.mydata;
	SET myLib.mydata;
	myMean=MEAN (OF q1-q4);
	myN=N (OF q1-q4);
	RUN;
	PROC MEANS;
	   VAR q1-q4 myMean myN;
	RUN;</pre>

SPSS programming statements	<pre>* SPSS Program for calculating N and Mean
	* across both variables and observations.
	* FuncProc.sps .
	CD 'C: \myRfolder '.
	GET FILE='mydata.sav '.
	* Functions work for each observation (row).
	COMPUTE myMean=Mean(q1 TO q4).
	COMPUTE mySum=Sum(q1 TO q4).
	COMPUTE myN=mySum/myMean.
	* Procedures work for all observations (column).
	DESCRIPTIVES VARIABLES=q1 q2 q3 q4 myMean myN.</pre>

R programming statements	<pre># R Program for calculating N and Mean
	# across both variables and observations.
	# FuncProc.R
	setwd("/myRfolder")
	load(file="myWorkspace.RData")</pre>

Table 14.3 (continued)

```
mydata
attach(mydata)
# Create myQmatrix.
myQmatrix <- as.matrix( mydata[ ,3:6] )
myQmatrix
# Get mean of whole matrix.
mean( myQmatrix )
mean( myQmatrix,na.rm=TRUE )
# Get mean of matrix columns
apply(myQmatrix,2,mean,na.rm=TRUE)
# Get mean of matrix rows.
apply(myQmatrix,1,mean,na.rm=TRUE)
rowMeans (myQmatrix,na.rm=TRUE )

# Add row means to mydata.
mydata$meanQ <- rowMeans(myQmatrix,na.rm=TRUE)
mydata$meanQ <- rowMeans(myQmatrix,na.rm=TRUE)
mydata <- transform (mydata,
  meanQ=rowMeans (myQmatrix, na.rm=TRUE )
) mydata
# Means of data frames & their vectors.
mean(mydata, na.rm=TRUE )
lapply(mydata[ ,3:6] , mean, na.rm=TRUE)
sapply(mydata[ ,3:6] , mean, na.rm=TRUE )
mean (
sapply(mydata[ ,3:6] , mean, na.rm=TRUE)
)
# Length of data frames & their vectors.
length(mydata[ , "q3"] )
is.na( mydata[ , "q3"] )
!is.na(mydata[ , "q3"] )
sum( !is.na( mydata[ , "q3"] ) )
# Like the SAS /SPSS n from stat procedures.
library("prettyR")
sapply(mydata, valid.n )
# Like the SAS /SPSS n function.
apply(myMatrix, 1, valid.n )
mydata$myQn <- apply (myMatrix, 1, valid.n )
mydata
```

14.3 Conditional Transformations

Conditional transformations apply different formulas to various subgroups of
the data. For example, the formulas for recommended daily allowances of
vitamins differ for males and females. The ifelse function does conditional
transformations in a way that is virtually identical to what SAS and SPSS do.
The general form of the function is:

```
ifelse (test, yes, no)
```

Where "test" is a logical condition, "yes" is the value to return when the logic is true, and "no" is the value to return when the logic is false. For example, to create a variable that has a value of 1 for people who *strongly agree* with question 4 on our survey, we could use:

```
> mydata$q4Sagree <- ifelse( q4 == 5, 1,0 )

> mydata$q4Sagree


[1] 0 0 0 0 0 1 0 1
```

This is such a simple outcome that we can also do this using:

```
mydata$q4Sagree <- as.numeric( q4 == 5 )
```

However, this approach only allows the outcomes of 1 and 0 while the first version allows for any value. The statement q4==5 will result in a vector of logical TRUE/FALSE values. The as.numeric function converts it into zeros and ones.

R uses some different symbols for logical comparisons, such as the " = = " for logical equality. Table 14.4 shows the different symbols used by each package.

If we want a variable to indicate when people agree with question 4, that is, they responded with agree or strongly agree, we can use:

```
> mydata$q4agree <- ifelse( q4 >= 4, 1,0)

> mydata$q4agree


[1] 0 0 0 0 1 1 1 1
```

Table 14.4 Logical operators. See also help (Logic) and help (Syntax)

	SAS	SPSS	R
Equals	= or EQ	= or EQ	==
Less than	< or LT	< or LT	<
Greater than	> or GT	> or GT	>
Less than or equal	<= or LE	<= or LE	<=
Greater than or equal	>= or GE	>= or GE	>=
Not equal	^=, <> or NE	~= or NE	!=
And	& or AND	& or AND	&
Or	| or OR	| or OR	|
0<=x<=1	0 <=x <=1	(X >=) 0 AND (X <=1)	(x >= 0) & (x <=1)
Missing value size	Missing is less than all numbers	Comparisons with missing are set to NA	Comparisons with missing are set to NA
Testing for missing values	x = . Can be true.	x =SYSMIS Cannot be true. Also MISSING	is.na (x) Can be true. Bad!: x = =NA

The logical condition can be as complicated as you like. Here is one that creates a score of 1 when people took workshop 1 (abbreviated ws1) and agreed that it was good:

```
> mydata$ws1agree <- ifelse( workshop == 1 & q4 >=4 , 1,0)
> mydata$ws1agree
[1] 0 0 0 0 1 0 1 0
```

We can fill in equations that will supply values under the two conditions. The equations below for males and females are a bit silly but they make the point obvious. SAS users might think that if gender were missing, the second equation would apply. Luckily, R is more like SPSS in this regard and if gender is missing, it sets the result to missing.

```
> mydata$score <- ifelse( gender=="f", (2*q1)+q2, (3*q1)+q2 )
```

And here is our resulting data frame:

```
> mydata
```

	workshop	gender	q1	q2	q3	q4	q4Sagree	q4agree	ws1agree	score
1	1	f	1	1	5	1	0	0	0	3
2	2	f	2	1	4	1	0	0	0	5
3	1	f	2	2	4	3	0	0	0	6
4	2	<NA>	3	1	NA	3	0	0	0	NA
5	1	m	4	5	2	4	0	1	1	17
6	2	m	5	4	5	5	1	1	0	19
7	1	m	5	3	4	4	0	1	1	18
8	2	m	4	5	5	5	1	1	0	17

Example programs in all three languages are shown in Table 14.5.

Table 14.5 Example programs for conditional transformations

SAS programming statements

```
* SAS Program for Conditional Transformations;
* TranformIF.sas ;
LIBNAME myLib 'C:\myRfolder';
DATA myLib.mydata;
SET myLib.mydata;
  If q4= 5 then x1=1; else x1=0;
  If q4>=4 then x2=1; else x2=0;
  If workshop=1 & q4>=5 then  x3=1; else x3=0;
  If gender="f" then scoreA=2*q1+q2;
            Else scoreA=3*q1+q2;
  If workshop=1 and q4>=5 then scoreB=2*q1+q2;
            Else scoreB=3*q1+q2;
RUN;
```

SPSS programming statements

```
* SPSS Program for Conditional Transformations.
* TransformIF.sps
```

Table 14.5 (continued)

```
CD 'C:\myRfolder' ;
GET FILE=='mydata.sav' .
DO IF (Q4 eq 5).
+ COMPUTE X1=1.
ELSE.
+ COMPUTE X1=0.
END IF.
DO IF (Q4 GE 4).
+ COMPUTE X2=1.
ELSE.
+ COMPUTE X2=0.
END IF.
DO IF (workshop=1 AND q4 GE 4).
+ COMPUTE X3=1.
ELSE.
+ COMPUTE X3=0.
END IF.
DO IF (gender=='f').
+ COMPUTE scoreA=2*q1+q2.
ELSE.
+ COMPUTE scoreA=3*q1+q2.
END IF.
DO IF (workshop EQ 1 AND q4 GE 5).
+ COMPUTE scoreB=2*q1+q2.
ELSE.
+ COMPUTE scoreB=3*q1+q2.
END IF.
EXECUTE.
```

R programming
statements

```
# R Program for Conditional transformations.
# TransformIF.R
setwd("/myRfolder")
load(file="myWorkspace.RData")
mydata
attach(mydata)
# Create a series of dichotomous 0/1 variables
# The new variable q4SAgree will be 1 if q4 equals 5,
# (Strongly agree) otherwise zero.
mydata$q4Sagree <- ifelse( q4 == 5, 1,0)
mydata$q4Sagree
# This does the same as above.
mydata$q4Sagree <- as.numeric( q4 == 5 )
mydata$q4Sagree
# Create a score for people who agree with q4.
mydata$q4agree <- ifelse( q4 >= 4, 1,0)
mydata$q4agree
# Find the people only in workshop1 agree to item 5.
mydata$ws1agree <- ifelse( workshop == 1 & q4 >=4 , 1,0)
mydata$ws1agree
# Use equations to calculate values.
mydata$score <- ifelse( gender=="f", (2*q1)+q2, (3*q1)+q2 )
mydata
```

14.4 Multiple Conditional Transformations

Conditional transformations apply different sets of formulas to different subsets of the data. If you have only one formula to apply to each subgroup, read
Sect. 14.3. While SAS and SPSS have distinctly different approaches to multiple
conditional transformations, R's approach is the same as for single conditional
transformations. It does however let us look at some interesting variations in R.
   The simplest approach is to use ifelse a few times:

```
mydata$score1 <- ifelse( gender=="f", (2*q1)+q2, (20*q1)+q2 )
mydata$score2 <- ifelse( gender=="f", (3*q1)+q2, (30*q1)+q2 )
```

Above, the data frame gets two new (and rather silly) scores cleverly named
score1 and score2. These are calculated different ways for females and males. As
before, the ifelse functions have three arguments:

1. The gender=="f" argument tests the condition.
2. The first formula applies to the TRUE condition, the females.
3. The second formula applies to the FALSE condition, the males.

We can do the same thing using the index approach, but it is a bit trickier.
First, let us add the new score names to our data frame so that we can refer to
the columns by name:

```
> mydata <- data.frame( mydata, score1=0, score2=0 )

> mydata

  workshop gender q1 q2 q3 q4 score1 score2
1        1      f  1  1  5  1      0      0
2        2      f  2  1  4  1      0      0
3        1      f  2  2  4  3      0      0
4        2   <NA>  3  1 NA  3      0      0
5        1      m  4  5  2  4      0      0
6        2      m  5  4  5  5      0      0
7        1      m  5  3  4  4      0      0
8        2      m  4  5  5  5      0      0
```

Next, we want to differentiate between the genders. We can use the form
gender=="f" but we do not want to use it directly as indices to our data
frame because gender has a missing value. What should R do with mydata[
NA,]? Luckily the which function only cares about TRUE values, so we will
use that to locate the observations we want:

```
> gals <- which( gender=="f" )
> gals

[1] 1 2 3

> guys <- which( gender=="m" )
```

```
> guys
```

```
[1] 5 6 7 8
```

We can now use the gals and guys variables (American slang for women and men) to make the actual formula with the needed indices much shorter:

```
> mydata[gals,"score1"] <-  2*q1[gals] + q2[gals]
> mydata[gals,"score2"] <-  3*q1[gals] + q2[gals]
> mydata[guys,"score1"] <- 20*q1[guys] + q2[guys]
> mydata[guys,"score2"] <- 30*q1[guys] + q2[guys]
> mydata
```

	workshop	gender	q1	q2	q3	q4	score1	score2
1	1	f	1	1	5	1	3	4
2	2	f	2	1	4	1	5	7
3	1	f	2	2	4	3	6	8
4	2	<NA>	3	1	NA	3	0	0
5	1	m	4	5	2	4	85	125
6	2	m	5	4	5	5	104	154
7	1	m	5	3	4	4	103	153
8	2	m	4	5	5	5	85	125

We no longer need the guys and gals variables, so we can delete them with

```
rm(guys,gals)
```

For example programs the demonstrate these topics in SAS, SPSS and R, see Table 14.6.

Table 14.6 Example programs for multiple conditional transformations

SAS programming statements

```
* SAS Program for Multiple Conditional Transformations;
* TranformIF2.sas ;
LIBNAME myLib 'C:\myRfolder' ;
DATA myLib.mydata;
SET  myLib.mydata;
IF gender="m" THEN DO;
   score1 = (1.1*q1)+q2;
   score2 = (1.2*q1)+q2;
END;
ELSE IF gender="f" THEN DO;
   score1 = (2.1*q1)+q2;
   score2 = (2.2*q1)+q2;
END;
RUN;
```

Table 14.6 (continued)

SPSS programming statements

```
* SPSS Program for Multiple Conditional Transformations.
* TranformIF2.sps .
CD ' C:\myRfolder' .
GET FILE='mydata.sav' .
DO IF (gender EQ 'm' ).
+   COMPUTE score1 = (2*q1)+q2.
+   COMPUTE score2 = (3*q1)+q2.
ELSE IF (gender EQ 'f' ).
+   COMPUTE  score1 = (20*q1)+q2.
+   COMPUTE  score2 = (30*q1)+q2.
END IF.
EXECUTE.
```

R programming statements

```
# R Program for Multiple Conditional Transformations.
# TransformIF2.R
# Read the file into a data frame and print it.
setwd ("/myRfolder")
load(file="myWorkspace.RData")
attach(mydata)
mydata
# Using the ifelse approach.
mydata$score1 <-
    ifelse( gender=="f",(2*q1)+q2,(20*q1)+q2 )
mydata$score2 <-
    ifelse( gender=="f",(3*q1)+q2,(30*q1)+q2 )
mydata
# Using the index approach.
load(file="myWorkspace.RData")
# Create names in data frame.
mydata <- data.frame( mydata, score1=0, score2=0 )
attach(mydata)
mydata
# Find which are males and females.
gals <- which( gender=="f" )
gals
guys <- which( gender=="m" )
guys
mydata[gals,"score1"] <-  2*q1[gals] + q2[gals]
mydata[gals,"score2"] <-  3*q1[gals] + q2[gals]
mydata[guys,"score1"] <- 20*q1[guys] + 2[guys]
mydata[guys,"score2"] <- 30*q1[guys] + 2[guys]
mydata
# Clean up.
rm(guys, gals)
```

14.5 Missing Values

We have discussed missing values briefly in several previous chapters. Let us bring those various topics together to review and expand upon. R represents missing values with NA, for Not Available. It also uses NaN (Not a Number) to represent missing values for impossible calculations, such as the square root of a negative number. The letters NA and NaN are also objects in R that you can use to assign missing values. Unlike SAS, the value used to store the NA is not the smallest number your computer can store, so logical comparisons such as $x < 0$ will result in NA when x is missing. SAS would give a result of TRUE instead while the SPSS result would be missing.

When importing numeric data, R reads blanks as missing (except when blanks are delimiters). R reads the string NA as missing for both numeric and character variables.

When importing a text file, both SAS and SPSS would recognize a period as a missing value for numeric variables. R will instead *read the whole variable as a character vector!* If you have control of the source of the data, it is best not to write them out that way. If not, you can use a text editor to replace the periods with NA, but you have to be careful to do so in a way that does not also replace valid decimal places. Some editors make that easier than others. A safer method would be to fix it in R, which we do below.

When other values represent missing, you will of course have to tell R about them. The read.table function provides an argument, na.strings, which allows you to provide a set of missing values. However, it applies that value to every variable, so its usefulness is limited. Here is a dataset that we will use to demonstrate the various ways to set missing values. The data frame we use, mydataNA, is the same as mydata in our other examples, except that it uses several missing value codes:

```
> mydataNA <- read.table ("mydataNA.txt")

> mydataNA
```

	workshop	gender	q1	q2	q3	q4
1	1	f	1	1	5	1
2	2	f	2	1	4	99
3	.	f	9	2	4	3
4	2	.	3	9	99	3
5	1	m	4	5	2	4
6	.	m	9	9	5	5
7	1	.	5	3	99	4
8	2	m	4	5	5	99

In the data we see that workshop and gender have periods as missing values, q1 and q2 have 9s and q3 and q4 have 99s. Do not be fooled by the periods in workshop and gender, they are not already set to missing! If so, they would have appeared as NA instead. R has seen the periods and converted both variables to character (string) variables. Since read.table converts string variables to factors unless the as.is = TRUE argument is added, both workshop and gender are now factors.

We can set all three codes to missing by simply adding the na.strings argument to the read.table function:

```
>   mydataNA   <-   read.table ("mydataNA.txt" ,
+      na.strings = c (".", "9", "99")    )

>   mydataNA

      workshop gender q1 q2 q3 q4
1            1      f  1  1  5  1
2            2      f  2  1  4 NA
3           NA      f NA  2  4  3
4            2   <NA>  3 NA NA  3
5            1      m  4  5  2  4
6           NA      m NA NA  5  5
7            1   <NA>  5  3 NA  4
8            2      m  4  5  5 NA
```

If the data did not come from a text file, we could still easily scan every variable for 9 and 99 to replace with missing values using:

```
mydataNA[ mydataNA == 9  |  mydataNA == 99]   <-   NA
```

Both of the above approaches treat all variables alike. If some variables, like age, had a valid value of 99, see Sect. 14.5.3. Of course "." never has meaning by itself, so getting rid of them with na.strings="." is usually fine.

14.5.1 Substituting Means for Missing Values

There are several methods for replacing missing values with estimates of what they would have been. These methods include simple mean substitution, regression and the gold standard, multiple imputation. We will just do mean substitution. See the packages listed in Appendix B: Comparison of SAS and SPSS Products with R Packages and Functions under *Missing Value Analysis* for a list of alternatives.

Any logical comparison on NAs results in an NA outcome, so q1==NA will *never* be TRUE, even when q1 is indeed NA. Therefore, if you wanted to substitute another value such as the mean, you would need to use the is.na function. Its output is TRUE when a value is NA. Here is how you can use it to substitute missing values.

```
mydataNA$q1[is.na(q1)] <-
  mean( mydataNA$q1, na.rm=TRUE )
```

On the left-hand side, the statement above selects mydataNA$q1 as a vector and then finds its missing elements with is.na(mydata$q1). On the right, it calculates the mean of mydataNa$q1 across all observations to assign to those NA values on the left.

14.5.2 Finding Complete Observations

You can omit all observations that contain *any* missing values with the na.omit function. The new data frame, myNoMissing contains no missing values for any variables.

```
> myNoMissing  <-  na.omit(mydataNA)
> myNoMissing
  workshop gender q1 q2 q3 q4
1        1         f  1  1  5  1
5        1         m  4  5  2  4
```

Yikes! We do not have much data left. Thank goodness this is not our dissertation data.

The complete.cases function returns a value of TRUE when a case is complete, that is, when an observation has no missing values:

```
> complete.cases(mydataNA)
[1]   TRUE FALSE FALSE FALSE  TRUE FALSE FALSE FALSE
```

Therefore, we can use this to get the cases that have no missing values (the same result as the na.omit function) by doing:

```
> myNoMissing  <-  mydataNA[ complete.cases(mydataNA) ,  ]
> myNoMissing
  workshop gender q1 q2 q3 q4
1        1         f  1  1  5  1
5        1         m  4  5  2  4
```

Since we already saw na.omit do that, it is of more interest to do the reverse. If we want to see which observations contain *any* missing values we can use "!" for NOT:

```
> myIncomplete <- mydataNA[ !complete.cases(mydataNA) , ]
> myIncomplete
  workshop gender   q1  q2  q3  q4
2        2       f    2   1   4  NA
3       NA       f   NA   2   4   3
4        2    <NA>    3  NA  NA   3
6       NA       m   NA  NA   5   5
7        1    <NA>    5   3  NA   4
8        2       m    4   5   5  NA
```

14.5.3 When "99" Has Meaning

Occasionally, datasets use different missing values for different sets of variables. In that case, the methods described above would not work because they assume *every* missing value code applies to *all* variables.

Variables often have several missing value codes to represent things like, "not applicable," "do not know," "refused to answer." Early statistics programs used to read blanks as zeros, so researchers got used to filling their fields with as many 9s as would fit. For example, a two-column variable such as years of education, would use 99 represent missing. It might also have a variable like age, for which 99 is a valid value. Age, requiring three columns would have a missing value of 999. Data archives like the Interuniversity Consortium of Political and Social Research (ICPSR) have many datasets coded with multiple values for missing.

We will use conditional transformations, covered in the previous chapter, to address this problem. Let us read the file again, and put NAs in for the values 9 and 99 independently:

```
> mydataNA <- read.table("mydataNA.txt", na.strings=".")

> attach(mydataNA)

> mydataNA$q1[q1==9]  <- NA
> mydataNA$q2[q2==9]  <- NA
> mydataNA$q3[q3==99] <- NA
> mydataNA$q4[q4==99] <- NA

> mydataNA
```

	workshop	gender	q1	q2	q3	q4
1	1	f	1	1	5	1
2	2	f	2	1	4	NA
3	NA	f	NA	2	4	3
4	2	<NA>	3	NA	NA	3
5	1	m	4	5	2	4
6	NA	m	NA	NA	5	5
7	1	<NA>	5	3	NA	4
8	2	m	4	5	5	NA

That approach can handle any values we might have and assign NAs only where appropriate, but it would be quite tedious with hundreds of variables. We have used the apply family of functions to execute the same function across sets of variables. We can use that method here. First, we need to create some function, letting x represent each variable:

```
my9isNA   <- function(x) {  x[x==9 ] <- NA; x}
my99isNA  <- function(x) {  x[x==99 ] <- NA; x}
```

These functions will return a value of NA when x = = 9 or x = = 99, and will return a value of just x (the original value the variable contained) if they are false. If you leave off that last "...x}" above, what will the functions return when the conditions are false? That would be undefined, so *every* value would become NA!

Now we need to apply each function where it is appropriate, using the lapply function.

```
> mydataNA <- read.table("mydataNA.txt", na.strings=".")
> attach(mydataNA)
>
> mydataNA[3:4] <- lapply( mydataNA[3:4], my9isNA )
> mydataNA[5:6] <- lapply( mydataNA[5:6], my99isNA )
> mydataNA

  workshop gender   q1   q2   q3   q4
1        1      f    1    f    1    f
2        2      f    2    f    2    f
3     <NA>      f <NA>    f <NA>    f
4        2   <NA>    2 <NA>    2 <NA>
5        1      m    1    m    1    m
6     <NA>      m <NA>    m <NA>    m
7        1   <NA>    1 <NA>    1 <NA>
8        2      m    2    m    2    m
```

With our small data frame, this has not saved much effort. However, to handle thousands of variables, all we would need to change are the indices above from 3:4 and 5:6 to perhaps 3:4000, 4001:6000. Example programs than demonstrate these topics in SAS, SPSS and R are contained in Table 14.7.

Table 14.7 Example programs to assign missing values

SAS programming statements

```
* SAS Program to Assign Missing Values;
* MissingValues.sas ;
LIBNAME myLib 'C:\myRfolder' ;
DATA myLib.mydata;
SET  myLib.mydata;
*Convert 9 to missing, one at a time.
IF q1=9 THEN q1=.;
IF q2=9 THEN q2=.;
IF q3=99 THEN q2=.;
IF q4=99 THEN q4=.;
* Same thing but is quicker for lots of vars;
ARRAY q9 q1-q2;
DO OVER q9;
   IF q9=9 THEN q=.;
END;
ARRAY q99 q1-q2;
DO OVER q99;
   IF q=99 THEN q99=.;
END;
```

Table 14.7 (continued)

SPSS programming statements

```
* SPSS Program to Assign Missing Values.
* MissingValues.sps .
CD 'C:\myRfolder' .
GET FILE=('mydata.sav')
MISSING q1 TO q2 (9) q3 TO q4 (99).
SAVE OUTFILE='mydata.sav' .
```

R programming statements

```
# R Program to Assign Missing Values.
# MissingValues.R
setwd("/myRfolder")
# Read the data to see what it looks like.
mydataNA <- read.table("mydataNA.txt")
mydataNA
# Now read it so that ".", 9, 99 are
# converted to missing.
mydataNA <- read.table("mydataNA.txt",
   na.strings=c(".", "9", "99") )
mydataNA
# Convert 9 and 99 manually
mydataNA <- read.table("mydataNA.txt",
   na.strings=".")
mydataNA[mydataNA==9 | mydataNA==99] <- NA
mydataNA
# Substitute the mean for missing values.
mydataNA$q1[is.na(mydataNA$q1)] <-
   mean(mydataNA$q1, na.rm=TRUE)
mydataNA
# Eliminate observations with any NAs.
myNoMissing <- na.omit(mydataNA)
myNoMissing
# Test to see if each case is complete.
complete.cases(mydataNA)
# Use that result to select compete cases.
myNoMissing <- mydataNA[ complete.cases(mydataNA), ]
myNoMissing
# Use that result to select incomplete cases.
myIncomplete <- mydataNA[ !complete.cases(mydataNA), ]
myIncomplete
# When "99" Has Meaning...
# Now read it and set missing values
# one variable at a time.
mydataNA <- read.table("mydataNA.txt", na.strings=".")
mydataNA
attach(mydataNA)
# Assign missing values for q variables.
mydataNA$q1[q1==9]  <- NA
mydataNA$q2[q2==9]  <- NA
mydataNA$q3[q3==99] <- NA
mydataNA$q4[q4==99] <- NA
mydataNA
```

Table 14.7 (continued)

```
detach(mydataNA)
# Read file again, this time use functions.
mydataNA <- read.table("mydataNA.txt",na.strings=".")
mydataNA
attach(mydataNA)
#Create a functions that replaces 9, 99 with NAs.
my9isNA  <- function(x){ x[ x==9 ] <- NA; x}
my99isNA <- function(x){ x[ x==99] <- A; x}
# Now apply our functions to the data frame using lapply.
mydataNA[ 3:4] <- lapply( mydataNA[ 3:4] , my9isNA )
mydataNA[ 5:6] <- lapply( mydataNA[ 5:6] , my99isNA )
mydataNA
```

14.6 Renaming Variables (... and Observations)

In SAS and SPSS, you do not know where variable names are stored or how. You just know they are in the dataset somewhere. Renaming is simply a matter of matching the new name to the old name with a RENAME statement. In R however, both row and column names are stored in attributes, essentially character vectors, within the data frame. In essence, they are just another form of variable that you can manipulate.

If you use Microsoft Windows, you can see the names in the data editor, and changing them there by hand is a very easy way to rename them. The command, fix(mydata), brings up the data editor. Clicking on the name of a variable opens a box that enables you to change its name. In Fig. 14.1, is an image that shows this process. I am in the midst of changing the name of the variable q1 (see the name in the spreadsheet) to x1.

Closing the Variable Name box, then the Data Editor completes your changes.

If you use Macintosh or Linux, the fix function does not work this way. However, on any operating system you can use functions to change variable names.

The programming approach to changing names that feels the closest to SAS or SPSS is the rename function in Hadley Wickham's reshape package [22]. To use it, install the package, then load it with the library function and create a character vector whose values are the new names. The names of the vector elements are the names of the old variables you want to change. This approach offers the distinct advantage of allowing you to rename only a subset of your variables.

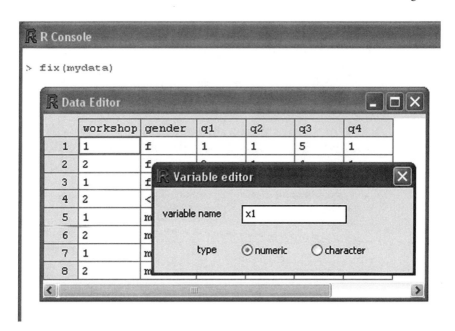

Fig. 14.1 Renaming a variable using R's data editor

```
> library("reshape")

> myChanges <- c(q1="x1", q2="x2", q3="x3", q4="x4")

> myChanges

   q1    q2    q3    q4
 "x1"  "x2"  "x3"  "x4"
```

Now it is very easy to change names with:

```
> mydata <- rename(mydata, myChanges)
> mydata

  workshop gender x1 x2 x3 x4
1        1      f  1  1  5  1
2        2      f  2  1  4  1
3        1      f  2  2  4  3
4        2   <NA>  3  1 NA  3
5        1      m  4  5  2  4
6        2      m  5  4  5  5
7        1      m  5  3  4  4
8        2      m  4  5  5  5
```

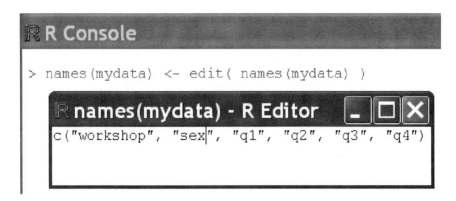

Fig. 14.2 Renaming variables using the `edit` function

R's built-in programming approach to renaming variables uses the `names` function. Simply entering `names(mydata)` causes R to print out the names vector.

```
> names(mydata)

[1] "group"  "gender" "q1"     "q2"     "q3"     "q4"
```

You can also assign a character vector of equal length to that function, which renames the variables. With this approach, you can supply a name for every variable.

```
> names(mydata) <- c("group", "gender", "x1", "x2",
  "x3", "x4")

> mydata

  group gender x1 x2 x3 x4
1     1      f  1  1  5  1
2     2      f  2  1  4  1
...
```

You can also use subscripting for this type of renaming. Since `names (mydata)` is a character vector of names, and its second element is "gender", you can change that to "sex" like this.

```
names(mydata)[2] <- "sex"
```

The `edit` function, described in the *Data Editor* section, will generate a character vector of variable names, complete with the `c` function and parentheses. In Fig. 14.2, you can see the command I entered and the window that it opened, titled *names(mydata) – R Editor*. I have changed the name of the variable "gender" to "sex". When I finish my changes, closing the box will execute the command.

14.7 Renaming Variables – Advanced Examples

The methods shown above are often sufficient to rename your variables. You can view this section as either beating the topic to death, or as a wonderful opportunity to extend what you have learned about R into further examples. I think it is worthwhile because if you read the r-help listserv, you will see these methods used to rename variables. Section 14.7.3 can be a real time saver.

14.7.1 Renaming by Index

Let us extract the names of our variables using the names function.

```
> mynames <- names(mydata)

> mynames

[1] "group"  "gender" "q1"      "q2"      "q3"      "q4"
```

Now we have a character vector whose values we can change using the R techniques we have covered elsewhere. Print out the variable names with a counter so that we can see the index value of each:

```
> data.frame(mynames)

   mynames
1    group
2   gender
3       q1
4       q2
5       q3
6       q4
```

We see from the list above that q1 is the third name and q4 is the sixth. We can now use that information to enter new names directly into this vector, and print the result so that we can see if we had made errors:

```
> mynames[3] <- "x1"
> mynames[4] <- "x2"
> mynames[5] <- "x3"
> mynames[6] <- "x4"

> mynames
[1] "group"  "gender" "x1"      "x2"      "x3"      "x4"
```

That looks good, so let us place those new names into the names attribute of our data frame and look at the results:

```
> names(mydata) <- mynames

> mydata
```

```
   group gender x1 x2 x3 x4
1      1       f  1  1  5  1
2      2       f  2  1  4  1
...
```

As you will see in the program below, each time we do another method of name changes, we need to restore the old names to demonstrate the new techniques. We can accomplish that by either reloading our original data frame or by using this:

```
names(mydata) <- c("group", "gender", "q1", "q2",
   "q3", "q4")
```

14.7.2 Renaming by Column Name

If you prefer to use variable names instead of indices, that is easy to do. We will make another copy of mynames:

```
> mynames <- names(mydata)
> mynames
[1] "group"  "gender" "q1"      "q2"      "q3"      "q4"
```

Now we will make the same changes but using a logical match to find where mynames = = "q1" and so on, and assigning the new names to those locations.

```
> mynames[ mynames=="q1" ] <- "x1"
> mynames[ mynames=="q2" ] <- "x2"
> mynames[ mynames=="q3" ] <- "x3"
> mynames[ mynames=="q4" ] <- "x4"

> mynames

[1] "group"  "gender" "x1"      "x2"      "x3"      "x4"
```

Finally, we put the new set mynames into the names attribute of our data frame, mydata.

```
> names(mydata) <- mynames

> mydata

   group gender x1 x2 x3 x4
1      1       f  1  1  5  1
2      2       f  2  1  4  1
...
```

You can combine all these steps into one, but I find it *very* confusing to read.

```
names(mydata)[names(mydata)=="q1"] <- "x1"
names(mydata)[names(mydata)=="q2"] <- "x2"
```

```
names(mydata)[names(mydata)=="q3"] <- "x3"
names(mydata)[names(mydata)=="q4"] <- "x4"
```

14.7.3 Renaming Many Sequentially Numbered Variable Names

Our next example works well if you are changing many variable names, like a 100 variables named x1, x2, ... over to similar names like y1, y2, ... You occasionally have to make changes like this when you measure many variables at different times and you need to rename the variables in each dataset before joining them all.

We learned how the paste function can append sequential digits onto any string in Sect. 10.4. We will use it here to create the new variable names:

```
> myXs <- paste( "x", 1:4, sep="")

> myXs

[1] "x1" "x2" "x3" "x4"
```

Now we need to find out where to put the new names. We already know this of course, but we found that out in the previous example by listing all the variables. If we had thousands of variables, that would not be a very good method. We will use the method we covered previously in Sect. 10.11:

```
> myA <- which( names(mydata)=="q1" )

> myA

[1] 3

> myZ <- which( names(mydata)=="q4" )

> myZ

[1] 6
```

Now we know the indices of the variable names to replace, we can replace them with the following. Keep in mind that this assumes the variables are stored next to each other in the data set, and are also in numeric order. That is usually the case, but it bears keeping in mind.

```
> names(mydata)[myA:myZ] <- myXs

> mydata

  group gender x1 x2 x3 x4
1     1      f  1  1  5  1
2     2      f  2  1  4  1
...
```

14.7.4 Renaming Observations

R has row names that work much the same as variable names, but they apply to observations. These names must be unique and often come from an ID variable. When reading a text file using `read.table`, the `row.names` argument allows you to specify an ID variable. See *Reading Delimited Text Files* in Section 9.2 for details.

When renaming rows using a variable like id, you must select it so that it will pass as a vector. In the examples below, the first two select a variable named "id" as a vector so they work. The third approach looks almost like the first, but it selects id as a data frame, which will not fit in the row.names attribute.

```
> row.names(mydata) <- mydata[ ,"id"] #This works.

> row.names(mydata) <- mydata$id       #This works too.

> row.names(mydata) <- mydata[["id"]]  #This does too.

> row.names(mydata) <- mydata["id"]    #This does not.

Error in `row.names<-.data.frame`(`*tmp*`, value =
   list(id = c(1, 2, 3,   :
         invalid 'row.names' length
```

Example programs which demonstrate renaming variables in SAS, SPSS and R are found in Table 14.8.

Table 14.8 Example programs for renaming variables

SAS programming statements

```
* SAS Program for Renaming Variables;
* Rename.sas ;
LIBNAME myLib 'C:\myRfolder' ;
DATA myLib.mydata;
  RENAME q1-q4=x1-x4;
  *or;
  *RENAME q1=x1  q2=x2  q3=x3  q4=x4;
RUN;
```

SPSS programming statements

```
* SPSS Program for Renaming Variables.
* Rename.sps .
CD 'C:\myRfolder' .
GET FILE='mydata.sav' .
RENAME VARIABLES (Q1=X1)(Q2=X2)(Q3=X3)(Q4=X4) .
```

Table 14.8 (continued)

R programming statements

```
# R Program for Renaming Variables.
# Rename.R
setwd("/myRfolder")
load(file="myWorkspace.RData")
mydata
#---This uses the data editor.
# Make the changes by clicking on the names in the
spreadsheet, then closing it.
fix(mydata)
mydata
# Restore original names for next example.
names(mydata) <- c("group", "gender",
  "q1", "q2", "q3", "q4")
#---This method is most like SAS or SPSS.
# It is easy to understand but does not
# help you understand what R is actually doing.
# It requires the reshape package.
library("reshape")
myChanges <- c(q1="x1",q2="x2",q3="x3",q4="x4")
myChanges

mydata <- rename(mydata, myChanges)
mydata
# Restore original names for next example.
names(mydata) <- c("group", "gender",
  "q1", "q2", "q3", "q4")
#---Simplest renaming with no packages required.
# With this approach, you simply list every name
# in order, even those you don't need to change.
names(mydata) <- c("group", "gender",
  "x1", "x2", "x3", "x4")
mydata
# Restore original names for next example.
names(mydata) <- c("group", "gender",
  "q1", "q2", "q3", "q4")
#---The edit function actually generates
# a list of names for you. If you edit them
# and close the window, they will change.
names(mydata) <- edit( names(mydata) )
mydata
# Restore original names for next example.
names(mydata) <- c ("workshop", "gender",
    "q1", "q2", "q3", "q4")
#---This method uses the row index numbers.
# First, extract the names to work on.
mynames <- names(mydata)
# Now print the names. Data.frame adds index numbers.
data.frame(mynames))
mynames[3] <- "q1"
mynames[4] <- "q2"
mynames[5] <- "q3"
mynames[6] <- "q4"
```

Table 14.8 (continued)

```
names (mydata) <- mynames #Put new names into data frame.
mydata

# Restore original names for next example.
names (mydata) <- c ("group", "gender",
  "q1", "q2", "q3", "q4")
#---Here's the exact same example, but now you
# do not need to know the order of each variable
# i.e. that q1 is column[3] .
mynames <- names (mydata) #Make a copy to work on
mynames
mynames[ mynames=="q1" ] <- "x1"
mynames[ mynames=="q2" ] <- "x2"
mynames[ mynames=="q3" ] <- "x3"
mynames[ mynames=="q4" ] <- "x4"
mynames
# Finally replace the names with the new ones.
names (mydata) <- mynames
mydata
# Restore original names for next example.
names (mydata) <- c ("group", "gender",
  "q1", "q2", "q3", "q4")
#---You can do the steps above without working
# on a copy, but I find it VERY confusing to read.
names (mydata)[ names (mydata)=="q1"] <- "x1"
names (mydata)[ names (mydata)=="q2"] <- "x2"
names (mydata)[ names (mydata)=="q3"] <- "x3"
names (mydata)[ names (mydata)=="q4"] <- "x4"
print (mydata)
# Restore original names for next example.
names (mydata) <- c ("group", "gender",
  "q1", "q2", "q3", "q4")
#---This approach works well for lots of numbered
# variable names names like x1,x2...

# First we'll see what the names look like.
names (mydata)
# Next we'll generate x1,x2,x3,x4
#               to replace q1,q2,q3,q4.
myXs <- paste ( "x", 1:4, sep="")
myXs
# Now we want to find out where to put the new names.
myA <- which ( names (mydata)=="q1" )
myA
myZ <- which ( names (mydata)=="q4" )
myZ
# Replace q1 thru q4 at index values A thru Z with
# the character vector of new names.
names (mydata)[ myA:myZ] <- myXs (mydata)
#remove the unneeded objects.
rm (myXs, myA, myZ)
```

14.8 Recoding Variables

Recoding is just a simpler way of doing a set of similar IF/THEN conditional transformations. Survey researchers often collapse five-point Likert-scale items to simpler three-point Disagree/Neutral/Agree scales to summarize results. They also often recode values to reverse the scale of negatively worded items so that a large numeric value has the same meaning across all items. It is easier to reverse scales by subtracting each score from 6 as in

```
qr1 <- 6-q1
```

That results in $6 - 5 = 1$, $6 - 1 = 5$, and so on.

SAS does not have a separate recode procedure as SPSS does, but it does offers a similar capability using its value label formats. That has the useful feature of applying the formats in categorical analyses and ignoring them otherwise. For example, PROC UNIVARIATE will ignore it. You can also recode the data with a series of IF/THEN statements. For simplicity, I leave the value labels out of the SPSS and R programs. I cover those in Chap. 15.

For recoding continuous variables into categorical, see the cut function in base R, and the cut2 function in the Hmisc package. For choosing optimal cut points with regard to a target variable, see the rpart function in rpart package [23], or the tree function in the Hmisc package.

It is wise to avoid modifying your original data, so that recoded variables are typically stored under new names. If you named your original variables q1, q2, ... then you might name the recoded ones qr1, qr2, ..., with "r" representing recoded.

14.8.1 Recoding a Few Variables

We will work with the recode function from the car package, which you will have to install before running this. See *Installing R and Add-on Packages* in Section 5.1 for details. We will apply it below to collapse our five-point scale down to a three-point one representing just disagree, neutral, agree.

```
> library("car")
> mydata$qr1 <- recode(q1, "1=2; 5=4")
> mydata$qr2 <- recode(q2, "1=2; 5=4")
> mydata$qr3 <- recode(q3, "1=2; 5=4")
> mydata$qr4 <- recode(q4, "1=2; 5=4")
> mydata
```

	workshop	gender	q1	q2	q3	q4	qr1	qr2	qr3	qr4
1	1	f	1	1	5	1	2	2	4	2
2	2	f	2	1	4	1	2	2	4	2
3	1	f	2	2	4	3	2	2	4	3
4	2	<NA>	3	1	NA	3	3	2	NA	3

```
5            1        m  4  5  2  4    4    4    2    4
6            2        m  5  4  5  5    4    4    4    4
7            1        m  5  3  4  4    4    3    4    4
8            2        m  4  5  5  5    4    4    4    4
```

The recode function needs only two arguments, the variable you wish to recode and a string of values in the form "old1 = new1; old2 = new2; ...."

14.8.2 Recoding Many Variables

That approach worked fine with our tiny dataset, but in a more realistic situation, we would have many variables to recode. So let us scale this example up. We learned how to rename variables in the previous section, so we will use that knowledge here.

```
> myQnames <-  paste( "q",  1:4, sep="")
> myQnames

[1] "q1" "q2" "q3" "q4"

> myQRnames <- paste( "qr", 1:4, sep="")
> myQRnames

[1] "qr1" "qr2" "qr3" "qr4"
```

Now we will use the original names to extract the variables we want to recode to a separate data frame.

```
> myQRvars <- mydata[ ,myQnames]

> myQRvars

  q1 q2 q3 q4
1  1  1  5  1
2  2  1  4  1
3  2  2  4  3
4  3  1 NA  3
5  4  5  2  4
6  5  4  5  5
7  5  3  4  4
8  4  5  5  5
```

We will use our other set of variable names to rename the variables we just selected.

```
> names(myQRvars) <- myQRnames

> myQRvars

  qr1 qr2 qr3 qr4
1   1   1   5   1
2   2   1   4   1
```

```
3   2   2   4    3
4   3   1   NA   3
5   4   5   2    4
6   5   4   5    5
7   5   3   4    4
8   4   5   5    5
```

Now we need to create a function that will allow us to apply the recode function to each of the selected variables. Our function only has one argument, x, which will represent each of our variables.

```
myRecoder <- function(x) {  recode(x,"1=2;5=4")  }
```

Here is how we can use myRecoder on a single variable. Notice that the qr1 variable had a 1 for the first observation, which myRecoder made a 2. It also had values of 5 for the sixth and seventh observations, which became 4s.

```
> myQRvars$qr1

[1] 1 2 2 3 4 5 5 4

> myRecoder(myQRvars$qr1)

[1]  2 2 2 3 4 4 4 4
```

To apply this function to our whole data frame, myQRvars, we can use the sapply function.

```
> myQRvars <- sapply( myQRvars, myRecoder)

> myQRvars

      qr1 qr2 qr3 qr4
[1,]   2   2   4    2
[2,]   2   2   4    2
[3,]   2   2   4    3
[4,]   3   2  NA    3
[5,]   4   4   2    4
[6,]   4   4   4    4
[7,]   4   3   4    4
[8,]   4   4   4    4
```

The sapply function has converted our data frame to a matrix, but that is fine. We will use the cbind function to bind these columns to our original data frame.

```
> mydata <- cbind(mydata,myQRvars)

> mydata

  workshop gender q1 q2 q3 q4 qr1 qr2 qr3 qr4
1        1      f   1  1  5  1   2   2   4    2
2        2      f   2  1  4  1   2   2   4    2
```

3	1	f	2	2	4	3	2	2	4	3
4	2	\<NA\>	3	1	NA	3	3	2	NA	3
5	1	m	4	5	2	4	4	4	2	4
6	2	m	5	4	5	5	4	4	4	4
7	1	m	5	3	4	4	4	3	4	4
8	2	m	4	5	5	5	4	4	4	4

Now we can use either the original variables or their recoded counterparts in any analysis we choose.

In this simple case, it was not necessary to create the myRecoder function. We could have used the form:

```
sapply(myQRvars, myRecoder,  "1=2;5=4")
```

However, you can generalize the approach we took to far more situations. For example programs that demonstrate these topics in SAS, SPSS and R, see Table 14.9.

Table 14.9 Example programs for recoding variables

SAS programming statements

```
* SAS Program for Recoding Variables;
* Recode.sas ;
LIBNAME myLib ' C:\myRfolder' ;
DATA myLib.mydata;
INFILE '\myRfolder\mydata.csv' delimiter = ','
   MISSOVER DSD LRECL=32767 firstobs=2 ;
INPUT id workshop gender$q1 q2 q3 q4;
PROC PRINT; RUN;
PROC FORMAT;
   VALUE Agreement 1="Disagree" 2="Disagree"
                   3="Neutral"
                   4="Agree"    5="Agree"; run;
DATA myLib.mydata;
   SET myLib.mydata;
   ARRAY q q1-q4;
   ARRAY qr qr1-qr4; *r for recoded;
   DO i=1 to 4;
   qr{i} =q{i} ;
   if q{i} =1 then qr{i} =2;
   else
   if q{i} =5 then qr{i} =4;
  END;
  FORMAT q1-q4 q1-q4 Agreement.;
RUN;
* This will use the recoded formats automatically;
PROC FREQ; TABLES q1-q4; RUN;
* This will ignore the formats;
* Note high/low values are 1/5;
PROC UNIVARIATE; VAR q1-q4; RUN;
* This will use the 1-3 codings, not a good idea!;
* High/Low values are now 2/4;
PROC UNIVARIATE; VAR qr1-qr4;
   RUN;
```

Table 14.9 (continued)

SPSS programming statements

```
* SPSS Program for Recoding Variables.
* Recode.sps .
CD 'C:\myRfolder' .
GET FILE='mydata.sav' .
RECODE q1 to q4 (1=2) (5=4) .
SAVE OUTFILE='C:\myRfolder\myleft.sav' .
```

R programming statements

```
# R Program for Recoding Variables.
# Recode.R
setwd("/myRfolder")
load(file="myWorkspace.RData")
mydata
attach(mydata)
library("car")
mydata$qr1 <- recode(q1, "1=2; 5=4")
mydata$qr2 <- recode(q2, "1=2; 5=4")
mydata$qr3 <- recode(q3, "1=2; 5=4")
mydata$qr4 <- recode(q4, "1=2; 5=4")
mydata
# Do it again, stored in new variable names.
load(file="myWorkspace.RData")
attach(mydata)
# Generate two sets of var names to use.
myQnames <-  paste( "q",  1:4, sep="")
myQnames
myQRnames <- paste( "qr", 1:4, sep="")
myQRnames
# Extract the q variables to a separate data frame.
myQRvars <- mydata[ ,myQnames]
myQRvars
# Rename all the variables with R for Recoded.
names(myQRvars) <- myQRnames
myQRvars
# Create a function to apply the labels to lots of variables.
myRecoder <- function(x) {  recode(x,"1=2;5=4") }
# Here's how to use the function on one variable.
myQRvars$qr1
myRecoder(myQRvars$qr1)
#Apply it to all the variables.
myQRvars <- sapply( myQRvars, myRecoder)
myQRvars
# Save it back to mydata if you want.
mydata <- cbind(mydata,myQRvars)
mydata
summary(mydata)
```

14.9 Keeping and Dropping Variables

In SAS, you use the KEEP and DROP statements to determine which variables to save in your dataset. The SPSS equivalent is the DELETE VARIABLES statement. In R, the main methods to do this within a data frame are in the section on *Selecting Variables*. For example, if we want to keep variables on the left side of our data frame, workshop through q2, (variables 1 through 4) an easy way to do this is with:

```
myleft <- mydata[ ,1:4]
```

Those are the variables on the left side of our original data frame, hence the name, myleft. We will strip off the ones on the right in a future example on merging data frames.

Another way to drop variables is to assign the NULL object to the variable:

```
mydata$varname <- NULL
```

This has the advantage of removing a variable without having to make a copy of the data frame. That may come in handy with data frames so large that your workspace will not hold a copy, but it is usually much safer to work on copies when you can. Mistakes happen! You can apply NULL repeatedly with the form:

```
myleft <- mydata
myleft$q3 <- myleft$q4 <- NULL
```

NULL is only used to remove components from data frames and lists. You cannot use it to drop elements of a vector, nor can you use it to remove a vector by itself from your workspace.

Later in Chap. 19, we will discuss removing objects using the rm function. That function removes only whole objects from your workspace; it cannot remove variables from a data frame:

```
rm( mydata$varname )   #This does NOT work.
```

Example programs that demonstrate these topics in SAS, SPSS and R are in Table 14.10.

Table 14.10 Example programs for keeping and dropping variables

SAS programming statements	
| | ```
* SAS Program for Keeping and Dropping Variables;
* DropKeep.sas ;
LIBNAME myLib' C:\myRfolder' ;
DATA myleft; SET mydata;
  KEEP id workshop gender q1 q2;
PROC PRINT;
RUN;
``` |
```

Table 14.10 (continued)

```
* or equivalently;
DATA myleft; SET mydata;
   DROP q3 q4;
PROC PRINT;
RUN;
```

SPSS programming
statements

```
* SPSS Program for Keeping and Dropping Variables.
* DropKeep.sps ;
CD' C:\myRfolder' .
GET FILE='mydata.sav' .
DELETE VARIABLES q3 to q4.
LIST.
SAVE OUTFILE='myleft.sav' .
```

R programming
statements

```
* R Program for Keeping and Dropping Variables.
* DropKeep.R
setwd("/myRfolder")
load(file="myWorkspace.RData")
# Using variable selection.
myleft <- mydata[ ,1:4]
myleft
# Using NULL.
myleft <- mydata
myleft$q3 <- myleft$q4 <- NULL
myleft
```

14.10 Stacking/Concatenating/Adding Datasets

The examples below first split mydata into separate datasets for males and
females. They then show how to put them back together. SAS calls this con-
catenation and accomplishes this with the SET statement. SPSS calls it adding
cases and does it using the ADD FILES statement. R, with its row/column
orientation, calls it binding rows.

To demonstrate this, let us take our practice dataset and split it into separate
ones for females and males. Then we will bind the rows back together. A split
function exists to do this type of task, but it puts the resulting data frames into a
list, so we will use an alternate approach. First, let us get the females:

```
> females <- mydata[ which(gender=="f"), ]

> females
```

```
  workshop gender q1 q2 q3 q4
1        1      f  1  1  5  1
2        2      f  2  1  4  1
3        1      f  2  2  4  3
```

And now we get the males:

```
> males <- mydata[ which(gender=="m"), ]

> males
```

```
  workshop gender q1 q2 q3 q4
5        1      m  4  5  2  4
6        2      m  5  4  5  5
7        1      m  5  3  4  4
8        2      m  4  5  5  5
```

We can put them right back together by binding their rows with the rbind function.

```
> both <- rbind(females, males)

> both
```

```
  workshop gender q1 q2 q3 q4
1        1      f  1  1  5  1
2        2      f  2  1  4  1
3        1      f  2  2  4  3
5        1      m  4  5  2  4
6        2      m  5  4  5  5
7        1      m  5  3  4  4
8        2      m  4  5  5  5
```

This works fine when the two data frames share the exact same variables. Often the data frames you will need to bind have a few variables missing. We will drop variable q2 in the males data frame to create such a mismatch.

```
> males$q2 <- NULL

> males
```

```
  workshop gender q1 q3 q4
5        1      m  4  2  4
6        2      m  5  5  5
7        1      m  5  4  4
8        2      m  4  5  5
```

Note that variable q2 is indeed gone. Now let us try to put the two data frames together again.

```
> both <- rbind(females, males)
```

```
Error in match.names(clabs, names(xi)) : names do not
   match previous names
```

It fails because the rbind function needs both data frames to have the exact same variable names. Luckily, Hadley Wickham's reshape package has a function rbind.fill that binds whichever variables it finds that match, and then fills in missing values for those that do not. This next example assumes you have installed the reshape package. See Section 5.1 for details.

```
> library("reshape")
> both <- rbind.fill(females, males)
> both
  workshop gender q1   q2 q3 q4
1        1      f  1    1  5  1
2        2      f  2    1  4  1
3        1      f  2    2  4  3
5        1      m  4   NA  2  4
6        2      m  5   NA  5  5
7        1      m  5   NA  4  4
8        2      m  4   NA  5  5
```

We can do the same thing with the built-in rbind function, but we have to first determine which variables we need to add, and then add them manually with the data.frame function and set them to NA.

```
> males <- data.frame( males, q2=NA )
> males
  workshop gender q1 q3 q4 q2
5        1      m  4  2  4 NA
6        2      m  5  5  5 NA
7        1      m  5  4  4 NA
8        2      m  4  5  5 NA
```

The males data frame now has a variable q2 again and so we can bind the two data frames using rbind. The fact that q2 is now at the end will not matter. The data frame you list first on the rbind function will determine the order of the final data frame. However, if you use index values to refer to your variables, you need to be aware of the difference.

```
> both <- rbind(females, males)
> both
  workshop gender q1 q2 q3 q4
1        1      f  1  1  5  1
2        2      f  2  1  4  1
3        1      f  2  2  4  3
5        1      m  4 NA  2  4
```

```
6            2          m   5 NA   5   5
7            1          m   5 NA   4   4
8            2          m   4 NA   5   5
```

This was an easy way to do it with such a tiny data frame. In situations that are more realistic, rbind.fill is usually a great time saver. Example programs that demonstrate these topics in SAS, SPSS and R are located in Table 14.11.

Table 14.11 Example programs for stacking/concatenating/adding datasets

| SAS programming statements | ```
* SAS Program for
    Stacking/Concatenating/Adding Data Sets;
* Stack.sas ;
LIBNAME myLib 'C:\myRfolder' ;
DATA males;
  SET mydata;
  WHERE gender='m' ;
  RUN;
DATA females;
  SET mydata;
  WHERE gender='f' ;
  RUN;
DATA both;
  SET males females;
RUN;
``` |
| SPSS programming statements | ```
* SPSS Program for
* Stacking/Concatenating/Adding Data Sets.
* Stack.sps .
CD ' C:\myRfolder' .
GET FILE='mydata.sav' .
SELECT IF(gender = "f").
SAVE OUTFILE='females.sav' .
EXECUTE .
GET FILE='mydata.sav' .
SELECT IF(gender = "m").
SAVE OUTFILE='males.sav' .
EXECUTE .
GET FILE='females.sav' .
ADD FILES /FILE=*
 /FILE='males.sav' .
``` |
| R programming statements | ```
# R Program for Stacking/Concatenating/Adding Data Sets.
# Stack.R
setwd("/myRfolder")
load(file="myWorkspace.RData")
mydata
``` |
```

Table 14.11 (continued)

```
attach(mydata)
# Create female data frame.
females <- mydata[ which(gender=="f"),]
females
# Create male data frame.
males <- mydata[ which(gender=="m"),]
males
#Bind their rows together with the rbind function.
both <- rbind(females, males)
both
# Drop q2 to see what happens.
males$q2 <- NULL
males
# See that row bind will not work.
both <- rbind(females, males)
# Use reshape's rbind.fill.
library("reshape")
both <- rbind.fill(females, males)
both
# Add a q2 variable to males.
males <- data.frame( males, q2=NA )
males
# Now rbind can handle it.
both <- rbind(females,males)
both
```

14.11 Joining/Merging Data Frames

One of the most frequently used data manipulation methods is joining or merging two datasets. If you have a one-to-many join, it will create a row for every possible match. A common example is a short data frame containing household-level information such as household income joined to a longer dataset of individual family member variables. A complete record of each family member along with his or her household income will result. Duplicates in more than one data frame are possible, but you should study them carefully for errors.

So that we will have an ID variable to work with, let us read our practice data without the row.names argument. That will keep our ID variable as is and fill in row names with 1,2,3,....

```
> mydata <- read.table("mydata.csv",header=TRUE,sep=
  ",",na.strings=" ")

> mydata

  id workshop gender q1 q2 q3 q4
1 1         1      f  1  1  5  1
2 2         2      f  2  1  4  1
3 3         1      f  2  2  4  3
4 4         2   <NA>  3  1 NA  3
```

```
5  5          1      m   4  5  2  4
6  6          2      m   5  4  5  5
7  7          1      m   5  3  4  4
8  8          2      m   4  5  5  5
```

Now we will split the left half of the data frame into one called myleft:

```
> myleft <- mydata[ c("id","workshop","gender",
  "q1","q2") ]

> myleft

  id workshop gender q1 q2
1  1          1      f   1  1
2  2          2      f   2  1
3  3          1      f   2  2
4  4          2   <NA>  3  1
5  5          1      m   4  5
6  6          2      m   5  4
7  7          1      m   5  3
8  8          2      m   4  5
```

And then do the same for the variables on the right, but we will keep id to match on later.

```
> myright <- mydata[ c("id","q3","q4") ]

> myright

  id  q3 q4
1  1   5  1
2  2   4  1
3  3   4  3
4  4  NA  3
5  5   2  4
6  6   5  5
7  7   4  4
8  8   5  5
```

Now we can use the merge function to put the two data frames back together.

```
> both <- merge(myleft, myright, by="id")

> both

  id workshop.x gender q1 q2 workshop.y q3 q4
1  1          1      f   1  1          1   5  1
2  2          2      f   2  1          2   4  1
3  3          1      f   2  2          1   4  3
4  4          2   <NA>  3  1          2  NA  3
```

```
5   5              1      m   4   5              1   2   4
6   6              2      m   5   4              2   5   5
7   7              1      m   5   3              1   4   4
8   8              2      m   4   5              2   5   5
```

This `merge` function call has three arguments.

1. The first data frame to merge.
2. The second data frame to merge.
3. The `by` argument that has either a single variable name in quotes, or a character vector of names.

If you leave out the by argument, it will match by all variables with common names! That is quite unlike SAS or SPSS, which would simply match the two row-by-row. That is what the R `cbind` function will do. It is much safer to match on some sort of id variable(s) though. You never know when rows do not match up as well as you think.

Sometimes the same variable has two different names in the data frames you need to merge. For example, one may have "id" and another "subject". If you have such a situation, you can use the `by.x` argument to identify the first variable or set of variables and the `by.y` argument to identify the second. The merge function will match them up in order and do the proper merge. In this next example, I do that but of course the variables have the same name. It still works though.

```
> both <- merge(myleft,       myright,
+                      by.x="id", by.y="id")

> both
  id workshop gender q1 q2 q3 q4
1  1        1      f  1  1  5  1
2  2        2      f  2  1  4  1
3  3        1      f  2  2  4  3
4  4        2   <NA>  3  1 NA  3
5  5        1      m  4  5  2  4
6  6        2      m  5  4  5  5
7  7        1      m  5  3  4  4
8  8        2      m  4  5  5  5
```

If you have multiple variables in common but you only want to match on a subset of them, you can use the form:

```
both <- merge( myleft,myright,
                by=c("id","workshop")  )
```

If each file had variables with slightly different names, you could use the form:

```
both <- merge( myleft,myright,
  by.x=c("id", "workshop")
  by.y=c("subject","shortCourse")
)
```

By default, SAS and SPSS keep all records regardless of whether or not they match (a full outer join). For observations that do not have matches in the other file, the `merge` function will fill in with missing values. R takes the opposite approach, keeping only those that have record in both (an inner join). To get `merge` to keep all records, use the argument `all = TRUE`. You can also use `all.x = TRUE` to keep all of the records in the first file regardless of whether or not they have matches in the second. The `all.y = TRUE` argument does the reverse.

While SAS and SPSS can merge any number of files at once, base R can only do two at a time. To do more, you can use the `merge_all` function in the `reshape` package [22]. Example programs that merge files in SAS, SPSS and R are in Table 14.12.

Table 14.12 Example programs for joining/merging data frames

SAS programming statements

```
* SAS Program for Joining/Merging Data Sets.
* Merge.sas ;
LIBNAME myLib ' C:\myRfolder' ;
DATA  myLib.myleft;
  SET mylib.mydata;
  KEEP id workshop gender q1 q2;
PROC SORT; BY id workshop; RUN;
DATA  myLib.myright;
  SET myLib.mydata;
  KEEP id q3 q4;
PROC SORT; BY id workshop; RUN;
DATA  myLib.both;
  MERGE myLib.myleft myLib.myright;
  BY id workshop;
RUN;
```

SPSS programming statements

```
* SPSS Program for Joining/Merging Data Sets.
* Merge.sps  .
CD 'C:\myRfolder' .
GET FILE='mydata.sav' .
DELETE VARIABLES q3 to q4.
SAVE OUTFILE='myleft.sav' .
GET FILE='mydata.sav' .
```

Table 14.12 (continued)

```
DELETE VARIABLES workshop to q2.
SAVE OUTFILE='myright.sav'.
GET FILE='myleft.sav'.
MATCH FILES /FILE=*
/FILE='myright.sav'
/BY id.
```

R programming statements

```
# R Program for Joining/Merging Data Sets.
# Merge.R
setwd("/myRfolder")
# Read data keeping ID as a variable.
mydata <- read.table("mydata.csv",
  header=TRUE,sep=",",na.strings=" ")
mydata
# Create a data frame keeping the left two q variables.
myleft <- mydata[ c("id","workshop","gender","q1","q2") ]
myleft
# Create a data frame keeping the right two q variables.
myright <- mydata[ c("id","workshop","q3","q4") ]
myright
#Merge the two dataframes by ID.
both <- merge(myleft,myright,by="id")
both
#Merge the two dataframes by ID.
both <- merge(myleft,  myright,
       by.x="id", by.y="id" )
#Merge dataframes by both ID and workshop.
both <- merge(myleft,myright,by=c("id","workshop"))
both
#Merge dataframes by both ID and workshop,
#while allowing them to have different names.
both <- merge(myleft,
       myright,
       by.x=c("id","workshop"),
       by.y=c("id","workshop") )
both
```

14.12 Creating Summarized or Aggregated Datasets

We often have to work on data that is a summarization of other data. For
example, you might work on household-level data that you aggregated from
a dataset that had each family member as its own observation. SAS calls
this *summarization* and performs it with the SUMMARY procedure. SPSS
calls this process *aggregation* and performs it using the AGGREGATE
procedure.

R has three distinct advantages over SAS and SPSS regarding aggregation. First, it is possible to perform multi-level calculations and selections in a single step. We will perform some of each below. Second, R can aggregate with *every function it has and any function you write yourself*, not just the few that SAS and SPSS build into SUMMARY and AGGREGATE. Third, R has data structures designed to hold aggregate results so that other functions can take advantage of that structure.

14.12.1 The `aggregate` Function

We will use the `aggregate` function to calculate the mean of the q1 variable by gender and save it to a new (very small!) data frame.

```
> attach(mydata)

> myAgg1 <- aggregate(q1,
+    by=data.frame(gender),
+    mean, na.rm=TRUE)

> myAgg1

  gender        x
1      f  1.666667
2      m  4.500000
```

The `aggregate` function call above has four arguments.

1. The variable you wish to aggregate.
2. One or more grouping factors. Unlike SAS, the data does not have to be sorted by these factors. This must be in the form of a list (or data frame, which is a list). Recall that single subscripting of a data frame creates a list. So `mydata[ "gender"]` and `mydata[2]` work. Adding the comma to either one will prevent them from working. Therefore, `mydata [ ,"gender"]` or `mydata[ ,2]` will not work. If you have attached the data frame, `data. frame(gender)` will work. The function call `list(gender)` will also work, but it loses track of the grouping variable names.
3. The function that you wish to apply, in this case the `mean` function. An important limitation of the `aggregate` function is that it can apply only functions that return a single value. If you need to apply a function that returns multiple values, you can use the `tapply` function. That is covered in the next section.
4. Arguments to pass to the function applied. Here `na.rm = TRUE` is passed to the `mean` function to remove missing, or NA, values.

Next we will aggregate by two variables, workshop and gender. To keep our by-factors in the form of a list (or data frame) we can use any one of these forms:

```
mydata[ c("workshop","gender")]
```
or
```
mydata[ c(2,3) ]
```
or, if you have attached the data frame,
```
data.frame( workshop, gender)
```
```
> myAgg2 <- aggregate(q1,
+    by=data.frame(workshop, gender),
+    mean, na.rm=TRUE)
```
```
> myAgg2
```
```
  workshop gender     x
1        R      f   1.5
2      SAS      f   2.0
3        R      m   4.5
4      SAS      m   4.5
```

Now let us use the mode and class functions to see the type of object the aggregate function creates are data frames. It is small but ready for further analysis.

```
> mode(myAgg2)
[1] "list"
```
```
> class(myAgg2)
[1] "data.frame"
```

14.12.2 The tapply Function

In the last section we discussed the aggregate function. That function has a significant limitation: you can only use it with functions that return single values. The tapply function works very similarly to the aggregate function, but can perform aggregation using any R function. To gain this ability, it has to abandon the convenience of creating a data frame. Instead, its output is in the form of a matrix or an array.

Let us first duplicate the last example from the section above using tapply.

```
> myAgg2 <- tapply(q1,
+    data.frame(workshop,gender),
+    mean, na.rm=TRUE)
```
```
> myAgg2
```
```
            gender
workshop     Female        Male
R          3.571429    3.000000
SAS        4.090909    2.692308
```

```
SPSS        4.230769    2.833333
Stata       4.111111    3.500000
```

The `tapply` example above uses four arguments.

1. The variable to aggregate.
2. One or more grouping factors. Unlike SAS, the data does not have to be sorted by these factors. This must be in the form of a list (or data frame, which is a list). Recall that single subscripting of a data frame creates a list. So `mydata[ "gender"]` and `mydata[ 2]` work. Adding the comma to either one will prevent them from working. Therefore, `mydata[ ,"gender"]` or `mydata[ ,2]` will not work. If you have attached the data frame, `data.frame(gender)` will work. The function call `list(gender)` will also work, but it loses track of the grouping variable names.
3. The function to apply, in this case the `mean` function. This function can return any result, not just single values.
4. Any additional parameters to pass to the applied function. In this case, `na.rm = TRUE` is used by the mean function to remove NA or missing values.

The actual means are, of course, the same as we got before using the `aggregate` function. However, the result is now a numeric matrix rather than a data frame.

```
> class(myAgg2)
[1] "matrix"

> mode(myAgg2)
[1] "numeric"
```

Now let us do an example that the `aggregate` function could not perform. The `range` function returns two values, the minimum and maximum for each variable.

```
> myAgg2 <- tapply(q1,
+    data.frame(workshop,gender),
+    range, na.rm=TRUE)

> myAgg2
                  gender
workshop      Female         Male
R             Numeric,2    Numeric,2
SAS           Numeric,2    Numeric,2
SPSS          Numeric,2    Numeric,2
Stata         Numeric,2    Numeric,2
```

This output looks quite odd! It is certainly not formatted for communicating results to others. Let us see how it is stored.

```
> mode(myAgg2)
[1] "list"
```

```
> class(myAgg2)
[1] "data.frame"
```

It is a list, so let us look at its first element. We see that the q1 responses of
females that took the R workshop range from 2 to 5.

```
> myAgg2[[1]]
[1] 2 5
```

14.12.3 *Merging Aggregates with Original Data*

It is often useful to add aggregate values back to the original data frame. This
allows you to perform multi-level transformations that involve both individual-
level and aggregate-level values. A common example of such a calculation is
a Z score, which subtracts a variable's mean and then divides by its standard
deviation (the scale function performs this calculation more easily).

Another important use for merging aggregates with original data is to
perform multi-level selections of observations. To select individual-level observations based on aggregate-level values requires access to both at once. For
example, we could create a subset of subjects who fall below their group's mean
value.

This is an area in which R has a distinct advantage over SAS and SPSS. R's
greater flexibility allows it to do both multi-level transformations and selections
in a single step. For example, you can calculate a Z score for variable q1 with the
single following statement. Note that we are specifying the long form of the
name for our new variable, mydata$Zq1, so that it will go into our data frame.
The data frame is attached so we can use the short variable names on the right
side of the assignment.

```
> mydata$Zq1 <- (q1 - mean(q1) ) / sd(q1)

> mydata

  workshop gender q1 q2 q3 q4         Zq1
1        R      f  1  1  5  1  -1.5120484
2      SAS      f  2  1  4  1  -0.8400269
3        R      f  2  2  4  3  -0.8400269
4      SAS   <NA>  3  1 NA  3  -0.1680054
5        R      m  4  5  2  4   0.5040161
6      SAS      m  5  4  5  5   1.1760376
7        R      m  5  3  4  4   1.1760376
8      SAS      m  4  5  5  5   0.5040161
```

You can also select the observations that were below average with this single
statement.

```
> mySubset <- mydata[ q1 < mean(q1), ]

> mySubset
```

```
  workshop gender q1 q2 q3 q4         Zq1
1        R        f  1  1  5  1  -1.5120484
2      SAS        f  2  1  4  1  -0.8400269
3        R        f  2  2  4  3  -0.8400269
4      SAS     <NA>  3  1 NA  3  -0.1680054
```

SAS and SPSS cannot perform such calculations in one step. You would
have to create the aggregate-level data and then merge it back into the individual-level dataset. R can use this approach too, and as the number of levels you
consider increases, it becomes more reasonable to do so.

So let us now merge myAgg2, created in Sect. 14.12.1, to mydata. To do that,
we will rename the mean of q1 from x to mean.q1. We will use the rename
function from the reshape package. If you do not have that installed, see the
section 5.1.

```
> library("reshape")

> myAgg3 <- rename(myAgg2, c(x="mean.q1"))

> myAgg3
```

```
  workshop gender  mean.q1
1        R      f      1.5
2      SAS      f      2.0
3        R      m      4.5
4      SAS      m      4.5
```

Now we merge the mean onto each of the original observations.

```
> mydata2 <- merge(mydata, myAgg3,
+    by=c("workshop","gender") )

> mydata2
  workshop gender q1 q2 q3 q4         Zq1  mean.q1
1        R      f  1  1  5  1  -1.5120484      1.5
2        R      f  2  2  4  3  -0.8400269      1.5
3        R      m  4  5  2  4   0.5040161      4.5
4        R      m  5  3  4  4   1.1760376      4.5
5      SAS      f  2  1  4  1  -0.8400269      2.0
6      SAS      m  5  4  5  5   1.1760376      4.5
7      SAS      m  4  5  5  5   0.5040161      4.5
```

The merge function call above has only two arguments.

1. The two data frames to merge. Unlike SAS and SPSS, which can merge
   many datasets at once, R can only do two at a time.

2. The `by` argument specifies the variables to match upon. In this case they have the same name in both data frames. They can however have different names. See the `merge` help files for details. While some other functions require by variables in list form, here you provide more than one variable in the form of a character vector.

We could now perform multi-level transformations or selections on mydata2.

14.12.4 Tabular Aggregation

The aim of table creation in SAS and SPSS is to communicate the results to people. You can create simple tables of frequencies and percents using the SAS FREQ procedure and SPSS CROSSTABS procedure. For more complex tables, SAS has PROC TABULATE, and SPSS has its CTABLES procedure. These two create complex tables with basic statistics in almost any form, and can perform some basic hypothesis tests. However, no other procedures are programmed to process these tables further automatically. You can analyze them further using the SAS ODS or SPSS OMS, but not as easily as in R.

R can create tables for presentation too, but it also creates tables and matrices that are optimized for further use by other functions. They are a different form of aggregated dataset. See Chap. 23 for other uses of tables.

Let us revisit simple frequencies using the `table` function. First, let us look at just workshop attendance (the data frame is attached so I am using short variable names).

```
> table(workshop)

workshop
    R SAS
    4   4
```

And now gender and workshop.

```
> table(gender,workshop)
        workshop

gender R SAS
     f 2   1
     m 2   2
```

Let us save this table to an object myCounts and check its mode and class.

```
> myCounts <- table(gender, workshop)

> mode(myCounts)
[1]  "numeric"
```

```
> class(myCounts)
[1]  "table"
```

We see that the mode of myCounts is *numeric* and its class is *table*. Other functions that exist to work with pre-summarized data know what to do with table objects. In Chap. 21, we will see the kinds of barplots we can make from tables. In Chap. 23, we will also work with table objects to calculate things like row and column percents.

Other functions prefer count data in the form of a data frame. This is the type of output created by the SAS SUMMARY procedure or the SPSS AGGREGATE procedure. The as.data.frame function makes quick work of this conversion.

```
> myCountsDF <- as.data.frame(myCounts)

> myCountsDF

  gender workshop Freq
1      f        R    2
2      m        R    2
3      f      SAS    1
4      m      SAS    2

> class(myCountsDF)

[1]  "data.frame"
```

14.12.5 The Reshape Package

If you perform a lot of aggregation, you will want to learn how to use Hadley Wickham's powerful reshape package [22] documented at http://cran.r-project.org/doc/packages/reshape.pdf. Programs that demonstrate this section's topics are found in Table 14.13.

Table 14.13 Example programs for aggregating/summarizing data

SAS programming statements	
	`* SAS Program for Aggregating/Summarizing Data;`
	`* Aggregate.sas ;`
	`LIBNAME myLib 'C:\myRfolder' ;`
	`* Get means of q1 for each gender;`
	`PROC SUMMARY DATA=myLib.mydata MEAN NWAY;`
	`  CLASS GENDER;`
	`  VAR q1;`
	`  OUTPUT OUT=myLib.myAgg;`
	`  RUN;`

Table 14.13 (continued)

```
                     PROC PRINT; RUN;
                     DATA myLib.myAgg;
                       SET myLib.myAgg;
                       WHERE _STAT_='MEAN' ;
                       KEEP workshop gender q1;
                       RENAME q1=meanQ1;
                     RUN;
                     PROC PRINT; RUN;
                     *Get means of q1 by workshop and gender;
                     PROC SUMMARY DATA=myLib.mydata MEAN NWAY;
                     CLASS WORKSHOP GENDER;
                     VAR Q1;
                     OUTPUT OUT=myLib.myAgg;RUN;
                     PROC PRINT; RUN;
                     *Strip out just the mean and matching variables;
                     DATA myLib.myAgg;
                       SET myLib.myAgg;
                       WHERE _STAT_='MEAN' ;
                       KEEP workshop gender q1;
                       RENAME q1=meanQ1;
                     RUN;
                     PROC PRINT; RUN;
                     *Now merge aggregated data back into mydata;
                     PROC SORT DATA=myLib.mydata;
                      BY workshop gender; RUN:
                     PROC SORT DATA=myLib.myAgg;
                      BY workshop gender; RUN:
                     DATA myLib.mydata2;
                      MERGE myLib.mydata myLib.myAgg;
                      BY workshop gender;
                     PROC PRINT; RUN;
```

SPSS programming
statements

```
                     * SPSS Program for Aggregating/Summarizing Data.
                     * Get mean of q1 by gender.
                     * Aggregate.sps .
                     CD 'C:\myRfolder' .
                     GET FILE='mydata.sav'
                     AGGREGATE
                       /OUTFILE='myAgg.sav'
                       /BREAK=gender
                       /q1_mean = MEAN(q1) .
                     GET FILE='myAgg.sav' .

                     LIST.
                     * Get mean of q1 by workshop and gender.
                     GET FILE='\myRfolder\mydata.sav' .
                     AGGREGATE
                       /OUTFILE='C:\myRfolder\myAgg.sav'
                       /BREAK=workshop gender
                       /q1_mean = MEAN(q1) .
```

Table 14.13 (continued)

```
GET FILE='\myRfolder\myAgg.sav' .
LIST.
* Merge aggregated data back into mydata.
 GET FILE='\myRfolder\mydata.sav' .
SORT CASES BY workshop (A) gender (A) .
MATCH FILES /FILE=*
    /TABLE='\myAgg.sav'
    /BY workshop gender.
SAVE OUTFILE= 'mydata.sav' .
```

R programming
statements

```
# R Program for Aggregating/Summarizing Data.
# Aggregate.R
setwd("/myRfolder")
load(file="myWorkspace.RData")
attach(mydata)
mydata
# The aggregate Function.
# Means by gender.
myAgg1 <- aggre gate(q1,
  by=data.frame(gender),
  mean, na.rm=TRUE)
myAgg1
# Now by workshop and gender.
myAgg2 <- aggregate(q1,
  by=data.frame(workshop, gender),
  mean, na.rm=TRUE)
myAgg2
mode(myAgg2)
class(myAgg2)
# Aggregation with tapply.
myAgg2 <- tapply(q1,
  data.frame(workshop,gender),
  mean, na.rm=TRUE)
myAgg2
class(myAgg2)
mode(myAgg2)
myAgg2 <- tapply(q1,
  data.frame(workshop,gender),
  range, na.rm=TRUE)
myAgg2
mode(myAgg1)
class(myAgg2)
myAgg2[[1]]
# Example multi-level transformation.
mydata$Zq1 <- (q1 - mean(q1) ) / sd(q1)
mydata
mySubset <- mydata [ q1 < mean(q1),    ]
mySubset
# Rename x to be mean.q1.
library("reshape")
```

Table 14.13 (continued)

```
myAgg3 <- rename(myAgg2, c(x="mean.q1"))
myAgg3
# Now merge means back with mydata.
mydata2 <- merge(mydata,myAgg3,
   by=c("workshop","gender") )
mydata2
# Tables of Counts
table(workshop)
table(gender,workshop)
myCounts <- table(gender, workshop)
mode(myCounts)
class(myCounts)
# Counts in Summary/Aggregate style.
myCountsDF <- as.data.frame(myCounts)
myCountsDF
class(myCountsDF)
# Clean up
mydata["Zq1"] <- NULL
rm(myAgg1, myAgg2, myAgg3,
   myComplete, myMeans, myCounts, myCountsDF)
```

14.13 By or Split File Processing

When you want to repeat an analysis for every level of a categorical variable, you can use the BY statement in SAS, or the SPLIT FILE command in SPSS. SAS and SPSS require you to sort the data by the factor variable(s) first, but R does not.

R has a by function, which repeats analysis for levels of factors. Section 14.12, *Creating Summarized or Aggregated Datasets*, did similar things while creating summary datasets. Let us look at the by function first, and then discuss how it compares to similar functions.

We will use the by function to apply the mean function. First, let us use the mean function by itself just for review. To get the means of our q variables, we can use:

```
> mean( mydata[ c("q1","q2","q3","q4") ] ,
+         na.rm=TRUE)

    q1     q2     q3     q4
3.2500 2.7500 4.1429 3.7500
```

Now let us get means for the males and females using the by function:

```
> myBYout <- by( mydata[ c("q1","q2","q3","q4") ] ,
+      mydata["gender"],
+      mean,na.rm=TRUE)
```

```
> myBYout
gender: f
       q1        q2        q3        q4
1.666667 1.333333 4.333333 1.666667
---------------------------------------------------------
gender: m
  q1   q2   q3   q4
4.50 4.25 4.00 4.50
```

The by function call above has four parameters.

1. The data frame name and/or variables to analyze, mydata
   [c("q1","q2","q3","q4")] .
2. One or more grouping factors. Unlike in SAS or SPSS, the data does
   not have to be sorted by these factors. This must be in the form of a list
   (or data frame, which is a list). Recall that single subscripting of a data
   frame creates a list. So mydata ["gender"] and mydata [2] work.
   Adding comma to either one will prevent them from working. Therefore, mydata[,"gender"] or mydata [,2] will not work. If you
   have attached the data frame, data.frame(gender) will work. The
   function call list(gender) will also work, but it loses track of the
   grouping variable names.
3. The function to apply, in this case the mean. The by function can apply
   functions that calculate more than one value (unlike the aggregate
   function).
4. Any additional arguments are ignored by the by function and simply passed
   to the function. In this case, na.rm=TRUE is simply passed to the mean
   function.

Let us check to see what the mode and class are of the output object.

```
> mode(myBYout)
 [1] "list"

> class(myBYout)
 [1] "by"
```

It is a list, with a class of "by." If we would like to convert that to a data
frame, we can do so with the following. The as.table function gets the
data into a form that the as.data.frame function can then turn into a data
frame.

```
> myBYdata <- as.data.frame( (as.table(myBYout) ) )

> myBYdata

   gender   Freq.f Freq.m
q1      f 1.666667   4.50
q2      m 1.333333   4.25
```

```
q3         f 4.333333    4.00
q4         m 1.666667    4.50
```

Now let us break the mean down by both workshop and gender. To keep our by-factors in the form of a list (or data frame) we can use any one of these forms:

```
mydata[ c("workshop","gender")]
```

or

```
mydata[ c(2,3) ]
```

or, if you have attached the data frame,

```
data.frame( workshop, gender)
```

This starts looking messy, so let us put both our variable list and our factor list into character vectors:

```
myVars <- c("q1","q2","q3","q4")
myBys  <- mydata[ c("workshop","gender") ]
```

By substituting our character vectors into the by function, it is much easier to read. This time, let us use the range function to show that the by function can apply functions that return more than one value.

```
> myBYout <- by( mydata[myVars],
+    myBys, range, na.rm=TRUE )
> myBYout
workshop: R
gender: f
[1] 1 5
----------------------------------------------------------
workshop: SAS
gender: f
[1] 1 4
----------------------------------------------------------
workshop: R
gender: m
[1] 2 5
----------------------------------------------------------
workshop: SAS
gender: m
[1] 4 5
```

That output is quite readable. Recall that when we did this same analysis using the tapply function, the results were in a form that is optimized for further analysis rather than communication. However, we can save the data to a dataframe if we like. The approach it takes is most interesting. Let us see what type of object we have.

```
> mode(myBYout)
[1] "list"

> class(myBYout)
[1] "by"

> names(myBYout)
NULL
```

It is a list with a class of "by" and no names. Let us look at one of its components.

```
> myBYout[[1]]

[1] 1 5
```

This is the first set of ranges from the printout above. If we wanted to create a data frame from these, we could bind them into the rows of a matrix and then convert that to a data frame with:

```
> myBYdata <- data.frame(
+     rbind( myBYout[[1]], myBYout[[2]],
+              myBYout[[3]], myBYout[[4]] )
+ )

> myBYdata
  X1 X2
1  1  5
2  1  4
3  2  5
4  4  5
```

That approach is easy to understand but not much fun to use if we had many more factor levels! Luckily the do.call function can call a function you choose once, upon all the components of a list, just as if you had entered them individually. That is quite different from the lapply function, which applies the function you choose repeatedly on each separate component. All we have to do is give it the function to feed the components into, rbind function in this case, and the list name, myBydata.

```
> myBYdata  <- data.frame ( do.call (rbind, myBYout ) )

> myBYdata
  X1 X2
1  1  5
2  1  4
3  2  5
4  4  5
```

Table 14.14 Summarization function comparison table

Function	Input	Functions it can apply	Output
by	Data frame	Any function	List with class of "by". Easier to read but not as easy to program.
aggregate	Data frame	Functions that return single values	Data frame
tapply	Lists or data frames	Any function	List. Easy to access elements for programming. Not as nicely formatted for reading.
table	Factors	Does counting only	Table object. Easy to read and easy to analyze further with functions that understand that class of object.

14.13.1 Comparing Summarization Methods

So we have seen that the capability of the by function closely mirrors that of the aggregate and tapply functions. Table 14.14 can help you choose which to use.

14.13.2 Example Programs for By or Split File Processing

Table 14.15 shows the example programs for By or Split File processing.

Table 14.15 Example programs for *By* or *Split File* processing

SAS programming statements	
| | ```
* SAS Program for By or Split File Processing;
* By.sas ;
LIBNAME myLib 'C:\myRfolder' ;
PROC MEANS DATA=myLib.mydata;
  RUN;
PROC SORT DATA=myLib.mydata;
  BY gender;
  RUN;
PROC MEANS DATA=myLib.mydata;
  BY gender;
  RUN;
PROC SORT DATA=myLib.mydata;
  BY workshop gender;
  RUN;
PROC MEANS DATA=myLib.mydata;
  BY workshop gender;
  RUN;
``` |
```

Table 14.15 (continued)

| SPSS programming statements | ```
* SPSS Program for By or Split File Processing;
* By.sps .
CD 'C:\myRfolder' .
GET FILE='mydata.sav' .
DESCRIPTIVES
  VARIABLES=q1 q2 q3 q4
  /STATISTICS=MEAN STDDEV MIN MAX.
SORT CASES BY gender.
SPLIT FILE
  SEPARATE BY gender.
DESCRIPTIVES
  VARIABLES=q1 q2 q3 q4
  /STATISTICS=MEAN STDDEV MIN MAX.
SORT CASES BY workshop gender.
SPLIT FILE
  SEPARATE BY workshop gender.
DESCRIPTIVES
  VARIABLES=q1 q2 q3 q4
  /STATISTICS=MEAN STDDEV MIN MAX.
``` |
| R programming statements | ```
# R Program for By or Split File Processing.
# By.R
setwd("/myRfolder")
load(file="myWorkspace.RData")
attach(mydata)
options(width=64)
mydata
# Get means of q variables for all observations.
mean( mydata [c("q1","q2","q3","q4") ] ,
        na.rm=TRUE)
# Now get means by gender.
myBYout <- by( mydata[ c("q1","q2","q3","q4") ] ,
  mydata ["gender"],
  mean,na.rm=TRUE)
myBYout
mode(myBYout)
class(myBYout)
myBYdata <- as.data.frame( (as.table(myBYout) ) )
myBYdata
# Get range by workshop and gender
myVars <- c("q1","q2","q3","q4")
myBys  <- mydata [c("workshop","gender") ]
myBYout <- by( mydata[myVars],
  myBys, range, na.rm=TRUE )
myBYout
# Converting output to data frame.
mode(myBYout)
``` |

Table 14.15 (continued)

```
class(myBYout)
names(myBYout)
myBYout[[1]]
# A data frame the long way.myBYdata <- data.frame(
 rbind(myBYout[[1]], myBYout[[2]],
    myBYout[[3]], myBYout[[4]])
)
myBYdata
# A data frame using do.call.
myBYdata <- data.frame( do.call( rbind, myBYout) )
myBYdata
mode(myBYdata)
class(myBYdata)
```

14.14 Removing Duplicate Observations

Duplicate observations frequently creep into datasets, especially those that are merged from various other datasets. The SAS approach is to use PROC SORT NODUP or NODUPKEY to get rid of duplicates without examining them. The SPSS approach uses a dialog box to generate programming code that will identify and/or filter the observations. Of course SAS and SPSS are powerful enough to do either approach. We will use both the methods in R.

First, we will create a data frame that takes the top two observations from mydata and appends them to the bottom with the rbind function:

```
> myDuplicates <- rbind (mydata, mydata [1:2,])

> myDuplicates
```

```
   workshop gender q1 q2 q3 q4
1         R      f  1  1  5  1   <- We are copying
2       SAS      f  2  1  4  1   <- these two...
3         R      f  2  2  4  3
4       SAS   <NA>  3  1 NA  3
5         R      m  4  5  2  4
6       SAS      m  5  4  5  5
7         R      m  5  3  4  4
8       SAS      m  4  5  5  5
9         R      f  1  1  5  1   <- ... down here so we
10      SAS      f  2  1  4  1   <- have duplicates to
                                    remove.
```

Next we will use the `unique` function to find and delete them.

```
> myNoDuplicates  <- unique (myDuplicates )

> myNoDuplicates
```

```
  workshop gender q1 q2 q3 q4
1        R      f  1  1  5  1
2      SAS      f  2  1  4  1
3        R      f  2  2  4  3
4      SAS    <NA> 3  1 NA  3
5        R      m  4  5  2  4
6      SAS      m  5  4  5  5
7        R      m  5  3  4  4
8      SAS      m  4  5  5  5
```

We see above the duplicates are removed, but we never saw them; so we would know nothing about them. Knowing more about them might help us prevent them from creeping into our future analyses.

Let us put the duplicates back and use the duplicated function to create a logical variable named *duplicated*.

```
> myDuplicates   <- rbind (mydata, mydata [1:2, ] )

> myDuplicates$Duplicated <- duplicated (myDuplicates)

> myDuplicates
   workshop gender q1 q2 q3 q4 Duplicated
1         R      f  1  1  5  1      FALSE
2       SAS      f  2  1  4  1      FALSE
3         R      f  2  2  4  3      FALSE
4       SAS    <NA> 3  1 NA  3      FALSE
5         R      m  4  5  2  4      FALSE
6       SAS      m  5  4  5  5      FALSE
7         R      m  5  3  4  4      FALSE
8       SAS      m  4  5  5  5      FALSE
9         R      f  1  1  5  1       TRUE
10      SAS      f  2  1  4  1       TRUE
```

The TRUE values show us that R has indeed located the duplicate records. It is interesting to note that *now* we technically no longer have complete duplicates! We added a new variable that has changed the records, so the unique function would no longer get rid of the last two records, if we ran it the same way we did before! That is OK because now we will just get rid of those marked TRUE after we print a report of duplicate records. With a realistically sized data frame, we are unlikely to want to see the whole data frame.

```
> attach (myDuplicates )

> myDuplicates [Duplicated, ]
    workshop gender q1 q2 q3 q4 Duplicated
```

```
9            R       f  1  1  5  1           TRUE
10          SAS      f  2  1  4  1           TRUE
```

Finally, we will choose those not duplicated (i.e., !Duplicated) and drop the seventh variable, which is the TRUE/FALSE variable itself (Table 14.16).

```
> myNoDuplicates   <- myDuplicates [!Duplicated, -7 ]

> myNoDuplicates

  workshop gender q1 q2 q3 q4
1        R      f  1  1  5  1
2      SAS      f  2  1  4  1
3        R      f  2  2  4  3
4      SAS   <NA>  3  1 NA  3
5        R      m  4  5  2  4
6      SAS      m  5  4  5  5
7        R      m  5  3  4  4
8      SAS      m  4  5  5  5
```

For example programs that demonstrate these topics in SAS, SPSS and R, see Table 14.16.

Table 14.16 Example programs for removing duplicate observations

SAS	

programming statements	`* SAS Program for Removing Duplicate Observations;`
	`* Duplicates.sas ;`
	`LIBNAME myLib 'C:\myRfolder' ;`
	`PROC SORT NODUP data=mydata;`
	`  BY id workshop gender q1-q4;`
	`RUN;`
	`PROC PRINT;`
	`RUN;`

SPSS	

programming statements	`* SPSS Program for Removing Duplicate Observations.`
	`* Duplicates.sps.`
	`CD 'C:\myRfolder' .`
	`GET FILE='mydata.sav' .`
	`* Identify Duplicate Cases.`
	`SORT CASES BY workshop(A) gender(A)`
	`  q2(A) q1(A) q3(A) q4(A).`
	`MATCH FILES /FILE = *`
	`  /BY workshop gender q2 q1 q3 q4`
	`  /FIRST = PrimaryFirst`
	`  /LAST = PrimaryLast.`
	`DO IF (PrimaryFirst).`
	`+ COMPUTE MatchSequence = 1 - PrimaryLast.`
	`ELSE.`
	`+ COMPUTE MatchSequence = MatchSequence + 1.`
	`END IF.`

Table 14.16 (continued)

LEAVE MatchSequence.
FORMAT MatchSequence (f7).
COMPUTE InDupGrp = MatchSequence > 0.
SORT CASES InDupGrp(D).
MATCH FILES /FILE = *
/DROP = PrimaryFirst InDupGrp MatchSequence.
VARIABLE LABELS PrimaryLast' Indicator of each last matching case
as Primary' .
VALUE LABELS PrimaryLast
0' Duplicate Case'
1' Primary Case' .
VARIABLE LEVEL PrimaryLast (ORDINAL).
FREQUENCIES VARIABLES = PrimaryLast .

R programming
statements

```
# R Program for Removing Duplicate Observations.
# Duplicates.R
setwd("/myRfolder")
load("myWorkspace.RData")
mydata
# Create some duplicates.
myDuplicates <- rbind( mydata, mydata[1:2,] )
myDuplicates
# Get rid of duplicates without seeing them.
myNoDuplicates <- unique(myDuplicates)
myNoDuplicates
# This checks for location of duplicates
# before getting rid of them.
myDuplicates
<- rbind( mydata, mydata[1:2,] )
myDuplicates
myDuplicates$Duplicated <- duplicated(myDuplicates)
myDuplicates
# Print a report of just the duplicates.
myDuplicates[ Duplicated, ]
# Remove duplicates and Duplicated variable.
attach(myDuplicates)
myNoDuplicates <- myDuplicates[ !Duplicated, -7]
myNoDuplicates
```

14.15 Selecting First or Last Observations per Group

When a dataset contains groups, members within each group are often sorted in a useful order. For example, a company may have divisions divided into departments. Each department might have salary information for each person and a running total. So the last person's running total value would be the total for each department.

The SAS approach on this problem is quite flexible. Simply saying,

```
DATA mydata;
SET  mydata;
BY workshop gender;
```

creates four temporary variables, first.workshop, first.gender, last.workshop and last.gender. These all have the values of 1 when true and 0 when false. These variables vanish at the end of the data step unless you assign them to regular variables, but that is usually not necessary.

SPSS uses a very similar approach in the MATCH FILES procedure. Normally, you think of MATCH FILES as requiring two files to join, but you can use it in this case with only one file. It creates only a single FIRST or LAST variable that is saved to the dataset. Be careful with this approach as it subsets the main file, so you need to save it to a new name.

SPSS can also view this problem as an aggregation. Unlike SAS, its AGGREGATE procedure has FIRST and LAST functions. This works fine for just a few variables, but since it requires naming every variable you wish to save, it is not very useful for saving many variables. The example SPSS program below demonstrates both approaches.

The R approach to this problem demonstrates R's extreme flexibility. It does not have a function aimed directly at this problem. However, it is easy to create one using several other functions. We have seen the head function print the top few observations of a data frame and the tail function do the same for the last few. We have also used the by function to apply a function to groups within a data frame. We can use the by function to apply the head function to get the first observation in each group, or the tail function to get the last. Since the head and tail functions both have an n = argument, we can not only use n = 1 to get the single first or last, but also we could use n = 2 to get the first two or last two observations per group, and so on.

The last record per group is often of greatest interest since it contains the value of interest when you have a sum, like salaries, that is added cumulatively for each observation. So we will look at how to select the last observation per group. The idea readily extends to the first record(s) per group.

First, we will put our *by* variables into a data frame. By using workshop and then gender, we will soon be selecting the last male in each workshop.

```
myBys <- data.frame (mydata$workshop, mydata$gender)
```

Next, we use the by function to apply the tail function to mydata by workshop and gender. We are saving the result to mylast, which is in the form of a list.

```
mylastList  <- by( mydata, myBys, tail, n =1 )

mylastList
```

This is the resulting list:

```
mydata.workshop: 1
mydata.gender: f
  workshop gender q1 q2 q3 q4
3         1       f  2  2  4  3
-------------------------------------------------------
mydata.workshop: 2
mydata.gender: f
  workshop gender q1 q2 q3 q4
4         2       f  3  1 NA  3
-------------------------------------------------------
mydata.workshop: 1
mydata.gender: m
  workshop gender q1 q2 q3 q4
7         1       m  5  3  4  4
-------------------------------------------------------
mydata.workshop: 2
mydata.gender: m
  workshop gender q1 q2 q3 q4
8         2       m  4  5  5  9
```

We would like to put this into a data frame by combining all the components from that list in the form of rows. The do.call function does this. It essentially takes all the components of a list and feeds them into a single call to the function you choose. In this case, that is the as.list function.

```
mylastDF  <- do.call ("rbind", as.list(mylast ) )
```

And here are the records we desired.

```
mylastDF
  workshop gender q1 q2 q3 q4
3         1       f  2  2  4  3
4         2       f  3  1 NA  3
7         1       m  5  3  4  4
8         2       m  4  5  5  9
```

That single call to the do.call function does this for you:

```
mylastDF  <- rbind(as.list(mylastList )[[1]],
            as.list (mylastList )[[2]],
            as.list (mylastList )[[3]],
            as.list (mylastList )[[4]])
```

If we had hundreds of groups, the do.call function would be a big time saver! Example programs that demonstrate these topics in all three languages are in Table 14.17.

Table 14.17 Example programs for selecting first or last observations per group

SAS programming statements	

```
* SAS Program for Selecting Last Obs per Group.
* FirstLastObs.sas ;
LIBNAME myLib ' C:\myRfolder' ;
PROC SORT DATA=sasuser.mydata;
  BY workshop gender;
RUN;
DATA sasuser.mylast;
  SET  sasuser.mydata;
  BY   workshop gender;
  IF   last.gender;
RUN;
PROC PRINT; RUN;
```

SPSS programming statements

```
* SPSS Program for Selecting Last Obs per Group.
* FirstLastObs.sps .
CD ' C:\myRfolder
' .* Match files method.
GET FILE='mydata.sav' .
SORT CASES BY workshop gender.
MATCH FILES FILE=* /By workshop gender /LAST=lastgender.
SELECT IF lastgender.
LIST.
SAVE OUTFILE= 'mylast.sav' .
* Aggregation method.
SORT CASES BY workshop gender.
AGGREGATE /OUTFILE= 'C:\mylast.sav'
/BREAK workshop gender
/q1 = LAST(q1)
/q2 = LAST(q2)
/q3 = LAST(q3)
/q4 = LAST(q4) .
* Using LIST here would display original file.
GET FILE='mylast.sav' .
DATASET NAME DataSet5 WINDOW=FRONT.
LIST.
```

R programming statements

```
# R Program Selecting Last Obs per Group.
# FirstLastObs.R
setwd("/myRfolder")
load(file="myWorkspace.RData")
mydata
myBys <- data.frame(mydata$workshop,mydata$gender)
mylastList <- by( mydata,myBys,tail,n=1 )
mylastList
#Back into a data frame:
mylastDF <- do.call(rbind, mylastList)
```

Table 14.17 (continued)

```
mylastDF
# Another way to create the dataframe:
mylastDF <- rbind(mylastList[[1]],
                  mylastList[[2]],
                  mylastList[[3]],
                  mylastList[[4]])
mylastDF
```

14.16 Reshaping Variables to Observations and Back

A common data management problem is reshaping data from "wide" format to "long" and back. If we assume our variables q1, q2, q3, and q4 are the same item measured at four times, we will have the standard wide format for repeated measures data. Converting this to the long format consists of writing out four records, each of which has just one measure, we will call it Y, and a counter variable, often called time, that goes 1,2,3,4. So in the simplest case, just two variables, Y and time could replace dozens of variables. Going from long to wide is just the reverse.

SPSS makes this process very easy to do with their *Restructure Data Wizard*. It actually generated the SPSS program in Table 14.18. SAS can do this at least two ways but PROC TRANSPOSE is probably the easiest to use.

In R, Hadley Wickham's `reshape` package is quite powerful and easy to use. It uses the analogy of melting your data so that you can cast it into a different mold. In addition to reshaping, the package makes quick work of a wide range of aggregation problems.

We will need an ID variable. For this example, we will create one named subject using the colon operator:

```
> mydata$subject  <- 1:8

> print (mydata)

  workshop gender q1 q2 q3 q4 subject
1        1      f  1  1  5  1       1
2        2      f  2  1  4  1       2
3        1      f  2  2  4  3       3
4        2   <NA>  3  1 NA  3       4
5        1      m  4  5  2  4       5
6        2      m  5  4  5  5       6
7        1      m  5  3  4  4       7
8        2      m  4  5  5  5       8
```

Now we will load the `reshape` package, attach the data so that we can use short variable names, and "melt" it into the long form. The `melt` function needs only two arguments, the name of the data frame to reshape and the variables that identify each unique value.

```
> library("reshape")

> attach(mydata)

> mylong  <- melt(mydata, id =c("subject", "workshop",
"gender")  )

> mylong
```

	subject	workshop	gender	variable	value
1	1	1	f	q1	1
2	2	2	f	q1	2
3	3	1	f	q1	2
4	4	2	<NA>	q1	3
5	5	1	m	q1	4
6	6	2	m	q1	5
7	7	1	m	q1	5
8	8	2	m	q1	4
9	1	1	f	q2	1
10	2	2	f	q2	1
11	3	1	f	q2	2
12	4	2	<NA>	q2	1
13	5	1	m	q2	5
14	6	2	m	q2	4
15	7	1	m	q2	3
16	8	2	m	q2	5
17	1	1	f	q3	5
18	2	2	f	q3	4
19	3	1	f	q3	4
20	4	2	<NA>	q3	NA
21	5	1	m	q3	2
22	6	2	m	q3	5
23	7	1	m	q3	4
24	8	2	m	q3	5
25	1	1	f	q4	1
26	2	2	f	q4	1
27	3	1	f	q4	3
28	4	2	<NA>	q4	3
29	5	1	m	q4	4
30	6	2	m	q4	5
31	7	1	m	q4	4
32	8	2	m	q4	5

Now let us cast it back into the wide format.

```
> mywide <- cast(mylong, subject+workshop+gender ~ variable)

> mywide

  subject workshop gender q1 q2 q3 q4
1       1        1      f  1  1  5  1
2       2        2      f  2  1  4  1
3       3        1      f  2  2  4  3
4       4        2   <NA>  3  1 NA  3
5       5        1      m  4  5  2  4
6       6        2      m  5  4  5  5
7       7        1      m  5  3  4  4
8       8        2      m  4  5  5  5
```

The cast function needs only two arguments, the data to reshape and a formula. The formula has the identifying variables on the left side separated by plus signs, then a tilde, "~" and the variable that will contain the new variables' values.

The reshape package can also aggregate values into group means.

The ability of people to contribute packages adds to R's power but also occasionally leads to confusion among names. Note that R comes with a *function* named reshape, and the Hmisc package has one named reShape. They are totally different due to the capital "S"! They can both do the task at hand, but the reshape *package* is the one we are using. While the package is named reshape, the functions to do the reshaping are named melt and cast. The example programs that reshape data are in Table 14.18.

Table 14.18 Example programs for reshaping variables to observations and back

SAS programming statements	
| | ```
* SAS Program to Reshape Data.
* Reshape.sas ;
LIBNAME myLib 'C:\myRfolder' ;
* Wide to long;
PROC TRANSPOSE DATA=mylib.mydata
  OUT=myLib.mylong;
  VAR q1-q4;
  BY id workshop gender;
PROC PRINT;
RUN;
DATA mylib.mylong;
  SET mylib.mylong( rename=(COL1=value) );
  time=INPUT( SUBSTR( _NAME_, 2) , 1.);
  DROP _NAME_;
RUN;
PROC PRINT;
RUN;
* Long to wide;
``` |
```

Table 14.18 (continued)

```
PROC TRANSPOSE DATA=mylib.mylong
   OUT=myLib.mywide PREFIX=q;
   BY id workshop gender;
   ID time;
   VAR value;
RUN;
DATA mylib.mywide;
   SET mylib.mywide(DROP=_NAME_);
RUN;
PROC PRINT;
RUN;
```

SPSS programming
  statements

```
* SPSS Program to Reshape Data.
* Reshape.sps .
CD 'C:\myRfolder' .
GET FILE='mydata.sav' .
* Wide to long.
VARSTOCASES   /MAKE Y FROM q1 q2 q3 q4
   /INDEX = Question(4)
   /KEEP =   id workshop gender
   /NULL = KEEP.
LIST.
SAVE OUTFILE='mywide.sav' .
* Long to wide.
GET FILE='mywide.sav' .
CASESTOVARS
   /ID = id workshop gender
   /INDEX = Question
   /GROUPBY = VARIABLE.
LIST.
SAVE OUTFILE='mylong.sav' .
```

R programming
  statements

```
# R Program to Reshape Data.
# Reshape.R
setwd("/myRfolder")
load(file="myWorkspace.RData")
# Create an id variable.
mydata$subject <- 1:8
mydata
library("reshape")
attach(mydata)
# Melt data into "long" format.
mylong <- melt(mydata,
id=c("subject","workshop","gender") )
mylong
# Cast data back into "wide" format.
mywide <- cast(mylong,
   subject+workshop+gender ~ variable)
mywide
```

14.17 Sorting Data Frames

Sorting is one of the areas that R differs most from SAS and SPSS. In SAS and SPSS, sorting is critical prerequisite for three frequent tasks:

1. Doing the same analysis repeatedly for different groups. In SAS this is called BY processing. SPSS calls it SPLIT FILE processing.
2. Calculating summary statistics for each group in the SAS SUMMARY procedure (the similar SPSS AGGREGATE procedure does not require sorted data).
3. Merging files matched on the sorted variables such as id.

As we have seen, R does not need the data sorted for any of these tasks. Still, sorting is useful in a variety of contexts. R sorts in a very different way. It does not directly sort a data frame. Instead, it determines the order that the rows would be in if sorted, and then applies them to do the sort.

Consider the names Ann, Eve, Carla, Dave, and Bob. They are almost sorted in ascending order. Since the number of names is small, it is easy to determine the order that the names would require to sort them. We need the first name, Ann, followed by the fifth name, Bob, followed by the third name, Carla, the fourth name, Dave, and finally the second name, Eve. The order function would get those index values for us: 1, 5, 3, 4, 2.

To understand how these index values will help us sort, let us review briefly how data frame indexes work. One way to select rows from a data frame is to use the form mydata[rows, columns]. If you leave them all out, as in mydata[,] then you will get all rows and all columns. You can select the first 4 records with:

```
> mydata[ c(1,2,3,4 ),  ]

   id workshop gender q1 q2 q3 q4
1  1         1      f  1  1  5  1
2  2         2      f  2  1  4  1
3  3         1      f  2  2  4  3
4  4         2   <NA>  3  1 NA  3
```

We can select them in reverse order with:

```
> mydata[ c(4,3,2,1 ),  ]

   id workshop gender q1 q2 q3 q4
4  4         2   <NA>  3  1 NA  3
3  3         1      f  2  2  4  3
2  2         2      f  2  1  4  1
1  1         1      f  1  1  5  1
```

Now let us create a variable to store the order of the observations if sorted by workshop:

```
> myW  <- order( mydata$workshop  )
```

```
> myW
```

```
[1 ] 1 3 5 7 2 4 6 8
```

We can use this variable as the row index to mydata to see it sorted by workshop:

```
> mydata[ myW,  ]
```

```
  id workshop gender q1 q2 q3 q4
1  1        1      f  1  1  5  1
3  3        1      f  2  2  4  3
5  5        1      m  4  5  2  4
7  7        1      m  5  3  4  4
2  2        2      f  2  1  4  1
4  4        2   <NA>  3  1 NA  3
6  6        2      m  5  4  5  5
8  8        2      m  4  5  5  5
```

The order function is one of the few R functions that allow you to specify multiple variables without combining them into a vector with the c function. So we can create an order variable to sort the data by gender and then workshop within gender with the following. GW stands for Gender then Workshop.

```
> myGW   <- order ( mydata$gender, mydata$workshop  )
```

```
> mydata [ myGW,  ]
```

```
  id workshop gender q1 q2 q3 q4
1  1        1      f  1  1  5  1
3  3        1      f  2  2  4  3
2  2        2      f  2  1  4  1
5  5        1      m  4  5  2  4
7  7        1      m  5  3  4  4
6  6        2      m  5  4  5  5
8  8        2      m  4  5  5  5
4  4        2   <NA>  3  1 NA  3
```

The default order is ascending (small to large). To reverse this, place the minus sign before any variable. Therefore, this will sort by workshop in descending order and then within that, gender in ascending order:

```
> mydata$myWdG   <- order ( -mydata$workshop, mydata$gender )
> mydata$myWdG
```

```
[1 ] 2 6 8 4 1 3 5 7
```

The WdG part of myWdG is an acronym for Workshop, descending, then Gender. While SAS and SPSS view missing values as the smallest values to sort, R simply places them last. Recall from our discussion of missing values that R does not view NAs as large or small. You can use the argument na.last

=FALSE to cause R to place NAs first. You can also remove records with missing values by setting na.last =NA .

To see mydata in sorted order, we then use myWG in the row index position:

```
> myWdG   <- order ( -mydata$workshop, mydata$gender )

> mydata[ myWdG,   ]

   id workshop gender q1 q2 q3 q4
2  2         2      f  2  1  4  1
6  6         2      m  5  4  5  5
8  8         2      m  4  5  5  5
4  4         2   <NA>  3  1 NA  3
1  1         1      f  1  1  5  1
3  3         1      f  2  2  4  3
5  5         1      m  4  5  2  4
7  7         1      m  5  3  4  4
```

Since it is so easy to create a variable to store your various order indexes in, you do not need to store the whole data frame in sorted form to have easy access to it. However, if you want to, you can save the data frame in sorted form by using:

```
> mydataSorted   <- mydata[ myWdG,   ]

> mydataSorted

   id workshop gender q1 q2 q3 q4
2  2         2      f  2  1  4  1
6  6         2      m  5  4  5  5
8  8         2      m  4  5  5  5
4  4         2   <NA>  3  1 NA  3
1  1         1      f  1  1  5  1
3  3         1      f  2  2  4  3
5  5         1      m  4  5  2  4
7  7         1      m  5  3  4  4
```

Example programs for sorting are in Table 14.19.

Table 14.19 Example programs for sorting data frames

SAS programming statements	
| | ```
* SAS Program to Sort Data;
* Sort.sas ;
LIBNAME myLib 'C:\myRfolder';
PROC SORT  DATA=myLib.mydata;
  BY workshop;
RUN;
PROC PRINT DATA=myLib.mydata;
RUN;
``` |
```

Table 14.19 (continued)

```
                    PROC SORT   DATA=myLib.mydata;
                    BY gender workshop;
                    RUN;
                    PROC PRINT DATA=myLib.mydata;
                    RUN;
                    PROC SORT   DATA=myLib.mydata;
                      BY workshop descending gender;
                    RUN;
                    PROC PRINT DATA=myLib.mydata;
                    RUN;
```

SPSS programming
statements
```
                    * SPSS Program to Sort Data.
                    * Sort.sps .
                    CD 'C:\myRfolder' .
                    GET FILE='mydata.sav' .
                    SORT CASES BY workshop (A).
                    LIST.
                    SORT CASES BY gender (A) workshop (A).
                    LIST.
                    SORT CASES BY workshop (D) gender (A).
                    LIST.
```

R programming
statements
```
                    # R Program to Sort Data.
                    # Sort.R
                    setwd("/myRfolder")
                    load(file="myWorkspace.RData")
                      mydata
                      # Show first four observations in order.
                    mydata[ c(1,2,3,4), ]
                    # Show them in reverse order.
                    mydata[ c(4,3,2,1), ]
                    # Create order variable for workshop.
                    myW <- order( mydata$workshop )
                    myW
                    mydata[ myW, ]
                    # Create order variable for gender then workshop.
                    myGW <- order( mydata$gender, mydata$workshop )
                    myGW
                      mydata[ myGW, ]
                    # Create order variable for
                    # descending (-) workshop then gender
                    myWdG <- order( -mydata$workshop, mydata$gender )
                    myWdG
                    # Print data in WG order.
                    mydata[ myWdG, ]
                    # Save data in WdG order.
                    mydataSorted <- mydata[ myWdG, ]
                    mydataSorted
```

Chapter 15
Value Labels or Formats (and Measurement Level)

This section blends two topics because in R they are inseparable. In both SAS and SPSS, assigning labels to values is independent of the variable's measurement level. In R, you can assign value labels only to variables whose measurement level is *factor*. To be more precise, only objects whose class is factor can have label attributes.

In SAS, a variable's measurement level of nominal, ordinal, or interval is not stored. Instead, you list the variable on a specific statement, such as CLASS or BY, to tell SAS that you wish to view it as categorical.

SAS' use of value labels is a two-step process. First, PROC FORMAT creates a "format" for every unique set of labels and then the FORMAT statement assigns a format to each variable or set of variables (see example program below, Table 15.1). The formats are stored outside the dataset in a format library.

In SPSS, the VARIABLE LEVEL command sets measurement level but this is merely a convenience to help the GUI work well. Newer SPSS procedures take advantage of this information and do not show you nominal variables in a dialog box that it considers inappropriate. However, if you enter them into the same procedure using the programming language, it will accept them. As with SAS, special commands tell SPSS how you want to view the scale of the data. These include GROUPS and BY.

Independently, the VALUE LABEL command sets labels for each level and the labels are stored within the dataset itself as an attribute.

R has the measurement levels of *factor* for nominal data, *ordered factor* for ordinal data, and *numeric* for interval or scale data. You set these in advance and then the statistical and graphical procedures use them in an appropriate way automatically. When creating a factor, assigning labels is optional. If you do not use labels, the variable's original values are stored as character labels. R stores value labels in the factor itself.

R.A. Muenchen, *R for SAS and SPSS Users*, DOI: 10.1007/978-0-387-09418-2_15,   225
© Springer Science+Business Media, LLC 2009

15.1 Character Factors

Let us review how the `read.table` function deals with character variables. If we do not tell it what to do, it will convert all character data to factors.

```
> mydata <- read.table("mydata.tab")

> mydata

  workshop gender q1 q2 q3 q4
1        1      f  1  1  5  1
2        2      f  2  1  4  1
3        1      f  2  2  4  3
4        2   <NA>  3  1 NA  3
5        1      m  4  5  2  4
6        2      m  5  4  5  5
7        1      m  5  3  4  4
8        2      m  4  5  5  5
```

You cannot tell what gender is by looking at it, but the `class` function can tell us that gender is a factor.

```
> class( mydata [ ,"gender"] )

[1] "factor"
```

In our case, this is helpful. However, there are times when you want to leave character data as simply characters. When reading people's names or addresses from a database for example, you do not want to store them as factors. The argument `stringsAsFactors = FALSE` will tell `read.table` to leave such variables as character.

```
> mydata2 <- read.table("mydata.tab",
+     stringsAsFactors=FALSE)

> mydata2

  workshop gender q1 q2 q3 q4
1        1      f  1  1  5  1
2        2      f  2  1  4  1
3        1      f  2  2  4  3
4        2   <NA>  3  1 NA  3
5        1      m  4  5  2  4
6        2      m  5  4  5  5
7        1      m  5  3  4  4
8        2      m  4  5  5  5
```

This sets how *all* the character variables are read, so if some of them do need to be factors, you can convert those afterwards. The data looks just the same, but the `class` function can verify that gender is indeed now a character variable.

```
> class( mydata2[ ,"gender"] )

[1] "character"
```

Many functions will not do what you expect with character data. For example, we have seen the `summary` function count factor levels. But it will not count them in character form:

```
> summary( mydata2$gender )

   Length      Class        Mode
        8  character   character
```

We will focus on the first data frame that has gender as a factor. As `read.table` scans the data, it assigns the numeric values to the character values in alphabetical order. Therefore, gender gets 1 for "f" and 2 for "m" since "f" precedes "m" in the alphabet. For character data, those defaults are often sufficient. However, you can use the `factor` function to specify the order.

```
mydata$genderF <- factor(
  mydata$gender,
  levels=c ("m", "f"),
  labels=c ("Male", "Female") )
```

The `factor` function call above has three arguments.

1. The name of the factor.
2. The `levels` argument with the levels *in order*. Since "m" appears first, it will be associated with 1 and "f" with 2.
3. The `labels` argument provides labels in the same order as the `levels` argument. This example sets "Male" as 1, "Female" as 2 and uses the fully written out labels. If you use the `lables` argument with character data, the original character values ("m" and "f" in this case) are not stored at all.

There is a danger in setting value labels with character data that does not appear at all in SAS or SPSS. In the statement above, if we instead set

```
levels=c ("m", "F")
```

R would set the values for all females to missing (NA) because the actual values are lower case. There are no capital F's in the data! This danger applies of course to other more obvious misspellings.

15.2 Numeric Factors

Unlike character data, R has no way to identify numeric factors automatically, such as our workshop variable. Workshop is a categorical measure, but initially R assumes it is numeric because it is entered as 1 and 2. If we do any analysis on workshop, it will be as an integer variable.

```
> class ( mydata$workshop )
```

```
[1] "integer"
```

```
> summary ( mydata$workshop )
```

```
   Min. 1st Qu.   Median    Mean 3rd Qu.    Max.
    1.0     1.0      1.5     1.5     2.0     2.0
```

We can nest the call to the as.factor function into any analysis to overcome this problem.

```
> summary ( as.factor (mydata$workshop) )
```

```
1 2
4 4
```

So we see four people took workshops 1 and 2. Note here that the values 1 and 2 are merely labels at this point.

We can use the factor function to convert it to a factor and optionally assign labels. Factor labels in R are stored in the factor itself.

```
mydata$workshop <- factor(
  mydata$workshop,
  levels=c (1,2,3,4)
  labels=c ("R", "SAS", "SPSS","Stata"
)
```

Notice that we have assigned four labels to four values even though our data only contain the values 1 and 2. This can be very helpful in making consistent value and label regardless of the data we currently have on hand. If we collected another dataset in which people only took SPSS and Stata workshops, we do not want 1 = R in one dataset and 1 = SPSS in another!

Now let us convert our q variables to factors. Since we will need to specify the same levels repeatedly, let us put them in a variable.

```
> myQlevels <- c(1,2,3,4,5)
```

```
> myQlevels
```

```
[1] 1 2 3 4 5
```

Now we will do the same for the labels.

```
> myQlabels <- c("Strongly Disagree",
+                "Disagree",
+                "Neutral",
+                "Agree",
+                "Strongly Agree")
```

```
> myQlabels
```

```
[1] "Strongly Disagree" "Disagree"              "Neutral"
[4] "Agree"              "Strongly Agree"
```

Finally, we will use the `ordered` function to complete the process. It works just like the `factor` function but tells R that the data values have order. In statistical terms, it sets the variable's measurement level to ordinal. Ordered factors allow you to perform logical comparisons of *greater than* or *less than*. R's statistical modeling functions also try to treat ordinal data appropriately. However, some methods are influenced by the assumption that the ordered values are equally spaced, which may not be the case in your data.

We will put the factors into new variables with an "f" for factor in their names, like qf1 for q1. The f is just there to help us remember, it has no meaning to R. We will keep both sets of variables because people who do survey research often want to view this type of variable as numeric for some analyses and as categorical for others. It is not necessary to do this if you prefer converting back and forth on the fly.

```
> mydata$qf1 <- ordered( mydata$q1, myQlevels, myQlabels)
> mydata$qf2 <- ordered( mydata$q2, myQlevels, myQlabels)
> mydata$qf3 <- ordered( mydata$q3, myQlevels, myQlabels)
> mydata$qf4 <- ordered( mydata$q4, myQlevels, myQlabels)
```

Now we can use the `summary` function to get frequency tables on them, complete with value labels

```
> summary( mydata[ c("qf1", "qf2", "qf3", "qf4")])

                qf1                  qf2         qf3
Strongly               Strongly             Strongly
  Disagree     :1        Disagree     :3      Disagree :0
Disagree       :2      Disagree       :1    Disagree   :1
Neutral        :1      Neutral        :1    Neutral    :0
Agree          :2      Agree          :1    Agree      :3
Strongly Agree :2      Strongly Agree :2    Strongly
                                              Agree    :3
                                            NA' s      :1

                qf4
Strongly Disagree:3
Disagree         :0
Neutral          :2
Agree            :2
Strongly Agree   :1
```

15.3 Making Factors of Many Variables

The approach used above works fine for small numbers of variables. However, if you have hundreds, it is needlessly tedious. We do the same thing again, this time in a form that would handle any number of variables whose names follow the format string1 to stringN. Our practice data only has q1 to q4, but the same number of commands would handle q1 to q4000.

First, we will generate variable names to use. We will use qf to represent the q variables in factor form. This is an optional step, needed only if you want to keep the variables in both forms.

```
> myQnames <- paste ( "q",  1:4, sep="")

> myQnames

 [1] "q1" "q2" "q3" "q4"

> myQFnames <- paste ( "qf", 1:4, sep="")

> myQFnames

 [1] "qf1" "qf2" "qf3" "qf4"
```

Now we will use the myQnames character vector as column names to select from our data frame. We will store those in a separate data frame.

```
> myQFvars <- mydata[ ,myQnames]

> myQFvars

  q1 q2 q3 q4
1  1  1  5  1
2  2  1  4  1
3  2  2  4  3
4  3  1 NA  3
5  4  5  2  4
6  5  4  5  5
7  5  3  4  4
8  4  5  5  5
```

Next, we will use the myQFnames character vector to rename these variables.

```
> names (myQFvars) <- myQFnames

> myQFvars

  qf1 qf2 qf3 qf4
1   1   1   5   1
2   2   1   4   1
3   2   2   4   3
4   3   1  NA   3
5   4   5   2   4
6   5   4   5   5
7   5   3   4   4
8   4   5   5   5
```

Now we need to make up a function to apply to the variables.

```
myLabeler <- function (x) {
  ordered(x, myQlevels, myQlabels)
}
```

The myLabeler function will apply myQlevels and myQlabels (defined in the section directly above) to any variable, x, that we supply to it. Let us try it on a single variable.

```
> summary( myLabeler( myQFvars[," qf1"] ) )
```

Strongly Disagree	Disagree	Neutral	Agree
1	2	1	2

```
  Strongly Agree
        2
```

It is important to understand that the myLabeler function will work only on vectors, since the ordered function requires them. So simply removing the comma in the above command would select a data frame containing only qf1, and would not work.

```
> summary( myLabeler(myQFvars["qf1"]) ) # Doesn't work!
```

Strongly Disagree	Disagree	Neutral	Agree
0	0	0	0

Strongly Agree	NA' s
0	1

Now we will use the sapply function to apply myLabeler to myQFvars.

```
myQFvars <- data.frame( sapply( myQFvars, myLabeler ) )
```

The sapply function simplified the result to a matrix and the data.frame function converted that to a data frame. Now the summary function will count the values and display their labels.

```
> summary(myQFvars)
```

	qf1		qf2		qf3
Agree	:2	Agree	:1	Agree	:3
Disagree	:2	Disagree	:1	Disagree	:1
Neutral	:1	Neutral	:1	Strongly	
				Agree	:3
Strongly Agree	:2	Strongly Agree	:2	NA' s	:1
Strongly Disagree	:1	Strongly Disagree	:3		

	qf4
Agree	:2
Neutral	:2

```
Strongly Agree    :2
Strongly Disagree:2
```

If you care to, you can bind myQFvars to our original data frame.

```
mydata <- cbind(mydata, myQFvars)
```

15.4 Converting Factors into Numeric or Character Variables

R has functions for converting factors into numeric or character vectors (variables). To extract the numeric values from a factor like gender, we can use the as.numeric function.

```
> mydata$genderNums <- as.numeric( mydata$gender )

> mydata$genderNums

[1]  1  1  1 NA  2  2  2  2
```

If we want to extract the labels themselves to use in a character vector, we can do so with the as.character function.

```
> mydata$genderChars <- as.character( mydata$gender)

> mydata$genderChars

[1] "f" "f" " f" NA   "m" "m" "m" "m"
```

We can apply the same two functions to variable qf1. Since we were careful to set all the levels, even for those that did not appear in the data, this works fine. First, we will do it using as.numeric.

```
> mydata$qf1Nums <- as.numeric(mydata$qf1)

> mydata$qf1Nums

[1] 1 2 2 3 4 5 5 4
```

Now let us do it again using as.character.

```
> mydata$qf1Chars <- as.character(mydata$qf1)

> mydata$qf1Chars

[1] "Strongly Disagree" "Disagree"        "Disagree"
[4] "Neutral"           "Agree"           "Strongly Agree"
[7] "Strongly Agree"    "Agree"
```

Where you can run into trouble is when you do not specify the original numeric values, and they are not simply 1, 2, 3,... For example, let us create a variable whose original values are 10, 20, 30:

```
> x <- c (10,20,30)

> x

[1] 10 20 30

> xf <<- as.factor(x)

> xf

[1] 10 20 30

Levels: 10 20 30
```

So far, the factor xf looks fine. However, when we try to extract the original values with the as.numeric function, we get instead the levels 1, 2, 3!

```
> as.numeric(xf)

[1] 1 2 3
```

If we use as.character to get the values, there are no nice value labels, so we get character versions of the original values.

```
> as.character (xf)

[1] "10" "20" "30"
```

If we want those original values in a numeric vector like the one we began with, we can use as.numeric to convert them. To extract the original values and store them in a variable x10, we can use the following:

```
> x10 <- as.numeric( as.character (xf) )

> x10

[1] 10 20 30
```

The original values were automatically stored as value labels. If you had specified value labels of your own, like low, medium, and high, the original values would have been lost. You would then have to recode the values to get them back. See Sect. 14.8 for details.

15.5 Dropping Factor Levels

Earlier in this chapter, we created labels for factor levels that did not exist in our data. While it is not at all necessary, it is helpful if you were to enter more data or merge your data frame with others that have a full set of values. However, when using such variables in analysis, it is often helpful to get rid of such empty levels. To get rid of the unused levels, append [,drop = TRUE] to the variable reference. For example, if you want to include the empty levels, skip the drop argument.

```
summary ( workshop )
```

```
R    SAS   SPSS Stata
```

```
4    4     0     0
```

But when you need rid of empty levels, add it the `drop` argument.

```
> summary( workshop[ ,drop=TRUE] )
```

```
R SAS
```

```
4   4
```

For example programs that demostrate those topics, see Table 15.1.

Table 15. 1   Example programs for value labels or formats (& measurement level)

SAS programming statements	
| | ```
* SAS Program to Assign Value Labels (formats);
* ValueLabels.sas ;

LIBNAME myLib 'C:\myRfolder';

PROC FORMAT;
  VALUES workshop_f 1="Control" 2="Treatment"
  VALUES$gender_f "m"="Male" "f"="Female"
  VALUES agreement
    1='Strongly Disagree'
    2='Disagree'
    3='Neutral'
    4='Agree'
    5='Strongly Agree'.;
RUN;

DATA  myLib.mydata;
  SET myLib.mydata;
FORMAT workshop workshop_f. gender gender_f.
  q1-q4 agreement.;
RUN;
``` |
| SPSS programming statements | ```
* SPSS Program to Assign Value Labels.
* ValueLabels.sps

CD   'C:\myRfolder'.

GET FILE='mydata.sav'.
VARIABLE LEVEL workshop (NOMINAL)
 /q1 TO q4 (SCALE).
VALUE LABELS workshop 1 'Control' 2 'Treatment'
 /q1 TO q4
``` |

Table 15. 1 (continued)

	1 'Strongly Disagree' 2 'Disagree' 3 'Neutral' 4 'Agree' 5 'Strongly Agree' SAVE OUTFILE="mydata.sav".
R programming statements	(see code below)

```
# R Program to Assign Value Labels & Factor Status.
# ValueLabels.R

setwd("/myRfolder")

#Character Factors

#Read gender as factor.
mydata <- read.table("mydata.tab")
mydata
class( mydata[ ,"gender"] )

# Read gender as character.
mydata2 <- read.table("mydata.tab", as.is=TRUE)
mydata2
class ( mydata2[ ,"gender"] )
summary ( mydata2$gender)
rm (mydata2)

# Numeric Factors

class ( mydata$workshop)
summary ( mydata$workshop )

summary ( as.factor(mydata$workshop) )

# Now change workshop into a factor:
mydata$workshop <- factor ( mydata$workshop,
  levels=c (1,2,3,4),
  labels=c ("R", "SAS""SPSS", "Stata") )
mydata

# Now see that summary only counts workshop attendance.
summary (mydata$workshop)

# Making the Q Variables Factors

# Store levels to use repeatedly.
myQlevels <- c(1,2,3,4,5)
myQlevels

# Store labels to use repeatedly.
```

Table 15. 1 (continued)

```
myQlabels <- c ("Strongly Disagree",
                "Disagree",
                "Neutral",
                "Agree",
                "Strongly Agree")
myQlabels

# Now create a new set of variables as factors.
mydata$qf1 <- ordered ( mydata$q1, myQlevels,
  myQlabels)
mydata$qf2 <- ordered ( mydata$q2, myQlevels,
  myQlabels)
mydata$qf3 <- ordered ( mydata$< q3, myQlevels,
  myQlabels)

mydata$qf4 <- ordered ( mydata$q4, myQlevels,
  myQlabels)

# Get summary and see that workshops are now counted.
summary ( mydata[ c ("qf1", "qf2", "qf3", qf4") ] )

# Making Factors of Many Variables

# Generate two sets of var names to use.
myQnames   <- paste( "q", 1:4, sep="")
myQnames
myQFnames  <- paste( qf", 1:4, sep="")
myQFnames

# Extract the q variables to a separate data frame.
myQFvars <- mydata[ ,myQnames]
myQFvars

# Rename all the variables with F for Factor.
names (myQFvars) <- myQFnames
myQFvars

# Create a function to apply the labels to lots of
   variables.
myLabeler <- function(x) {
  ordered (x, myQlevels, myQlabels)
}

# Here' s how to use the function on one variable.

summary ( myLabeler (myQFvars[ ,"qf1"] ) )
summary ( myLabeler (myQFvars[ "qf1"] ) ) # Doesn' t
  work!
```

Table 15. 1 (continued)

```
# Apply it to all the variables.
myQFvars <- data.frame ( s apply ( myQFvars, myLabeler) )

# Get summary again, this time with labels.
summary(myQFvars)

# You can even join the new variables to mydata.
# (this gives us two labeled sets if you ran
# the example above too.)
mydata <- cbind(mydata,myQFvars)
mydata

#--- Converting Factors into Character or Numeric
   Variables

# Converting the gender factor, first with as.numeric.

mydata$genderNums <- as.numeric ( mydata$gender )
mydata$genderNums

# and again with as.character.

mydata$genderChars <- as.character ( mydata$gender)
mydata$genderChars

# Converting the qf1 factor.

mydata$qf1Nums <- as.numeric (mydata$qf1)
mydata$qf1Nums

mydata$qf1Chars <- as.character (mydata$qf1)
mydata$qf1Chars

# Example with bigger values.
x <- c(10,20,30)
x
xf <- factor (x)
xf
as.numeric (xf)
as.character (xf)
x10 <- as.numeric (as.character (xf) )
x10
```

Chapter 16
Variable Labels

Perhaps, the most fundamental feature missing from the main R distribution is support for variable labels. It does have a comment attribute that you can apply to each variable, but only the `comment` function itself will display it.

In SAS or SPSS, you might name a variable BP and want your publication-ready output to display, "Systolic Blood Pressure" instead. SAS does this using the LABEL statement and SPSS does it using the very similar VARIABLE LABELS command.

Survey researchers in particular rely on variable labels. They often name their variables Q1, Q2, and so on and assign labels as the full text of the survey items. R is the only statistics package that I am aware of which lacks such a feature.

It is a testament to R's openness and flexibility that a user can add such a fundamental feature. Frank Harrell's `Hmisc` package [6] does just that. It adds a "label" attribute to the data frame and stores the labels there, even converting them from SAS datasets automatically (but not SPSS datasets). As amazing as this addition is, the fact that variable labels were not included in the main distribution means that most procedures do not take advantage of what `Hmisc` adds. The many wonderful functions in the `Hmisc` package do, of course. `Hmisc` creates these labels using its `label` function.

```
library("Hmisc")
label (mydata$q1) <- "The instructor was well prepared."
label (mydata$q2) <- "The instructor communicated well."
label (mydata$q3) <- "The course materials were helpful."
label (mydata$q4) <- "Overall, I found this workshop
  useful."
```

Now the `Hmisc` `describe` function will take advantage of the labels.

```
> describe( mydata[ ,3:6] )
mydata [, 3:6]

  4  Variables      8  Observations
----------------------------------------------
```

R.A. Muenchen, *R for SAS and SPSS Users*, DOI: 10.1007/978-0-387-09418-2_16,   239
© Springer Science+Business Media, LLC 2009

```
q1 : The instructor was well prepared.
      n missing  unique      Mean
      8        0        5      3.25

              1  2  3  4  5
Frequency 1   2  1  2  2
%            12 25 12 25 25
------------------------------------------------
q2 : The instructor communicated well.
      n missing  unique      Mean
      8        0        5      2.75

              1  2  3  4  5
Frequency 3   1  1  1  2
%            38 12 12 12 25
------------------------------------------------
q3 : The course materials were helpful.
      n missing  unique      Mean
      7        1        3      4.143

2 (1, 14%), 4 (3, 43%), 5 (3, 43%)
------------------------------------------------
q4 :  Overall, I found this workshop useful.
      n missing  unique      Mean
      8        0        4      3.25

1 (2, 25%), 3 (2, 25%), 4 (2, 25%), 5 (2, 25%)
------------------------------------------------
```

Unfortunately built-in functions such as summary ignore the labels.

```
> summary( mydata[ ,3:6] )
      q1              q2              q3              q4
Min.   :1.00  Min.   :1.00  Min.   :2.00  Min.   :1.00
1st Qu.:2.00  1st Qu.:1.00  1st Qu.:4.00  1st Qu.:2.50
Median :3.50  Median :2.50  Median :4.00  Median :3.50
Mean   :3.25  Mean   :2.75  Mean   :4.14  Mean   :3.25
3rd Qu.:4.25  3rd Qu.:4.25  3rd Qu.:5.00  3rd Qu.:4.25
Max.   :5.00  Max.   :5.00  Max.   :5.00  Max.   :5.00
                            NA's   :1.00
```

A second approach to variable labels is to store them as a character variable. That is the approach used by the prettyR package [7].

Finally, you can use illegal variable names of any length by enclosing them in quotes. This has the advantage of working with most R functions.

```
> #Assign long variable names to act as variable labels.
> names(mydata) <- c("Workshop","Gender",
+    "The instructor was well prepared.",
+    "The instructor communicated well.",
+    "The course materials were helpful.",
+    "Overall, I found this workshop useful.")
```

Notice here that names like q1, q2,... do not exist now. The labels *are* the names.

```
> names(mydata)
```

```
[1] "Workshop"
[2] "Gender"
[3] "The instructor was well prepared."
[4] "The instructor communicated well."
[5] "The course materials were helpful."
[6] "Overall, I found this workshop useful."
```

Now many R functions, even those built-in, will use the labels.

```
> summary( mydata[ ,3:6] )
```

```
The instructor was well prepared.
Min.   :1.00
1st Qu.:2.00
Median :3.50
Mean   :3.25
3rd Qu.:4.25
Max.   :5.00
```

```
The instructor communicated well.
Min.   :1.00
1st Qu.:1.00
Median :2.50
Mean   :2.75
3rd Qu.:4.25
Max.   :5.00
...
```

You can still select variables by their names.

```
> summary( mydata["Overall, I found this workshop
useful."] )
```

```
Overall, I found this workshop useful.
Min.   :1.00
1st Qu.:2.50
Median :3.50
Mean   :3.25
```

```
3rd Qu.:4.25
Max.    :5.00
```

You can type out the long name or use the search function, grep, to find key words in your labels. The grep below finds the string 'instructor' in variables 3 and 4.

```
> myvars <- grep( 'instructor', names(mydata) )

> myvars

[1] 3 4

> summary ( mydata[myvars] )

 The instructor was well prepared.
 Min.    :1.00
 1st Qu.:2.00
 Median :3.50
 Mean    :3.25
 3rd Qu.:4.25
 Max.    :5.00

 The instructor communicated well.
 Min.    :1.00
 1st Qu.:1.00
 Median :2.50
 Mean    :2.75
 3rd Qu.:4.25
 Max.    :5.00
```

Some important R functions, such as data.frame convert the spaces in the labels to periods. If you use this, you must replace the names as we initially created them above.

```
> newdata <- data.frame( mydata )

> names ( newdata[ ,3:6] )

[1] "The.instructor.was.well.prepared."
[2] "The.instructor.communicated.well."
[3] "The.course.materials.were.helpful."
[4] "Overall..I.found.this.workshop.useful."
```

To avoid this change, add the check.names = FALSE argument to the data.frame function.

```
> newdata <- data.frame(mydata, check.names=FALSE)
> names( newdata[ ,3:6] )
[1] "The instructor was well prepared."
[2] "The instructor communicated well."
[3] "The course materials were helpful."
[4] "Overall, I found this workshop useful."
```

Example programs that demonstrate variable labels in SAS, SPSS and R are in Table 16.1.

Table 16.1 Example programs for variable labels

| SAS programming statements | ```
* SAS Program for Variable Labels;
* VarLabels.sas ;

LIBNAME myLib 'C:\myRfolder';

DATA myLib.mydata;
  SET myLib.mydata ;
  LABEL
    Q1="The instructor was well prepared"
    Q2="The instructor communicated well"
    Q3="The course materials were helpful"
    Q4="Overall, I found this workshop useful";
RUN;

PROC FREQ; TABLES q1-q4; RUN;
. RUN;
``` |
| SPSS programming statements | ```
* SPSS Program for Variable Labels.
* VarLabels.sps .

CD 'C:\myRfolder'.

GET FILE='mydata.sav'.
VARIABLE LABELS
  Q1 "The instructor was well prepared"
  Q2 "The instructor communicated well"
  Q3 "The course materials were helpful"
  Q4 "Overall, I found this workshop useful".

FREQUENCIES VARIABLES=q1 q2 q3 q4.
SAVE OUTFILE="mydata.sav".
``` |
| R programming statements | ```
# R Program for Variable Labels.
# VarLabels.R

setwd("/myRfolder")
load(file="myWorkspace.RData")
options(width=64)
mydata
``` |
```

Table 16.1 (continued)

```
# Using the Hmisc label attribute.
library("Hmisc")
label(mydata$q1) <- "The instructor was well prepared."
label(mydata$q2) <- "The instructor communicated well."
label(mydata$q3) <- "The course materials were helpful."
label(mydata$q4) <- "Overall, I found this workshop useful."
# Hmisc describe function uses the labels.
describe( mydata[ ,3:6] )

# Buit-in summary function ignores the labels.
summary ( mydata[ ,3:6] )

#Assign long variable names to act as variable labels.
names (mydata) <- c("Workshop","Gender",
  "The instructor was well prepared.",
  "The instructor communicated well.",
  "The course materials were helpful.",
  "Overall, I found this workshop useful.")

names (mydata)

# Now summary uses the long names.
summary( mydata[ ,3:6] )

# You can still select variables by name.
summary ( mydata[ "Overall, I found this workshop
  useful."] )

# Searching for strings in long variable names.
myvars <- grep('instructor',names(mydata))
myvars
summary ( mydata[ myvars] )

# Data.frame replaces spaces with periods.
newdata <- data.frame( mydata )
names ( newdata[ ,3:6] )

# Data.frame now keeps the spaces.
newdata <- data.frame( mydata, check.names=FALSE )
names ( newdata[ ,3:6] )
```

Chapter 17
Generating Data

Generating your own data is helpful in several ways. When you are designing an experiment, the levels of the experimental variables usually follow simple repetitive patterns. You can generate those and then add the measured outcome values to it manually. With such a nice neat dataset to start with, it is tempting to collect data in that order. However, it is important to collect it in random order whenever possible so that factors such as human fatigue or machine wear do not bias the results of your study.

Some of our generating data examples use R's random number generator. It will give a different result each time you use it unless you use the set.seed function *before each function that generates random numbers.*

Our workshop variable is an easy factor to generate. It repeats the pattern 1,2,1,2. . .

Generated data is also helpful for learning R and for submitting questions to the r-help e-mail list. That way you can send a working example of your problem without having to send a large or confidential dataset. Finally, you can easily generate patterns of values to use as column widths when reading fixed width text files (see Section 9.5).

17.1 Generating Numeric Sequences

SAS generates data using do loops in a data step. SPSS uses input programs to generate data. You can also use SPSS' interface to Python to generate data. R generates data using specialized functions. We have used the simplest one, the colon operator. We can generate a simple sequence with

```
> 1:10
 [1]  1  2  3  4  5  6  7  8  9 10
```

You can store the results of any of our data generation example in a vector using the assignment operator. So we can create the id variable we used with

R.A. Muenchen, *R for SAS and SPSS Users*, DOI: 10.1007/978-0-387-09418-2_17,   245
© Springer Science+Business Media, LLC 2009

```
> id<- 1:8
> id
[1] 1 2 3 4 5 6 7 8
```

The seq function generates sequences like this too, and it offers more flexibility. Here is an example:

```
> seq (from=1, to=10, by=1)
[1]  1  2  3  4  5  6  7  8  9 10
```

This seq function call has three arguments.

1. The from argument tells it where to begin the sequence.
2. The to argument tells it where to stop.
3. The by argument tells it the increments to use between each number.

Here is an example that goes from five to 50 in increments of 5.

```
> seq(from=5,to=50,by=5)
[1]   5 10 15 20 25 30 35 40 45 50
```

Of course, you do not need to name the parameters if you use them in order. So you can do the example above using this form too.

```
> seq(5,50,5)
[1]   5 10 15 20 25 30 35 40 45 50
```

17.2 Generating Factors

The gl function generates levels of factors. Here is an example that generates the series 1,2,1,2...

```
> gl(n=2, k=1, length=8)
[1] 1 2 1 2 1 2 1 2
Levels: 1 2
```

This gl function call has three arguments.

1. The n argument tells it how many levels your factor will have.
2. The k argument tells it how many of each level to repeat before incrementing to the next value.
3. The length argument is the total number of values generated. Although this would usually be divisible by n*k, it does not have to be.

To generate our gender variable, we just need to change k to be five.

```
> gl(n=2, k=5, length=8)
[1] 1 1 1 1 1 2 2 2
Levels: 1 2
```

There is also an optional `label` argument. Here we use it to generate workshop and gender, complete with value labels.

```
> workshop <- gl(n=2,k=1, length=8, label=c("R",
  "SAS") )
> workshop
[1] R     SAS R     SAS R     SAS R     SAS
Levels:  R  SAS
> gender <- gl(n=2,  k=4,
+ length=8,  label=c("f","m")  )
> gender
[1] f f f f m m m m
Levels:  f  m
```

17.3 Generating Repetitious Patterns (not factors)

When you need to generate repetitious sequences of values, you are often creating levels of factors, which are covered in the previous section. However, sometimes you need similar patterns that are numeric, not factors. The `rep` function generates these.

```
> gender <- rep(1:2, each=4, times=1)
> gender
[1] 1 1 1 1 2 2 2 2
```

The call to the `rep` function above has three simple arguments: the numbers to repeat, how often to repeat each, and how many times to repeat that combination. Note that while we are generating the gender variable as an easy example, gender is not a factor and to make it one, you would have to use the `factor` function.

Next, we generate the workshop variable by repeating each number in the 1:2 sequence only one time each but repeat that set four times.

```
> workshop <- rep(1:2, each=1, times=4)
> workshop
[1] 1 2 1 2 1 2 1 2
```

The `rep` function is particularly helpful for generating a vector of a single constant value. We use a variation of this in the chapter, *Traditional Graphics*, to generate a set of zeros to use on a histogram.

```
> myZeros <- rep(0, each=8)
> myZeros
[1] 0 0 0 0 0 0 0 0
```

17.4 Generating Integer Measures

R's ability to generate samples of integers is easy and not quite like anything in SAS or SPSS. You provide a list of possible values and then use the sample function to generate your data. First, we put the Likert scale values 1, 2, 3, 4, and 5 into myValues.

```
> myValues <- c(1, 2, 3, 4,5)
```

Next, we set the random number seed using the set.seed function, so you can see the same result when you run it, and then generate a random sample of 1000 from it.

```
> set.seed(1234)  # Set random number seed.
```

Finally, we generate a random sample of 1000 from my Values using the sample function and calculate its mean and standard deviation.

```
> q1 <- sample( myValues, size=1000, replace = TRUE)
> mean(q1)
[1] 3.029
> sd(q1)
[1] 1.412854
```

To generate a sample using the same numbers but with a roughly normal distribution, we can change the values to have more as you reach the center.

```
> myValues <- c(1, 2, 2, 3, 3, 3, 4, 4, 5)
> set.seed(1234)
> q2 <- sample( myValues, 1000, replace = TRUE)
> mean(q2)
[1] 3.012
> sd(q2)
[1] 1.169283
```

You can see from the barplot in Fig. 17.1 the distributions of both variables. Do not worry about how we created the plot; we will cover that in Chap. 21.

```
> barplot( table(q1) )
> barplot( table(q2) )
```

If you would like to generate two Likert-scale variables that have a mean difference, you can do so by providing them with different sets of values from which to sample. In the example below, I nest the call to the c function to generate a vector of values within the call to the sample function. Notice that the vector for q1 has no values greater than 3 and q2 has none less than 3. This difference will create the mean difference. Here I am only asking for a sample size of 8.

Fig. 17.1 Barplots showing distributions of generated variables

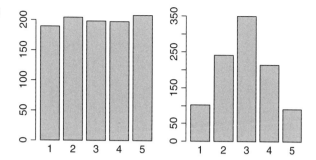

```
> set.seed(1234)
> q1 <- sample( c(1, 2, 2, 3), size=8, replace = TRUE)
> mean(q1)
[1] 1.75
> set.seed(1234)
> q2 <- sample( c(3, 4, 4, 5), size=8, replace = TRUE)
> mean(q2)
[1] 3.75
```

17.5 Generating Continuous Measures

You can generate continuous random values from a uniform distribution using the runif function

```
> set.seed(1234)
> x1 <- runif(n=1000)
> mean(x1)
[1] 0.5072735
> sd(x1)
[1] 0.2912082
```

where the *n* argument is the number of values to generate. You can also provide min and max arguments to set the lowest and highest possible values, respectively. So you might generate 1000 pseudo test scores that range from 60 to 100 with

```
> set.seed(1234)
> x2 <- runif(n=1000, min=60, max=100)
```

```
> mean(x2)
[1] 80.29094

> sd(x2)
[1] 11.64833
```

Normal distributions with a mean of zero and standard deviation of one have many uses. You can use the rnorm function to generate 1000 values from such a distribution with

```
> set.seed(1234)

> x3 <- rnorm(n=1000)

> mean(x3)
[1] -0.0265972

> sd(x3)
[1] 0.9973377
```

You can specify other means and standard deviations as in the following example.

```
> set.seed(1234)

> x4 <- rnorm(n=1000, mean=70, sd=5)

> mean(x4)
[1] 69.86701

> sd(x4)
[1] 4.986689
```

We can use the hist function to see what two of these distributions look like (Fig. 17.2).

```
> hist(x2)
> hist(x4)
```

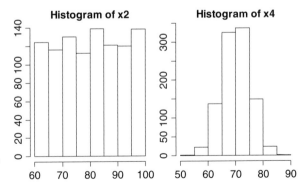

Fig. 17.2 Histograms showing the distributions of continuous generated data

17.6 Generating a Data Frame

Putting all the above ideas together, we can use the following commands to create a data frame similar to our practice dataset, with a couple of test scores added. I am not bothering to set the random number generator seed, so each time you run this, you will get different results.

```
> id <- 1:8
> workshop <- gl( n=2, k=1,
+   length=8, label=c("R","SAS") )
> gender <-    gl( n=2, k=4,
+   length=8, label=c("f","m")    )
> q1 <- sample( c(1, 2, 2, 3), 8, replace = TRUE)
> q2 <- sample( c(3, 4, 4, 5), 8, replace = TRUE)
> q3 <- sample( c(1, 2, 2, 3), 8, replace = TRUE)
> q4 <- sample( c(3, 4, 4, 5), 8, replace = TRUE)
> pretest   <- rnorm( n=8, mean=70, sd=5)
> posttest  <- rnorm( n=8, mean=80, sd=5)
>
> myGenerated <-  data.frame(id, gender, workshop,
+   q1, q2, q3, q4, pretest, posttest)
> myGenerated

id gender workshop q1 q2 q3 q4  pretest posttest
1  1       f         R   1  5  2   3 67.77482 77.95827
2  2       f       SAS   1  4  2   5 58.28944 78.11115
3  3       f         R   2  3  2   4 68.60809 86.64183
4  4       f       SAS   2  5  2   4 64.09098 83.58218
5  5       m         R   1  5  3   4 70.16563 78.83855
6  6       m       SAS   2  3  3   3 65.81141 73.86887
7  7       m         R   2  4  3   4 69.41194 85.23769
8  8       m       SAS   1  4  1   4 66.29239 72.81796
```

Example programs which domonstrate data generation in SAS, SPSS and R are in Table 17.1.

Table 17. 1 Example programs for generating data

| SAS programming statements | ```
* SAS Program for Generating Data;
* GenerateData.sas ;

DATA myID;
DO id=1 TO 8;
    OUTPUT;
END;
PROC PRINT; RUN;

DATA myWorkshop;
``` |
| --- | --- |
```

Table 17. 1 (continued)

```
Do i=1 to 4;
      DO workshop= 1 to 2;
         OUTPUT;
      END;
END;
DROP i;
RUN;
PROC PRINT; RUN;

DATA myGender;
DO i=1 to 2;
  DO j=1 to 4;
     gender=i;
     OUTPUT;
  END;
END;
DROP i j;
RUN;

PROC PRINT; RUN;

DATA myMeasures;
DO i=1 to 8;
   q1=round  (uniform(1234) * 5 );
   q2=round ((uniform(1234) * 5 ) -1 );
   q3=round (uniform(1234) * 5 );
   q4=round (uniform(1234) * 5 );
   test1=normal(1234)*5 + 70;
   test2=normal(1234)*5 + 80;
   OUTPUT;
END;
DROP i;
RUN;

PROC PRINT; RUN;

PROC FORMAT;
VALUES wLabels 1='R' 2='SAS';
VALUES gLabels 1='Female' 2='Male';
RUN;

* Merge and eliminate out of range values;
DATA myGenerated;
  MERGE myID myWorkshop myGender myMeasures;
  FORMAT workshop wLabels. gender gLabels. ;
  ARRAY q{ 4} q1-q4;
  DO i=1 to 4;
      IF q{ i} < 1 then q{ i} = 1;
    ELSE  IF  q{ i}  > 5 then q{ i} =5;
  END;
  RUN;

  PROC PRINT; RUN;
```

Table 17. 1 (continued)

SPSS programming statements

```
*SPSS Program for Generating Data.
* GenerateData.sps.

input program.
numeric id (F4.0).
string gender(a1) workshop(a4).
vector q(4).
loop #i = 1 to 8.
compute id = #i.
do if #i <= 5.
+ compute gender = 'f'.
else.
+ compute gender = 'm'.
end if.

compute #ws = mod(#i, 2).

if #ws = 0 workshop = 'SAS'.
if #ws = 1 workshop = 'R'.

do repeat #j = 1 to 4.
+ compute q(#j)=trunc(rv.uniform(1,5)).
end repeat.

compute pretest = rv.normal(70, 5).
compute posttest = rv.normal(80,5).
end case.
end loop.
end file.
end input program.

list.
```

R programming statements

```
# R Program for Generating Data.
# GenerateData.R

# Simple sequences.

1:10
seq(from=1,to=10,by=1)
    seq(from=5,to=50,by=5)
    seq(5,50,5)

    # Generating our ID variable
    id <- 1:8
    id

    # gl function Generates Levels.

    gl(n=2,k=1,length=8)
    gl(n=2,k=5,length=8)
```

Table 17. 1 (continued)

```
#Adding labels.

workshop <- gl(n=2,k=1,length=8, label=c("R", "SAS"))
workshop

gender <- gl( n=2,k=4,length=8, label=c("f","m") )
gender

# The rep function.

# Simple sequences.

 1:10
 seq(from=1,to=10,by=1)

seq(from=5,to=50,by=5)
seq(5,50,5)

# Generating our ID variable
id <- 1:8
id

# gl function Generates Levels.

gl(n=2,k=1,length=8)
gl(n=2,k=5,length=8)

#Adding labels.

workshop <- gl(n=2,k=1,length=8,label=c("R", "SAS"))
workshop

gender <- gl( n=2,k=4,length=8, label=c("f","m") )
gender

# Generating uniformly distributed Likert data

myValues <- c(1, 2, 3, 4, 5)
set.seed(1234)

# Simple sequences.

1:10
seq(from=1, to=10, by=1)
seq(from=5, to=50, by=5)
seq(5, 50, 5)

# Generating our ID variable
id <- 1:8
id
```

Table 17. 1 (continued)

```
# gl function Generates Levels.

gl(n=2,k=1,length=8)
gl(n=2,k=5,length=8)

#Adding labels.

workshop <- gl(n=2,k=1,length=8,label=c("R",  "SAS"))
workshop

gender <- gl( n=2,k=4,length=8,  label=c("f","m")  )
gender

# Generating repetitious Patterns (Not Factors).

gender <- rep(1:2, each=4, times=1)

gender

workshop<- rep(1:2, each=1, times=4)
workshop

myZeros <- rep(0, each=8)
myZeros

# Generating uniformly distributed Likert data

myValues <- c(1, 2, 3, 4, 5)
set.seed(1234)
q1 <- sample( myValues, size=1000, replace = TRUE)
mean(q1)
sd(q1)

# Generating normally distributed Likert data

myValues <- c(1, 2, 2, 3, 3, 3, 4, 4, 5)
set.seed(1234)
q2 #x003C;- sample( myValues , size=1000, replace = TRUE)
mean(q2)
sd(q2)


# Plot details in Traditional Graphics chapter.
par( mar=c(2, 2, 2, 1)+0.1 )
par( mfrow=c(1, 2) )
   barplot( table(q1) )
   barplot( table(q2) )
```

Table 17. 1 (continued)

```
# par ( mfrow =c (1, 1 )  )
#Sets back to 1 plot per page.
# par ( mar =c (5, 4, 4, 2 )+0.1 )

# Two Likert scales with mean difference set.seed (1234 )
q1 <- sample ( c (1, 2, 2, 3 ), size =8, replace = TRUE )
mean (q1 )

set.seed (1234 )
q2 <- sample ( c (3, 4, 4, 5 ), size =8, replace = TRUE )
mean (q2 )


# Generating continuous data
# From uniform distribution.
# mean =0.5
set.seed (1234 )
x1  <- runif (n =1000 )
mean (x1 )
sd (x1 )

# From a uniform distribution
# between 60 and 100

set.seed (1234 )
x2 <- runif (n =1000, min =60, max =100 )
mean (x2 )
sd (x2 )

# From a normal distribution.

set.seed (1234 )
x3 <- rnorm (n =1000 )
mean (x3 )
sd (x3 )

set.seed (1234 )
x4 <- rnorm (n =1000, mean =70, sd =5 )
mean (x4 )
sd (x4 )

# Plot details are in Traditional Graphics chapter.
par ( mar =c (2,2,2,1)+0.1 )
par ( mfrow =c (1,2) )
hist (x2 )
hist (x4 )
# par ( mfrow =c (1,1) ) #Sets back to 1 plot per page.
# par ( mar =c (5,4,4,2)+0.1 )

# Generating a Data Frame.
id <- 1:8
```

Table 17. 1 (continued)

```
workshop <- gl ( n =2, k =1,
    length =8, label =c ("R ", "SAS ") )
  gender <-  gl ( n =2, k =4,
    length =8, label =c ("f ", "m ")   )
  q1 <- sample ( c (1, 2, 2, 3 ), 8, replace = TRUE )
  q2 <- sample ( c (3, 4, 4, 5 ), 8, replace = TRUE )
  q3 <- sample ( c (1, 2, 2, 3 ), 8, replace = TRUE )
  q4 <- sample ( c (3, 4, 4, 5 ), 8, replace = TRUE )
  pretest   <- rnorm ( n =8, mean =70, sd =5 )
  posttest  <- rnorm ( n =8, mean =80, sd =5 )

myGenerated <- data.frame (id, gender, workshop,
  q1, q2, q3, q4, pretest, posttest )
myGenerated
```

Chapter 18
How R Stores Data

While in use, R stores its data in your computer's main memory rather than on its hard drive or tape drives. That means R cannot analyze huge datasets. An exception to this is Thomas Lumley's `biglm` package [24], which processes data in "chunks" for some linear and generalized linear models.

SAS and SPSS on the other hand, can store datasets on disk or tape, even across multiple tapes, and process essentially unlimited amounts of data. Of course, some methods of analysis require that all data reside in random access memory. No package can get around that type of limitation.

Given the low cost of memory today this is much less of a problem than you might think. R can handle hundreds of thousands of records on a computer with 2 gigabytes of memory available to R. That is the current memory limit for a single process or program in today's 32-bit operating systems. To have 2 gigabytes free just for R, you would want to have perhaps 3 gigabytes of total memory so that your operating system would have the room it needs. You can use virtual memory to go up to 3 gigabytes for a single process, but that slows things down considerably. Virtual memory uses the computer's hard drive to simulate memory.

Operating systems capable of 64-bit memory spaces are becoming more popular. The huge amounts of memory they can handle mitigate this problem.

Another way around the limitation is to store your data in a relational database and use its facilities to generate a sample to analyze. However, if you need to ensure that certain small groups are included in your sample, (e.g., those that got a rare disease, the small proportion that defaulted on a loan, etc.) then you end up taking a complex sample, which complicates your analysis considerably. R has specialized package to help analyze such samples, including: `pps`, `sampfling`, `sampling`, `spsurvey`, and `survey`.

Another alternative is to purchase S-PLUS, a commercial package that has an almost identical language with extensions to handle what it calls, "big data." It solves the problem in a way very similar to the SAS and SPSS approach.

Chapter 19
Managing Your Files and Workspace

When using SAS and SPSS, you manage your files with the same operating system commands that you use for your other software. SAS does have a few file management procedures such as DATASETS and CATALOG, but you can get by just fine without them for most purposes. R is quite different. It has a set of commands that replicate many operating system functions such as listing names of objects, deleting them, setting search paths, and so on. Learning how to use these commands is especially important because of the way R stores its data. You need to know how to make the most of your computer's memory.

19.1 Loading and Listing Objects

You can see what objects are in your workspace with the `ls` function. To list all objects such as data frames, vectors, and functions, use

```
ls ()
```

The `objects` function does the same thing and its name is more descriptive, but `ls` is more widely used since it is the same command UNIX, Linux, and MacOS X users can use to list the files in a particular directory or folder (without the parentheses).

When you first start R, using the `ls` function will tell you there is nothing in your workspace. How it does this is quite odd by SAS or SPSS standards. It tells you that the list of objects in memory is a character vector with zero values.

```
> ls ()
character (0)
```

After bringing our practice data frame into our workspace using the `load` function, `ls` will show us the data is now available.

```
> load ("myWorkspace.RData")
> ls ()
[1]"mydata" "mylist" "mymatrix"
```

R.A. Muenchen, *R for SAS and SPSS Users*, DOI: 10.1007/978-0-387-09418-2_19,   261
© Springer Science+Business Media, LLC 2009

You can use the `pattern` argument to search for any regular expression. Therefore, to get a list of all objects that begin with the letter x, you can use the following:

```
> ls (pattern="x")

[1] "x1" "x2" "x3"
```

The `ls` function does not look inside data frames to see what they contain, and it does not even tell you when an object is a data frame. You can use many of the commands we have already covered to determine what an object is and what it contains. To review, typing its name or using the `print` function will show you the whole object, or at least something about it. What print shows you depends upon the class of the object. The `head` and `tail` functions will show you the top or bottom few observations of vectors and data frames (again, depending upon the object's class). The `class` function will tell you if it is a data frame, list, or some other object. The `names` function will show you object names within a data frame or list.

The `attributes` function will display all the attributes that are stored in an object such as variable names, the object's class and any labels that it may contain.

```
> attributes (mydata)

$names
[1] "id"  "workshop"  "gender"  "q1"  "q2"  "q3"  "q4"

$class
[1] "data.frame"

$row.names
[1] 1 2 3 4 5 6 7 8
```

The `str` function displays the *str*ucture of any R object in a compact form.

```
> str (mydata)

'data.frame':    8 obs. of  6 variables:
 $ workshop: int  1 2 1 2 1 2 1 2
 $ gender  : Factor w/ 2 levels "f", "m": 1 1 1 NA 2 2 2 2
 $ q1      : int 1 2 2 3 4 5 5 4
 $ q2      : int 1 1 2 1 5 4 3 5
 $ q3      : int 5 4 4 NA 2 5 4 5
 $ q4      : int 1 1 3 3 4 5 4 5
```

The `str` function works on functions too. Here is the structure it shows for the `lm` function.

```
> str ( lm )

function (formula, data, subset, weights, na.action,
method = "qr", model = TRUE, x = FALSE, y = FALSE,
qr = TRUE, singular.ok = TRUE, contrasts = NULL,
  offset, ...)
```

The `ls.str` function applies the `str` function to every object in your workspace. It is essentially a combination of the `ls` function and the `str` function. Here is the structure of all the objects I had in my workspace as I wrote this paragraph.

```
> ls.str ()
myCounts :   'table' int [1:2, 1:2] 2 2 1 2
myCountsDF : 'data.frame':      4 obs. of  3 variables:
  $ gender  : Factor w/ 2 levels "f", "m": 1 2 1 2
  $ workshop: Factor w/ 2 levels "R", "SAS": 1 1 2 2
  $ Freq    : int  2 2 1 2
mydata : 'data.frame':  8 obs. of  6 variables:
  $ workshop: int  1 2 1 2 1 2 1 2
  $ gender  : Factor w/ 2 levels "f", "m": 1 1 1 NA 2 2 2  2
  $ q1      : int  1 2 2 3 4 5 5 4
  $ q2      : int  1 1 2 1 5 4 3 5
  $ q3      : int  5 4 4 NA 2 5 4 5
  $ q4      : int  1 1 3 3 4 5 4 5
```

The `Hmisc` package has a `contents` function modeled after the SAS CONTENTS procedure. It also lists names and other attributes as shown below. However, it works only with data frames.

```
> library ("Hmisc")
Attaching package: 'Hmisc'
...
> contents (mydata)
Data frame:mydata 8 observations and 7 variables  Maximum #
  NAs:1
```

	Levels	Storage	NAs
id		integer	0
workshop		integer	0
gender	2	integer	1
q1		integer	0
q2		integer	0
q3		integer	1
q4		integer	0

```
+---------+------+
| Variable | Levels|
+---------+------+
| gender  |  f,m  |
+---------+------+
```

19.2 Understanding Your Search Path

Once you have data in your workspace, where exactly is it? It is in an environment called *.GlobalEnv*. The `search` function will show us where that resides in R's *search path*.

```
> search ()
[1] ".GlobalEnv"        "package:stats" "package:graphics"
[4] "package:grDevices" "package:utils"  "package:datasets"
[7] "package:methods" "Autoloads"        "package:base"
```

Since our workspace (.GlobalEnv) is in position 1, R will search it first. By supplying no arguments to the `ls` function, we were asking for a listing of objects in the first position of the search path. Let us see what happens if we apply `ls` to different levels. We can either use the path position value, 1,2,3,... or their names.

```
> ls (1)  # This uses position number.
[1] "mydata"
> ls (".GlobalEnv")  # This does the same using name.
[1] "mydata"
```

The package:stats at level 2 contains some of R's built-in statistical functions. There are a lot of them, so let us use the `head` function to show us just the top few results.

```
> head (ls (2))
[1] "acf"          "acf2AR"       "add.scope"    "add1"
[5] "addmargins" "aggregate"
> head (ls ("package:stats"))  # Same result.
[1] "acf"          "acf2AR"       "add.scope"    "add1"
[5] "addmargins" "aggregate"
```

19.3 Attaching Data Frames

Understanding the search path is essential to understanding what the `attach` function really does. We will attach mydata and see what happens.

```
> attach (mydata)
> search ()
 [1] ".GlobalEnv"  "mydata"  "package:stats"
 [4] "package:graphics" "package:grDevices"
     "package:utils"
 [7] "package:datasets" "package:methods"    "Autoloads"
[10] "package:base"
```

```
> ls (2)
[1] "gender"    "id"        "q1"        "q2"        "q3"
[6] "q4"        "workshop"
```

You can see that attach has made virtual copies of the variables stored in mydata and placed them in search position 2. When we refer to just "gender" rather than "mydata$gender," R looks for it in position 1 first. It does find anything with just that short name even though mydata$gender is in that position. R then goes on to position 2 and finds it. This is the process that makes it so easy to refer to variables by their short names. It also makes them very confusing to work with if you create new variables. Let us say we want to take the square root of q4:

```
> q4 <- sqrt (q4)
> q4
[1] 1.000000  1.000000  1.732051  1.732051  2.000000
       2.236068
[7] 2.000000  2.236068
```

This looks like it worked fine. However, let us list the contents of search positions 1 and 2 to see what really happened:

```
> ls (1)
[1] "mydata" "q4"
> ls (2)
[1] "gender"    "id"        "q1"        "q2"        "q3"
[6] "q4"        "workshop"
```

It created the new version of q4 as a separate vector in our main workspace. The copy of q4 that the attach function put in position 2 was never changed! Since search position 1 dominates, asking for q4 will cause R to show us the one in our workspace. Asking for mydata$q4 will go inside the data frame and show us the original, untransformed values. There are two important lessons to learn from this:

1. Never use the attach function to create new variables.
2. Variables with the same name higher in the path order (i.e., those that appear toward the beginning of the search list) will always be the one used.

When the attach function places objects in position 2 of the search path (a position you can change but rarely need to) those objects will block, or *mask*, any others of the same name in lower positions (i.e., further toward the end of the search list). In the following example, I started with a fresh launch of R, loaded mydata, and attached it twice to see what happens.

```
> attach (mydata)
> attach (mydata)
The  following  object  (s)  are  masked  from  mydata
   (position 3)  :
```

```
   gender id q1 q2 q3 q4 workshop
> search ()
 [1] ".GlobalEnv"     "mydata"              "mydata"
 [4] "package:stats"  "package:graphics"  "package:grDevices"
 [7] "package:utils"  "package:datasets"  "package:methods"
[10] "Autoloads"            "package:base"
```

See that above "mydata" is now in search position 2 *and* 3. If you refer to any variable, R has to settle which one you mean (they do not need to be identical as in this example.) The message about masked objects in position 3 tells us that the second attach brought in variables with those names and now we cannot refer to them by their simple names. Those names are already in use somewhere higher in the search path. In this case, the variables from the first attach went to position 2 and they will be used. However, if objects with any of those names were in our main workspace (not in a data frame) *they* would be used. When we first learned about vectors, we created q1, q2, and so on as vectors and then formed them into a data frame. If we had left them as separate vectors in our main workspace, even the first attach would have given us a similar message. The vectors in position 1 would have blocked those in positions 2 and 3.

19.4 Attaching Files

So far we have only used the `attach` function on data frames. It can also be very useful with R data files. If you load a file, it brings all objects into your workspace. However, if you attach the file, you can bring in only what you need and then detach it. For example, let us create a variable x and then add only the vector q4 from the file myall.RData, a file that contains the objects we created in Chap. 8. Recall that in that section, we created each of our practice variables first as vectors, and then converted them to factors, a matrix, a data frame, and a list.

```
> x <- c (1,2,3,4,5,6,7,8)
> attach ("myall.RData")
> search ()
  [1] ".GlobalEnv"       "file:myall.RData" "package:stats"
  [4] "package:graphics" "package:grDevices" "package:utils"
  [7] "package:datasets" "package:methods"   "Autoloads"
 [10] "package:base"
> q4 <- q4
```

The last statement looks quite odd! What is going on? The `attach` function loaded myall.RData, but put it at position 2 in the search path. R will place any variables you create in your workspace (position 1) and the attach allows R to find q4 in position 2. So it copies it from there to your workspace. Let us look at what we now have in both places.

```
> ls (1)  # Your workspace.
[1] "q4" "x"
> ls (2)  # The attached file.
[1] "mystats"   "gender"   "mydata"   "mylist" "mymatrix"
[6] "q1"        "q2"       "q3"       "q4"     "workshop"
> detach (2)
```

19.5 Removing Objects from Your Workspace

To delete an object from your workspace, use the remove function or the equivalent rm function as in

```
rm (mydata)
```

The rm function is one of the few that will accept multiple objects separated by commas (i.e., the names do not have to be in a single character vector). So to delete the our q variables that we created outside our data frame, we can use

```
rm (q1,q2,q3,q4)
```

You can create a character vector of objects to delete:

```
> load (file="myall.RData")
> ls ()

  [1] "mystats" "gender"   "mydata"   "mylist"  "mymatrix"
  [6] "q1"      "q2"       "q3"       "q4"      "workshop"
```

Notice that the results are simply a character vector, and the q variables are in the sixth to ninth position. Therefore, we can save just those to a character vector of items to delete, and then delete them with the following commands:

```
> myDeleteItems <- ls ()[6:9]
> myDeleteItems
[1] "q1" "q2" "q3" "q4"
> rm (list=myDeleteItems)
> ls ()
[1] "myStats"        "gender"        "mydata"
[4] "myDeleteItems"  "mylist"        "mymatrix"
[7] "workshop"
```

If you are sure you want to remove all the objects in your workspace, you can do it with the following approach. Be careful, there is no "undo" for this radical step!

```
> myDeleteItems <- ls ()
> myDeleteItems
```

```
[1] "mystats"          "gender"      "mydata"
[4] "myDeleteItems"    "mylist"      "mymatrix"
[7] "workshop"
> rm (list=myDeleteItems)
> ls ()
character (0)
```

The last line that says character (0) is the response to the ls () command. It says the result of the object listing is a character vector with zero
entries. That is, all the objects are gone. It may appear that the logical thing to
do is

```
rm (myDeleteItems)
```

However, that would delete only the list of item names, not the items
themselves. That is why the rm function needs a list argument when dealing
with character vectors.

A much shorter way to delete all the objects in your workspace is below. It
combines the functions, but it looks quite cryptic at first. The steps above make
it much more obvious what is going on.

```
rm ( list=ls () )
```

19.6 Minimizing Your Workspace

Removing unneeded objects from your workspace is one important way to save
space. You can also use the cleanup.import function in the Hmisc package.
It automatically stores the variables in a data frame in their most compact form.
You use it as:

```
mydata  <- cleanup.import (mydata)
```

See *Installing R and Add-on Packages* for more details.

To conserve workspace by saving only the variables you need within a data
frame, see Keeping and Dropping Variables, Sect. 14.9. The rm function cannot
drop variables within a data frame.

19.7 Setting Your Working Directory

Your working directory is the location R uses to retrieve or store files, if you do
not specify the full path. The getwd function will tell you where that is:

```
> getwd ()
[1] "c: \Program Files\ R\ R 2.7.0\"
```

Fig. 19.1 Windows dialog box for setting working directory

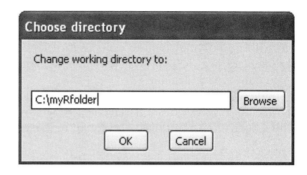

Windows users can see and/or change their working directory by choosing *File> Change dir*. . . .. The dialog box will appear as in Fig. 19.1.

On any operating system you can change the working directory with the setwd function. This is the equivalent to the SPSS CD command and somewhat similar to the SAS LIBNAME statement. Simply provide the full path between the quotes:

```
setwd ("/myRfolder")
```

We have discussed before that R uses the forward slash, "/" even on computers running Windows. That is because within strings, R uses "\t", "\n" and "\\" to represent the single character tab, newline, and backslash, respectively. In general, a backslash followed by another character may have a special meaning. So when using R on Windows, always specify the paths with either a single forward slash, or two backslashes in a row. This book uses the single forward slash because that works with R on all operating systems.

You can set your working directory automatically by putting it in your .Rprofile. For details, see *Appendix C*, Automating your Settings.

19.8 Saving Your Workspace

You exit R by choosing *File> Exit* or entering the command quit () or just q (). R will then offer to save your workspace. If you click yes, it stores it in a file named *.RData* in your working directory (see how to set that in Sect. 19.7). The next time you start R from the same working directory, it automatically loads that file back into memory and you can continue working.

The name .RData seems like an odd choice, because most operating systems hide files that begin with a period! That is true on Windows, Macintosh, and Linux/UNIX systems. Of course, you can tell your operating system to show you such files.

To get Windows XP to show you this file, in Explorer uncheck the option below and click *Apply to all folders*: *Tools> Folder Options> View> [ ]Hide*

extensions to known file types. In Windows Vista, use this selection: *Start>*
Control Panel> Folder Options> View> Show hidden files and folders. Note that
this will still not allow you to click on a filename like myProject.RData and
rename it to just .RData. The Windows Rename message box will tell you,
"You must type a filename."

Linux/UNIX users can see files named .RData with the command ls –a.

Macintosh users can see files named .RData by starting a terminal window
with Applications> Utilities> Terminal. In the terminal type

```
defaults write com.apple.finder AppleShowAllFiles TRUE
killall Finder
```

To revert back to normal file view, simply type the same thing but with
"FALSE" instead of "TRUE."

If you want to avoid using the *filename* .RData, you can at any time choose
File> Save Workspace and name it anything you like that has the *extension*
.RData. If you are a Windows user, R does not automatically append the
.RData extension, as do most Windows programs. Later, when you start R,
you can use *File> Load Workspace* to load it from the hard drive back into the
computer's memory.

You can also use the save.image function to save your workspace:

```
save.image(file="myWorkspace.RData")
```

This will save all your data and function objects. Later when you start R, you
can restore them using the load function.

```
load(file="myWorkspace.RData")
```

If you want to save only a subset of your workspace, the save function allows
you to list the objects to save, separated by commas, before the file argument:

```
save(mydata,file="myWorkspace.RData")
```

This is one of the few functions that can accept many objects separated by
commas, so might save three as in the example below.

```
save(mydata,mylist,mymatrix,file="myExamples.RData")
```

It also accepts a list argument that lets you specify a character vector of
objects to save.

If you are a Windows user and like using shortcuts, there is another way to
keep your various projects organized. You can create an R shortcut for each of
your analysis projects. Then you right-click the shortcut, choose *Properties* and
set the *Start in folder* to a unique folder. When you use that shortcut to start R,
upon exit it will store the .RData file for that project. Although neatly organized
into separate folders, each project workspace will still be in a file named .RData.
That means it is harder to find the particular file you need via search engines or
backup systems. If you accidentally moved an .RData file to another folder, you
would not know which project it contained without first loading it into R.

19.9 Saving Your Programs and Output

R users that prefer the GUI can easily save programs, called scripts, and output to files in the usual way. Just click anywhere on the window you wish to save and choose *File> Save as* and supply a name. The standard extension for R programs is ".R" and for output is simply ".txt". You can also save bits of text output to your word processor using the typical cut/paste steps.

Windows users will need to type the extension ".R" on their filenames. R does not append it automatically. Since the *File>Open Script* dialog box displays only files with the .R extension, it is easy for a file lacking it to appear to be lost.

R users that prefer to use the command line interface often use text editors such as Emacs, or the one in JGR, that will check their R syntax for errors. Those files are no different from any other file created in a given editor.

Windows and Macintosh users can cut and paste graphics output into their word processors or other applications. Users of any operating system can rerun graphs, directing their output to a file. See Chap. 20 for details.

19.10 Saving Your History (Journal)

The R console displays commands you enter (or that menus enter for you) and their output. It is a good idea to submit commands from a script window (program editor) but sometimes you enter them directly into the console and

Table 19.1 Workspace management functions

List names including hidden objects like .First and .Last	`ls(all.names=TRUE)` objects `(all.names=TRUE)`
List names of most objects	`ls ()`
	`objects ()`
List object attributes	`attributes (mydata)`
Load workspace	`load(file="myWorkspace.RData")`
Remove a variable from a data frame	`mydata$myvar <- NULL`
Remove all objects (non-hidden)	`rm ( list=ls () )`
Remove an object	`rm ( mydata )`
Remove several	`rm (mydata, myvars, myobs)`
Save all objects	`save.image (file="myWorkspace.RData")`
Save some	`save (mydata,x,y,z,file= "myExamples.RData)`
Show structure of all objects	`ls.str (all.names=TRUE)`
Show structure of data frame only (requires Hmisc)	`contents (mydata)`
Show structure of most objects	`ls.str ()`

Table 19.1 (continued)

Show structure of objects by name	`str (mydata, str(lm)`
Store data efficiently (requires `Hmisc`)	`mydata <- cleanup.import (mydata)`
Working directory, getting	`getwd ()`
Working directory, setting on any operating system	`setwd("/mypath/myfolder")`
	`setwdsetwd("C:/mypath/myfolder")`
Working directory, setting on Windows	`setwd("C:\\mypath\\myfolder")`
	`setwd("/mypath/myfolder")`

then later realize you need to save the program. You could save the input and output in the console window, but you would need to edit out the output to create a usable program.

R has a history file that saves all of the commands in a given session. This is just like the SPSS journal file. SAS has no equivalent. Unlike SPSS, the history file is not cumulative on Windows computers. It is cumulative on Linux and Macintosh.

You can save the current session's history to a file in your working directory (see Sect. 19.6) with the `savehistory` function.

```
savehistory(file="myHistory.Rhistory")
```

You can later recall it using the `loadhistory` function.

```
loadhistory(file="myHistory.Rhistory")
```

Note that the filename can be anything you like but the extension should be ".Rhistory". The entire default filename, if you do not provide one, is just ".Rhistory". I prefer to always save a cumulative history file automatically. For details, see Appendix C, Automating Your Settings. For a summary of the commands covered in this chapter, see Table 19.1.

Chapter 20
Graphics Overview

Graphics is perhaps the most difficult topic to compare across SAS, SPSS, and R. Each package contains at least two graphical approaches, each with dozens of options, each with entire books devoted to them. Therefore, we will focus on only two main approaches in R and will discuss many more examples in R than in SAS or SPSS. This chapter focuses on a broad, high-level comparison of the three. The next chapter focuses on R's traditional graphics. The one after that focuses just on the grammar of graphics approaches in R and SPSS.

Dynamic visualization allows you explore your data by interacting with plots. Selections you make in one graph, such as the females in a bar chart, are reflected in all graphs. SAS/INSIGHT and its successor, SAS Stat Studio, provide this capability in SAS. SPSS does not offer much interactivity. Although dynamic visualization is outside our scope, R has this capability through the `iplots` package [25] , documented at http://www.rosuda.org/iplots/. R also has a link to the excellent GGobi package available free at http://www.ggobi.org/ [26]. The `rggobi` package links R to GGobi and is available at http://www.ggobi.org/rggobi/ [47].

20.1 SAS/GRAPH

When you purchase Base SAS, its graphics procedures such as CHART and PLOT use only primitive printer characters. For example, lines are drawn using series of "-" and "|" characters, while plotting symbols are things like "*" or " + ". It cannot even draw diagonal lines! You have to purchase a separate add-on package, SAS/GRAPH, to get real graphics.

From its release in 1980 through 2008, SAS/Graph offered a limited set of popular graphic displays. It used short commands, so getting a basic plot was easy. However, its default settings were so poorly chosen that it took many commands to create a publication-quality graph. If you needed the same set of graphs in a periodic report, that was not too much of a problem. But if you tended to do frequent unique graphs, it was frustrating. If you needed a wider selection of graph styles, you had to switch to another package.

R.A. Muenchen, *R for SAS and SPSS Users*, DOI: 10.1007/978-0-387-09418-2_20,   273
© Springer Science+Business Media, LLC 2009

The release of SAS version 9.2 in 2008 finally corrected this problem. The SGPLOT, SGPANEL, and SGSCATTER procedures added modern plots with well-chosen default settings. In addition, ODS Graphics gave SAS the ability to automatically provide a standard set of graphics with many statistical analyses. This makes sense as people typically do the same core set of graphs for a given analysis. See http://www2.sas.com/proceedings/forum2007/193-2007.pdf and http://support.sas.com/rnd/base/topics/statgraph/192-31-updated.pdf for details. As nice as these improvements are, they still leave SAS behind both SPSS and R in graphics flexibility.

20.2 SPSS Graphics

SPSS Base includes three types of graphics: those based upon Graphics Production Language (GPL), and the two "legacy" systems of standard and interactive graphics. Their standard legacy graphics use short commands and reasonable default settings. They can also plot data stored in either wide or long formats without having to restructure the data, which is very convenient. The interactive legacy graphics offer a very limited range of interactivity and do not reflect selections you make in one graph in others, as do SAS/INSIGHTgraphics and GGobi graphics. As you can tell by the "legacy" label that the company applied to them in version 15, SPSS Inc. is likely to phase them out eventually.

SPSS' third graphics approach is its GPL. We will discuss this extremely flexible approach below in Sect. 20.4. SPSS' standard legacy commands typically take one statement per graph, while GPL takes over a dozen! That demonstrates the classic trade-off of simplicity versus power. The example SPSS programs in this chapter use standard legacy graphics, those in the next chapter use GPL.

20.3 R Graphics

R offers three main types of graphics: traditional, `lattice` [27], and `ggplot2` [28].

R's traditional graphics functions (also called base graphics) were written by one of the original R developers, Ross Ihaka, and they come with the main R installation [28 p. 5]. Traditional graphics include high-level functions such as barplots, histograms, and scatterplots. These functions are brief and the default settings are good. While SAS and SPSS have you specify your complete plot in advance, R's traditional graphics allow you to plot repeatedly on the same graph. For example, you might start with a scatterplot, and then add a regression line. As you add new features, each writes on top of the previous ones, so the order of commands is important. Once written, a particular feature cannot be changed without starting the plot over from the beginning.

Table 20. 1 A comparison of R's three main graphics packages

	Traditional (or Base)	`lattice`	`ggplot2`
Automatic output from many other functions	Yes	No	No
Automatic legends	No	Sometimes	Yes
Easily repeats plots for different groups	No	Yes	Yes
Easy to use multiple data sources	Yes	No	Yes
Allows you to build-up plots piece-by-piece	Yes	No	Yes
Allows you to replace pieces after creation	No	No	Yes
Level of control so fine it even extends beyond data graphics[1]	Yes	No	No
Consistent functions	No	No	Yes
Attractive defaults	Good	Good	Excellent
Mosaic plots	Yes	Yes	No
Underlying graphics system used	Traditional	Grid	Grid

[1] Since R is open source, you could extend any of these systems using other languages. This focuses on extensibility within the package's existing framework.

R's traditional graphics also includes low-level functions for drawing things like points, lines, and axes. These provide flexibility and control that people can use to invent new types of graphs. Their use has resulted in many add-on graphics packages for R. The level of control is so fine that you can even use it for graphics that have nothing to do with data. See Paul Murrell's book, *R Graphics* [29], for examples.

The second major graphics package added to R was `lattice`, written by Deepayan Sarkar [27]. It implements Cleveland's Trellis graphics system [30]. Now part of the base R distribution, it does an excellent job of creating multi-frame plots. A good book on that package is Deepayan Sarkar's *Lattice: Multi-variate Data Visualization with R* [31]. We will not examine `lattice` graphics, as the next system covers most of its abilities and offers additional advantages.

The third major package, `ggplot2` [28], written by Hadley Wickham, is based on the *Grammar of Graphics* described in the next section. Even though it is based on the same concepts, it does not use SPSS' GPL. It offers a GPL-like flexibility that can require many statements. However, unlike SPSS' GPL, it also has a short, often one-statement, version that makes it easier to use. We will cover `ggplot2` Chap. 22. Given our space limitations however, all of our `ggplot2` examples (except Fig. 20.1) could be created as well in the `lattice` package. A comparison table of these three main packages is given in Table 20.1.

20.4 The Grammar of Graphics

Leland Wilkinson's watershed work, *The Grammar of Graphics* [32], forms the foundation for both SPSS GPL and Hadley Wickham's `ggplot2` package. Wilkinson's key insight was the realization that there general principals that

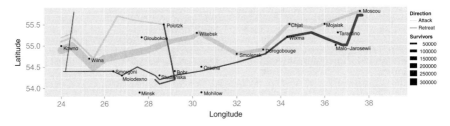

Fig. 20.1 Minard's plot of Napoleon's march created with the `ggplot2` package

form the foundation of all data graphics. Once you design a graphics language to follow those principals, the language should then be able to do any existing data graphics as well as variations that people had not previously considered.

An example is a stacked bar chart that shows how many students took each workshop (i.e., one divided bar). This is not a popular graph, but if you change its coordinate system from rectangular to circular (Cartesian to polar) it becomes a pie chart. So rather than requiring separate procedures, you can have one that includes changing coordinate systems. That type of generalization applies to various other graphical elements as well.

A much more interesting example is Charles Joseph Minard's famous plot of Napoleon's march to Moscow and back (Fig. 20.1). The light grey line shows the army's advance toward Moscow and the dark grey line shows its retreat. The thickness of the line reflects the size of the army. Given enough effort, you can get many graphics packages to create this graph, including R's traditional graphics. However, SPSS' GPL and R's `ggplot2` can do it easily using the same general concepts that it uses for any other plots. The data and `ggplot2` program to create this graph is included in the files you can download at http://RforSASandSPSSusers.com.

20.5 Other Graphics Packages

There are other `graphics` packages that we will not have space to examine in depth. These include the `maps` package for creating geographic maps and the `vcd` package for visualizing categorical data. The latter was written by David Meyer, Achim Zeileis, and Kurt Hornik and was inspired by Michael Friendly's book, *Visualizing Categorical Data* [33]. That book describes how to do its plots using SAS macros. The fact that plots initially implemented in SAS were also possible to add to R is testament to R's graphical flexibility.

There are comprehensive collections of graphs and the R programs to make them at the R Graph Gallery http://addictedtor.free.fr/graphiques/ and the R Graphical Manual at http://bg9.imslab.co.jp/Rhelp/. You can also use the demo capability in R to have it show you a sample of what it can do. Enter the `demo` function calls below in the R console.

```
demo(graphics)
demo(persp)
#Although we will not use the lattice package
library("lattice")
demo(lattice)
```

20.6 Graphics Procedures Versus Graphics Systems

In SAS/GRAPH and SPSS Base, there are high-level *graphics procedures* such as PROC GPLOT in SAS or GRAPH /SCATTER in SPSS. R has many add-on packages that contain similar high-level graphics functions.

At a lower level are *graphics systems*. This is where a command to do something like a scatterplot is turned into the precise combinations of lines and points needed. Controlling the graphics system allows you to change settings that affect all graphs, like text fonts, fill patterns, and line or point types. In SAS, you control these settings using the GOPTIONS command. In SPSS, you use the menu *File> Options> Edit> Charts*. SPSS also allows you to control these settings when you create a template to apply to future graphs.

Graphics *systems* are more visible in R because there are two of them. The traditional graphics system controls the high-level traditional graphics functions such as plot, barchart, and pie as well as some add-on packages, such as maps. It also controls low-level traditional graphics functions that do things like draw lines or rectangles. The packages lattice and ggplot2 instead use the grid graphics system. The implications of this will become much more obvious as we look at examples. The grid graphics system was written by Paul Murrell, and is documented in his book, *R Graphics* [29]. That book is an excellent reference for both graphics systems, as well as high-level traditional graphics procedures. It also gives an introduction to lattice graphics.

20.7 Graphics Devices

At the lowest level of graphics control is the graphics device itself. Regardless of the graphics package or system in R you use, the way you see or save the result is the same: the graphics *device*.

SAS chooses the device with the GOPTIONS statement. For example to create a postscript file you might use:

GOPTIONS GSFNAME = myfile GSFMODE = Replace;

SPSS, being more menu oriented, offers to save graphics for various devices via its File> Export menu.

By default R writes to your screen, each plot taking the place of the previous one. Windows users can buffer the plots for viewing with the Page Up/Page Down keys, choosing *History > Recording* in your plot window.

R has functions named for the various graphics devices. So to write a plot to a postscript file, you could use:

```
> postscript( file="myPlot.ps") # Opens the device.
> plot(pretest,posttest)        # Does the plot.
> dev.off()                     # Closes the device.
```

The last command is optional since R will close the device when you exit R. However, any additional plots you do will go to that file until you either use dev.off or explicitly choose to use the screen device again. Screen device functions are windows(), quartz(), and x11() for Microsoft Windows, Macintosh OS X, and UNIX or Linux, respectively. Other popular device functions include jpeg, pdf, pictex (for LATEX), png, and win.metafile for Microsoft Windows. The Portable Document Format (PDF) is particularly important as it is the only high-resolution format that supports transparency. Transparency is a feature we will use in Chap. 22. If your word processing software does not support PDF files, the Portable Network Graphics (PNG) format is a good alternative for most data graphics. See the help files for details.

Multiple plots can go to the file in the above example because postscript (and PDF) support multiple pages. Other formats, like Microsoft Windows Metafile do not. For that type of format, you can send multiple plots to separate files using the following form. The part of the filename "%03d" tells R to append 1, 2,... to the end of each filename. The numbers are padded by blanks or zeros depending on the operating system.

```
win.metafile( file="myPlot%03d.wmf" )
barplot( table(workshop) ) # 1st plot goes to myPlot  1.wmf
hist(posttest)             # 2nd plot goes to myPlot  2.wmf
plot( pretest,posttest)    # 3rd plot goes to myPlot  3.wmf
dev.off()
```

The ggplot2 package has its own ggsave function to save its plots to files. See Chap. 22 for details.

```
ggsave(file = ``march.pdf'' , width=16, height=4)
```

For much greater depth on these and many other graphics subjects, see *R Graphics* [29].

20.8 Practice Data: Mydata100

The following examples use a longer version of our practice dataset called mydata100. It has 100 observations, includes people who have taken all four workshops and adds test scores taken before and after the workshops (pretest and posttest). The file also contains full variable and value labels. It is also attached, so all the examples use short variable names. Here is the top part of the file.

```
> head(mydata100)
  gender workshop q1 q2 q3 q4 pretest posttest
1 Female        R  4  3  4  5      77       76
2   Male     SPSS  3  4  3  4      75       78
3 Female     SPSS  3  2  3  3      79       81
4 Female     SPSS  5  4  5  3      85       85
5 Female    Stata  4  4  3  4      80       84
6 Female     SPSS  5  4  3  5      77       80
```

Chapter 21
Traditional Graphics

In the previous chapter, we discussed the various graphics packages in R, SAS, and SPSS. Now we will delve into R's traditional, or base, graphics.

21.1 Barplots

Barplots using R's traditional graphics are easy to do, but they use a method quite different from SAS and SPSS. While SAS and SPSS assume your data need summarizing and require options to tell them when it is pre-summarized, many of R's base graphics functions assume just the opposite. We will first look at counts, then grouped counts, and finally various types of barplots for means.

21.1.1 Barplots of Counts

The `barplot` function call below makes a chart with just those two bars (Fig. 21.1). They could be counts, means, or any other measure. The main point to see here is that we get a bar for *every observation*. We are ignoring options for the moment, so the plot lacks labels.

```
barplot( c(40,60)  )
```

If we apply the same command to variable q1, we get a bar representing each observation in the data frame (Fig. 21.2). Notice that the y-axis is labeled 1 through 5. It is displaying the raw values for every observation rather than summarizing them. That is not a very helpful plot!

Recall that the table function gets frequency counts.

```
> table(q4)

q4
 1  2  3  4  5
 6 14 34 26 20
```

R.A. Muenchen, *R for SAS and SPSS Users*, DOI: 10.1007/978-0-387-09418-2_21,   281
© Springer Science+Business Media, LLC 2009

Fig. 21.1 Unlabeled barplot
using traditional graphics

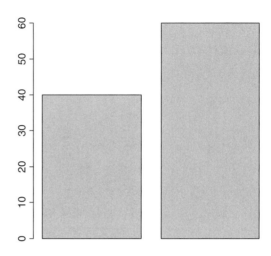

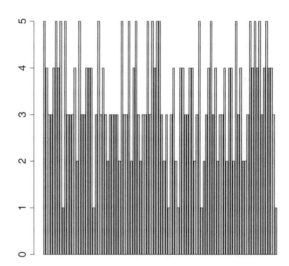

Fig. 21.2 A bad barplot on
un-summarized variable q4

Since the table function gets the data in the form we need for a barplot, we can simply nest one function within the other to finally get a reasonable plot (Fig. 21.3).

```
> barplot( table(q4)   )
```

When we make the same mistake with a barplot on gender, we see a different message.

Fig. 21.3 A more reasonable barplot for variable q4, this time summarized first

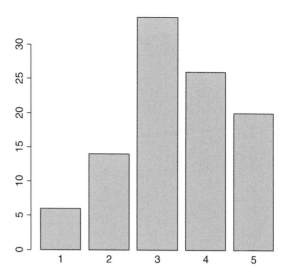

```
> barplot (gender)

Error in barplot.default(gender) :  'height' must be a
    vector or a matrix
```

If gender was coded 1 or 2, and was not stored as a factor, it would have created one bar for every subject, each with a height of 1 or 2. However, because gender *is* a factor, the message is saying barplot accepts only a vector or matrix. The solution is the same as before: count the genders before plotting (Fig. 21.4).

```
barplot ( table (gender) )
```

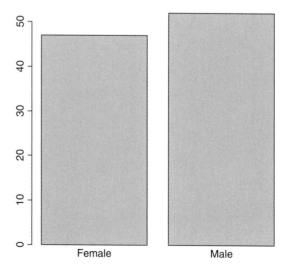

Fig. 21.4 A barplot for gender

Fig. 21.5 A horizontal
barplot of workshop flipped
using horiz = TRUE

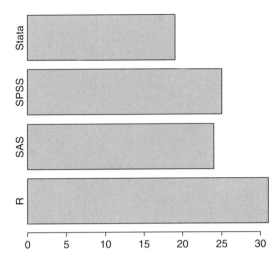

We can use the horiz = TRUE argument to turn the graph sideways
(Fig. 21.5).

```
> barplot( table(workshop), horiz=TRUE)
```

If we are interested in viewing groups as a proportion of the total, we can
stack the bars into a single one (Fig. 21.6). As we will see in the next chapter, this
is essentially a pie chart in rectangular Cartesian coordinates rather than
circular polar coordinates. To do this we use the as.matrix function to
convert the table into the form we need and use the beside = FALSE argument
to prevent the bars from appearing beside each other.

```
barplot( as.matrix( table(workshop)  ),
    beside  = FALSE)
```

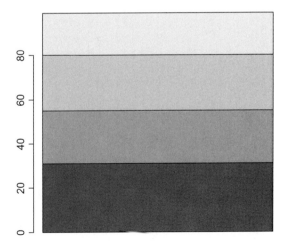

Fig. 21.6 An unlabeled
stacked barplot of workshop

21.1.2 Barplots for Subgroups of Counts

Recall that the `table` function can handle two variables. Nested within the `barplot` function, this results in a clustered bar chart (Fig. 21.7).

```
> barplot( table(gender,workshop)  )
```

The `plot` function is generic, so it does something different depending on the variables you give it. If we give it two factors, it will create a plot similar to Figure 21.8.

```
> plot( workshop,gender )
```

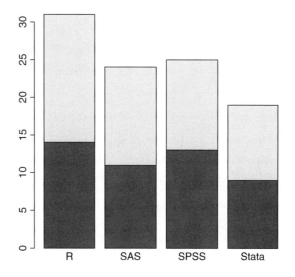

Fig. 21.7 A stacked barplot of workshop with bars split by gender (lacking legend)

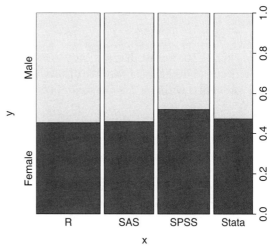

Fig. 21.8 A mosaic plot of workshop by gender, done using the `plot` function. Gender is labeled automatically

Fig. 21.9 A mosaic plot of workshop by gender, done using the `mosaicplot` function

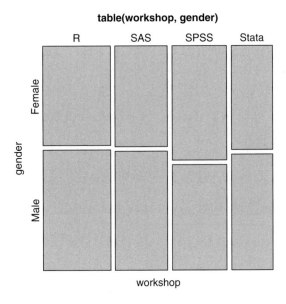

The difference is that the bars fill the `plot` vertically so the shading gives us proportions instead of counts. Also, the width of each bar varies, reflecting the marginal proportion of observations in each workshop. This is called a spine plot. The `mosaicplot` function does something similar (Fig. 21.9).

```
> mosaicplot( table(workshop,gender)   )
```

However, the `mosaicplot` function can handle the complexity of a third factor. We do not have one so let us use an example of Titanic survivors from the `mosaic` function help file.

```
> mosaicplot( ~ Sex + Age + Survived,
+    data  = Titanic, color  = TRUE)
```

In Fig. 21.10, we see that not only are the marginal proportions of sex and age reflected, but the third variable of survival is reflected as well. It is essentially four bar charts within a 2 × 2 cross-tabulation.

21.1.3 Barplots of Means

The `table` function counted for us. Now let us use the `mean` function, along with the `tapply` function to get a similar table of means. To make it easier to read, we will store the table of means in myMeans and then plot it (Fig. 21.11).

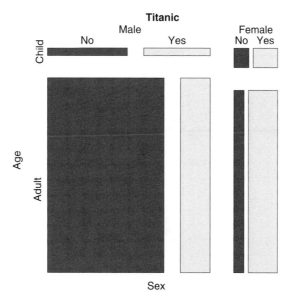

Fig. 21.10 A mosaic plot of Titanic survivors demonstrating the display of three factors at once

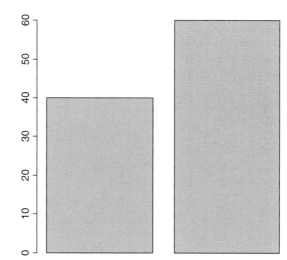

Fig. 21.11 A barplot of means

```
> myMeans  <- tapply(q1, gender, mean, na.rm=TRUE)
> barplot(myMeans )
```

Adding workshop to the tapply function is easy but you must combine gender and workshop into a list first, using the list function (Fig. 21.12).

```
> myMeans <- tapply(q1, list(workshop,gender),
  mean,na.rm=TRUE)
> barplot(myMeans, beside=TRUE)
```

Fig. 21.12 A barplot of
q1 means by workshop
(unlabeled) and gender

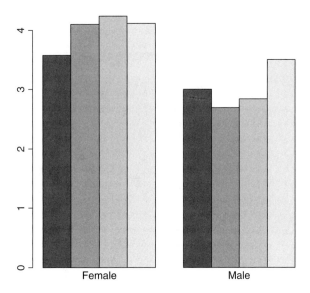

Many of the variations we used with barplots of counts will work with means of course. For example, the `horiz = TRUE` argument will flip any of the above examples on their sides.

21.2 Adding Titles, Labels, Colors, and Legends

So far our graphs have been fairly bland. Worse than that, without a legend, some of the bar charts above are essentially worthless. Let us now polish them up. Although we are using barplots, these steps apply to many of R's traditional graphics functions (Fig. 21.13).

```
> barplot( table(gender,workshop),
+    beside=TRUE,
+    col =c("gray90", "gray60"),
+    main="Number of males and females  \nin each work
    shop ")
> legend(  "topright",
+    c("Female", "Male"),
+    fill=c("gray90", "gray60")  )
```

The `barplot` function call above has four arguments.

1. The `table (gender,workshop)` argument generates a two-way table of counts.

Fig. 21.13 A barplot with a manually added title, legend, and shades of gray

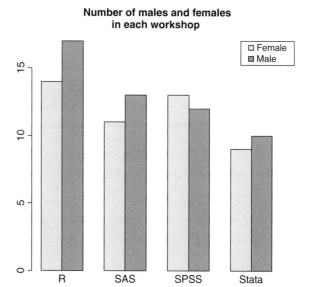

2. The beside = TRUE argument places the workshop bars side-by-side instead of stacking them on top of each other. That is better for perceiving the count per workshop. Leaving it off would result in a stacked bar chart that might be better for estimating relative total attendance of each workshop rather than absolute counts.
3. The col = c ("gray90","gray60") argument supplies the *col*ors (well, shades of gray in this case) for the bars in order of the factor levels in gender. To get a complete list of colors, enter the function colors().

Until the legend function ran, the plot had no legend. It appeared after the fact. The legend function call above has three arguments.

1. The first argument positions the legend itself. This can be the values "topright," "topleft," "bottomright," "bottomleft," "right," and "center." It can also be a pair of *x*, *y* values such as 10, 15. The 15 is a value of the *y*-axis, but the value of 10 for the *x*-axis was determined by the barplot function. You can query the settings with par ("usr"), which will return the start and end of the *x*-axis followed by the same figures for the *y*-axis. Those figures for this graph are 0.56 12.44 –0.17 17.00. So we see that the *x*-axis goes from 0.56 to 12.44, and the *y*-axis goes from –0.17 to 17.00.
2. The fill = c ("Female ", "Male ") argument supplies the value labels to print in the legend. It is up to you to match them to the values of the factor gender, *so be careful!*
3. The fill = c ("gray90 ", "gray60 ") argument supplies colors to match the col argument in the barplot function. Again, it is up to you to make these match the labels as well as the graph itself! The ggplot2 package covered in the next chapter does this for you automatically.

This might seem like an absurd amount of work for a legend, but as with SPSS' GPL, it trades off some ease-of-use for power. We are only skimming the surface of R's traditional graphics flexibility.

21.3 Graphics Parameters and Multiple Plots on a Page

R's traditional graphics make it very easy to place multiple graphs on a page. You could also use your word processor to create a table and insert graphs into the cells, but then you would have to do extra work to make them all the proper size and position, especially if their axes should line up for comparison. You still have to specify the axes' ranges to ensure compatibility but then R would size and position them properly. The built-in lattice package and the ggplot2 package will also standardize the axes automatically.

Another problem with the word processing table method is that text can shrink so small as to be illegible. Using R to create multiframe plots will solve these problems.

There are three different approaches to creating multiframe plots in R. We will discuss the approach that uses the par function. It is easy to use but is limited to equal-sized plots. If you need to create more complex multiframe plots, see the help files for either the layout function or the split.screen function.

R also has some functions, such as coplot and pairs, that create multi-frame plots themselves. Those plots cannot be one piece of an even larger multiframe plot. Even if R could do it, it would be quite a mess!

In traditional graphics, you use the par function to set or query graphic *par*ameter settings. This is equivalent to the SAS GOPTIONS statement. Entering simply par() will display all 71 parameters and how they are set. That is a lot, so we will use the head function to print just the top few parameters. In a later section, you will see a table of all the graphics parameters and functions we use.

```
> head( par()   )
$xlog
  [1] FALSE

$ylog
  [1] FALSE

$adj
  [1] 0.5

$ann
  [1] TRUE

$ask
  [1] FALSE

$bg
  [1] "transparent"
```

We can see above that the `xlog` parameter is set to FALSE, meaning the *x*-axis will not be scaled via the logarithm. The `ask` parameter is also FALSE telling us that if R wants to display multiple graph pages, it will not pause to ask you to click your mouse (or Enter key) to continue. Entering the following will change that setting. However, if you are already running graphs one at a time, this setting can get irritating. Setting it back to FALSE will turn it off and plots will automatically replace one another again.

```
> par( ask=TRUE )
```

Notice that there is no verification of this for the moment. If we wish to query the setting of any particular parameter, we can enter it in quotes, as we do below.

```
> par(  "mfrow"  )

  [1] 1 1
```

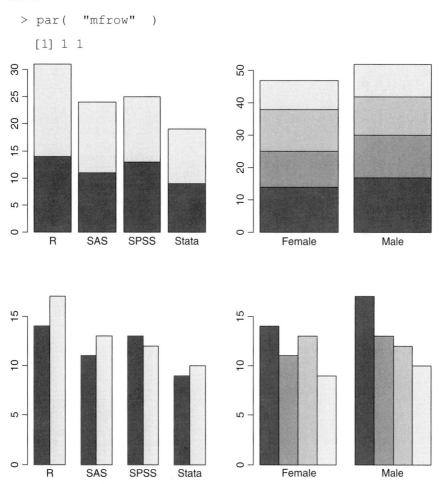

Fig. 21.14 Barplots of counts by workshop and gender. The top two use the default argument, beside = FALSE; the bottom two specify beside = TRUE. The left two tabulated (gender, workshop); the right two tabulated (workshop,gender)

The mfrow parameter sets how many rows and columns of graphs appear on one *multi*frame plot. The setting 1,1 means that only one graph will appear (one row and one column). Let us change that to 2,2 so we can see four graphs on a page. We will create four different bar charts of counts to see the impact of the argument, beside = TRUE (the default value is FALSE). The graphs will appear as we read words: left to right and top to bottom (Fig. 21.14).

```
> par( mfrow=c(2,2) ) #set to 2 rows, 2 columns of
  graphs.
>
> barplot( table(gender,workshop)  )   #top left
> barplot( table(workshop,gender)  )   #top right
> barplot( table(gender,workshop), beside=TRUE )
  #bottom left
> barplot( table(workshop,gender), beside=TRUE )
  #bottom right
>
> par( mfrow=c(1,1) )   #set back to 1 graph per page.
```

21.4 Pie Charts

As the R help file for the pie function says, "Pie charts are a very bad way of displaying information. The eye is good at judging linear measures and bad at judging relative areas. A bar chart or dotchart is a preferable way of displaying this type of data."

The pie function works much in the same way as the barchart function (Fig. 21.15).

Piechart of Workshop Attendance

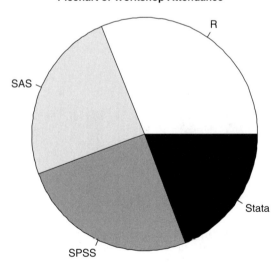

Fig. 21.15 Pie chart of workshop

```
> pie( table(workshop ),
+    col=c("white", "gray90", "gray60", "black" ),
+    main="Piechart of Workshop Attendance"  )
```

21.5 Dotcharts

William Cleveland popularized the dotchart in his book, *Visualizing Data* [30]. Based on research that showed people excel at determining the length of a line, he reduced the bar chart to just dots on lines. R makes it very easy to do (Fig. 21.16). The arguments to the dotchart function are essentially the same as those for the bar chart function. The dots in the dotchart do not show up well in the small image below, so I have added cex = 1.5 for character expansion.

```
> dotchart( table(workshop,gender),
+    main="Dotchart of Workshop Attendance",
+    cex=1.5 )
```

21.6 Histograms

Many statistical methods make assumptions about the distribution of your data. As long as you have enough data, say 30 or more data points, histograms are a good way to examine those assumptions. R's hist function makes it quite easy to do. We will start with basic histograms and will then overlay them.

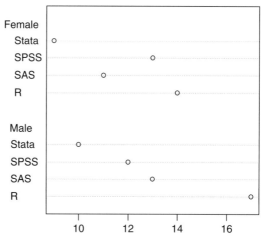

Fig. 21.16 Dotchart of workshop

21.6.1 Basic Histograms

To get a histogram of our posttest score, all we need to do is enter a call to the hist function (Fig. 21.17).

```
> hist(posttest)
```

One of the problems with histograms is that they break continuous data into artificial bins. Trying a different number of bins to see how that changes the view of the distribution is a good idea. In Fig. 21.18, we use the breaks = 20 argument to get far more bars than we saw in the default plot. The argument probability = TRUE causes the y-axis to display probability instead of counts. That does not change the overall look of the histogram but it does allow us to add a kernel density fit with a combination of the lines function and the density function.

```
> hist(posttest, breaks=20, probability=TRUE)
```

The lines function draws the smooth kernel density calculated by the call to the density function. You can vary the amount of smoothness in this function with the adjust argument. See the help file for details.

```
> lines( density(posttest)   )
```

Finally, we can add tick marks to the x-axis to show the exact data points. We can do that with the points function, which adds a series of x, y points anywhere on a plot. We already have the x values in our posttest variable. If we want ticks on the y-axis, we need a vector of zeros that is the same length as our posttest score. The rep function, discussed in Chap. 17, makes quick work of it by repeating the value zero to match the length of posttest.

Histogram of posttest

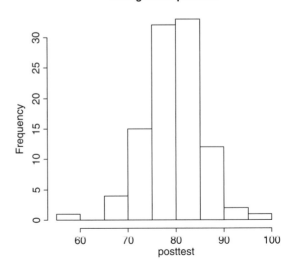

Fig. 21.17 Histogram of posttest

Histogram with Density and Points

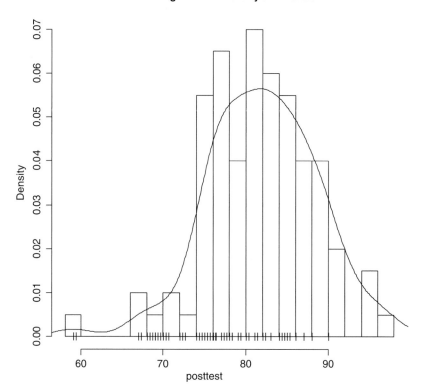

Fig. 21.18 Histogram of posttest with breaks = 20 and a kernel density added

```
> myZeros   <- rep(0, length(posttest)   )
> myZeros
  [1] 0 0 0 0 0 0 0 0 0 0 0 0 0 0 0 0 0 0 0 0 0 0 0 0 0 0 0 0 0
      0 0 0
 [30] 0 0 0 0 0 0 0 0 0 0 0 0 0 0 0 0 0 0 0 0 0 0 0 0 0 0 0 0 0
      0 0 0
 [59] 0 0 0 0 0 0 0 0 0 0 0 0 0 0 0 0 0 0 0 0 0 0 0 0 0 0 0 0 0
      0 0 0
 [88] 0 0 0 0 0 0 0 0 0 0 0 0 0 0 0
> points( posttest, myZeros, pch="|" )
```

Now let us get a histogram of just the males. Recall from Chap. 11, *Selecting Observations* that posttest[gender=="Male"] will make the selection we need. We will also add the argument col = gray60 to give the bars a "color" of gray (Fig. 21.19).

```
> hist( posttest[which (gender=="Male")],+ col="gray60")
```

Fig. 21.19 Histogram of
posttest scores for the males
only

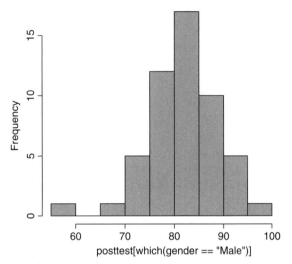

If we want to compare these two plots more directly, we can put them
onto a multiframe plot (Fig. 21.20) with the graphic parameter command
par (mfrow = c(1, 2)). To make them more comparable, we will ensure
they break the data into bars at the same spots using the breaks argument.

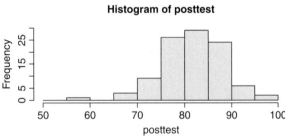
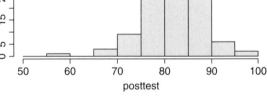

Fig. 21.20 Multiframe plot
of posttest histograms for
whole dataset (top) and just
the males (bottom)

I have written the breakpoints out to make it clear for beginners. Once you get more used to R, it would be much easier to specify `breaks = seq (50, 100, by = 5)`.

```
> par(mfrow=c(2,1) )
> hist(posttest, col="gray90",
+     breaks=c(50,55,60,65,70,75,80,85,90,95,100) )
> hist(posttest[ which (gender=="Male") ],
+     col="gray60",
+     breaks=c(50,55,60,65,70,75,80,85,90,95,100) )
> par(mfrow=c(1,1) )
```

21.6.2 Histograms Overlaid

That looks nice, but we could get a bit fancier and plot the two graphs right on top of one another. The next few examples start slow and end up rather complicated compared to similar plots in SAS or SPSS. In the next chapter, the same plot will be much simpler. However, this is an important example because it helps you learn what type of information is held inside a graphics object.

Since the males are a subset, their bars will be shorter and the overlay will work fine. The `add = TRUE` argument is what tells the `hist` function to add the second histogram on top of the first. Notice below that we also use the `seq` function to generate the numbers 50, 55 ... 100 without having to write them all out as we did before (Fig. 21.21).

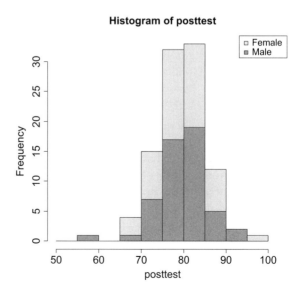

Fig. 21.21 Histogram of posttest for all data (tall bars) overlaid with just the males. The difference between the two represents females

```
> hist(posttest[ which(gender=="Male")],
+     col="gray60",
+     breaks=seq(from=50, to=100, by=5),
+     add=TRUE   )
> legend(  "topright", c("Female", "Male"),
+     fill=c("gray90", "gray60")   )
```

This looks good, but we did have to manually decide what the breaks should be. In a more general-purpose program, we may want R to choose the break-points in the first plot and then apply them to the second automatically. We can do that by saving the first graph to an object called, say, myHistogram.

```
> myHistogram   <- hist(posttest, col ="gray90")
```

Now let us use the names function to see the names of its components.

```
> names(myHistogram)
 [1] "breaks"    "counts"    "intensities"   "density"
 [5] "mids"      "xname"     "equidist"
```

One part of this object, myHistogram$breaks, stores the breakpoints that we will use in the second histogram. The graph of all the data appears at this point and we can print the contents of myHistogram$breaks. Notice that R has decided that our manually selected breakpoint of 50 was not needed.

```
> myHistogram$breaks

 [1]  55  60  65  70  75  80  85  90  95 100
```

Let us now do the histogram for males again, but this time with the argument, breaks = myHistogram $breaks so the breakpoints for males will be the same as those automatically chosen for the whole sample (Fig. 21.22).

```
> hist(posttest[ which(gender=="Male")],
+     col='gray60',
+     add=TRUE, breaks=myHistogram$breaks)
> legend(  "topright", c("Female","Male"),
+     fill=c("gray90","gray60"))
```

This is essentially the same as the previous graph, but the axis fits better by not extending all the way down to 50. Of course, we could have noticed that and fixed it manually if we had wanted. To see what else a histogram class object contains, simply enter its name. You see the breaks listed as its first element. You can change any of these histogram parameters to adjust your plot.

```
> myHistogram

$breaks
 [1]  55  60  65  70  75  80  85  90  95 100

$counts
```

Fig. 21.22 Same histogram
as previous one but now bar
breakpoints were chosen
from whole dataset and
then applied to males. This
showed the *x*-axis did not
need to go all the way to 50

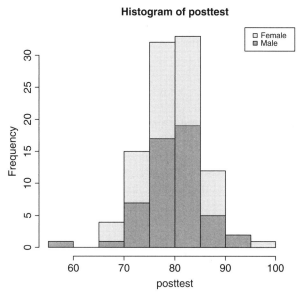

```
[1]  1  0  3  9 26 29 24  6  2

$intensities
[1] 0.002 0.000 0.006 0.018 0.052 0.058 0.048 0.012
  0.004

$density
[1] 0.002 0.000 0.006 0.018 0.052 0.058 0.048 0.012
  0.004

$mids
[1] 57.5 62.5 67.5 72.5 77.5 82.5 87.5 92.5 97.5

$xname
[1]  "posttest"

$equidist
[1]  TRUE

attr (, "class")
[1]  "histogram"
```

21.7 Normal QQ Plots

A normal QQ plot plots the quantiles of each data point against the quantiles
that each point would have if the data were normally distributed. If these points
fall on a straight line, they are likely to be from a normal distribution.

Histograms give you a nice view of a variable's distribution, but if you
have fewer than 30 or so points, the resulting histogram is often impossible to

interpret. Another problem with histograms is that they break the data into artificial bins, so unless you fiddle with bin size, you might miss an interesting pattern in your data. A QQ plot has none of these limitations.

So why use histograms at all? Because they are easy to interpret. At a statistical meeting I attended, the speaker displayed a QQ plot and asked the audience, all statisticians, what the distribution looked like. It was clearly not normal and people offered quite an amusing array of responses! When the shape is not a straight line, it takes time to learn how the line's shape reflects the underlying distribution. To make matters worse, some software reverses the roles of the two axes! The plot shown in Fig. 21.23, created using the qq.plot function from John Fox's car package, has the theoretical quantiles on the *x*-axis. This is how SAS displays it but SPSS reverses the axes.

```
> library("car")

> qq.plot(posttest,
+     labels=row.names(mydata100),
+     col="black"   )

> detach("package:car")
```

The call to the qq.plot function call above has three arguments.

1. The variable to plot.
2. labels = row.names (mydata100) allows you to interactively identify any point you wish and label it according by the values you request. Your cursor will become a cross-hair and when you click on (or near) a point, its label will appear. The escape key will end this interactive mode.
3. col="black" sets the color, which is red by default.

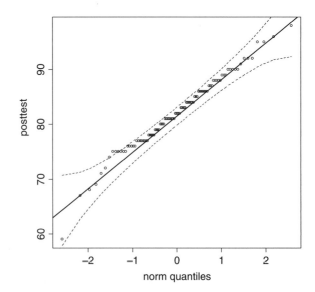

Fig. 21.23 A normal quantile plot of posttest using qq.plot from the car package

R also has a built-in function for qqplots called qqnorm. However, it lacks confidence intervals. It also does not let you identify points without calling the identify function (graph not shown). See section 21.9 for an example using the identify function.

```
myQQ   <- qqnorm(posttest)
identify(myQQ)
```

21.8 Strip Charts

Strip charts are scatterplots for one variable. Since they plot each data point, you might see clusters of points that would be lost in a boxplot or errorbar plot. The first stripchart function call below uses "jitter" or random noise added to the vertical axis to help you see points that would otherwise be obscured by falling on top of other point(s) at the same location. The second one uses method="stack" to stack the points instead (Fig.21.24).

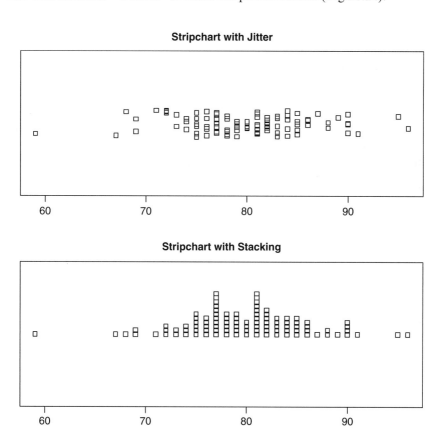

Fig. 21.24 Strip chart demonstrating the methods jitter and stack

```
> par(mfrow=c(2,1)  )
> stripchart(posttest, method="jitter",
+   main="Stripchart with Jitter")
> stripchart(posttest, method="stack",
+   main="Stripchart with Stacking")
> par( mfrow=c (1,1)  )
```

Let us now compare groups using strip charts (Fig. 21.25). Notice the use of the formula posttest ~ workshop to compare the workshop groups. You can reverse the order of those two variables to flip the scatter vertically, but you would lose the automated labels for the factor levels.

```
> par( las=2, mar=c(4,8,4,1)+0.1 )

> stripchart(posttest  ~ workshop, method ="jitter")
```

The par function here is optional. I use it just to angle the workshop value labels so that they are easier to read. Turned that way, they need more space on the left of the chart. The par function call above uses two arguments to accomplish this.

1. The las = 2 argument changes the angle of the text to be perpendicular to the *y*-axis. For more settings, see Table 21.3.
2. The mar = c (4,8,4,1) +0.1 argument sets the size of the margins at the bottom, left, top, and right sides, respectively. The point of this was to make the left side much wider so that we could fit the labels turned on their sides. For details, see Table 21.2.

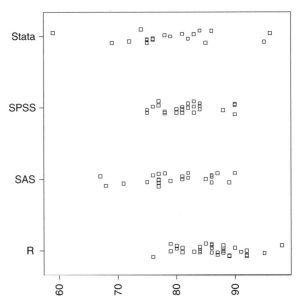

Fig. 21.25 Strip chart of posttest scores by workshop

The `stripchart` function call above contains two arguments.

1. The formula `posttest~workshop`, which asks for a strip chart of post-test for every value of workshop.
2. The `method="jitter"` argument that tells it to add random noise to help us see the points that would otherwise be plotted on top others.

21.9 Scatterplots

Scatterplots are helpful in many ways. They show the nature of a relationship between two variables. Is it a line? A curve? Is the variability in one variable the same at different levels of the other? Is there an outlier now visible that did not show up when checking minimum and maximum values one variable at a time? A scatterplot can answer all these questions.

The `plot` function takes advantage of R's object orientation by being generic. That is, it looks at the class of objects you provide it and takes the appropriate action. For example, when you give it two continuous variables, it does a scatterplot (Fig. 21.26).

```
> plot(pretest, posttest)
```

Note the "92" on the lower left point. That did not appear on the graph at first. I wanted to find which observation that was, so I used the `identify` function.

```
> identify(pretest,posttest)
 [1] 92
```

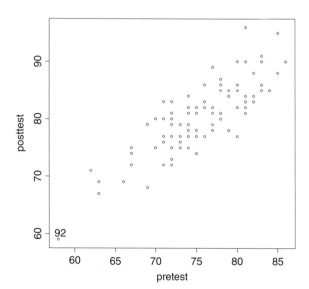

Fig. 21.26 Plot of pretest and posttest. Observation 92 was identified after the plot was created

The `identify` function lets you use your mouse to label points by clicking on them. You list the *x* and *y* variables in your plot and optionally provide a `label` = argument to specify an ID variable, with the row names as the default. The function will label the point you click on or near. Unfortunately, it does not let you draw a box around a set of points. You must click on each one, which can get a bit tedious. When finished, users can right-click and choose "stop", (or press the Esc key on Windows). It will then print out the values you chose.

If you assigned the result to a vector as in

```
myPoints   <- identify(pretest,posttest)
```

then that vector would contain all your selected points. You could then use logic like

```
summary( mydata100[!myPoints,  ]  )
```

(not myPoints) to exclude the points and see how it changes your analysis. Below, I use a logical selection to verify that it is indeed observation 92 that has the low score.

```
> mydata100 [pretest <60,  ]

    gender workshop q1 q2 q3 q4 pretest posttest
92    Male     Stata  3  4  4  4      58       59
```

Back to the scatterplot, you can specify the `type` argument to change how the points are displayed. The values use the letter "p" for *p*oints, the letter "l" for *l*ines, "b" for *b*oth points and lines, and "h" for *h*istogram-like lines that rise vertically from the *x*-axis. Connecting the points using either "l" or "b" makes sense only when the points are collected in a certain order, such as time series data. As you can see in Fig.21.27, that is not the case with our data, so those appear as a jumbled nest of lines.

```
> par(mfrow=c(2,2)   )

> plot( pretest, posttest, type="p", main="type=p" )
> plot( pretest, posttest, type="l", main="type=l" )
> plot( pretest, posttest, type="b", main="type=b" )
> plot( pretest, posttest, type="h", main="type=h" )

> par(mfrow =c (1,1)   )
```

21.9.1 Scatterplots with Jitter

The more points you have in a scatterplot, the more likely you are to have them overlap, potentially hiding the true structure of the data. This is a particularly bad problem with Likert-scale data since it only uses the values 1 through 5. This data is typically averaged into scales that are more continuous, but we will look at an example with just two Likert measures, q1 and q4. Jitter is simply some random variation added to the data to prevent overlap. You will see the

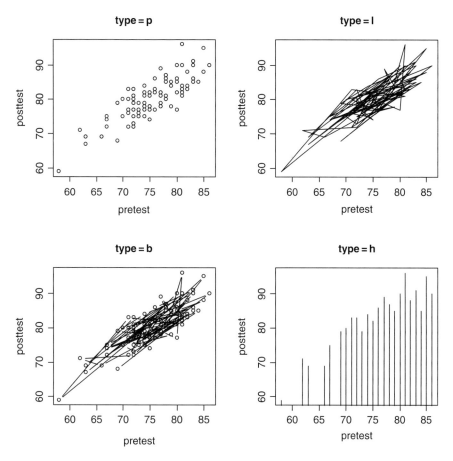

Fig. 21.27 Various scatterplots demonstrating the effect of the `type` argument

jitter function in the second plot in Fig. 21.28. Its arguments are simply the variable to jitter and a value "3" for the amount of jitter. That was derived from trial and error. The bigger the number, the greater the jitter.

```
> par( mfrow=c(1,2)   )
> plot( q1,q4,
+    main="Likert Scale Without Jitter")
> plot( jitter(q1,3), jitter(q4,3),
+    main="Likert Scale With Jitter")
```

21.9.2 Scatterplots with Large Datasets

The larger your dataset, it is more likely that some points will be hidden by others. That makes jitter, described in the previous section, particularly useful.

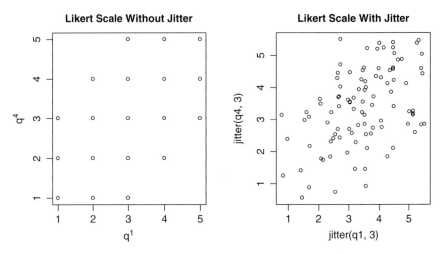

Fig. 21.28 Scatterplots demonstrating the impact of jitter on five-point Likert-scale data

It is also helpful to use the smallest point character with `pch="."`. Here is an example using 5000 points (Fig. 21.29).

```
> plot( jitter(pretest2,4), jitter(posttest2,4),
  pch=".",
+   main="5,000 Points Using pch='.' \nand Jitter")
```

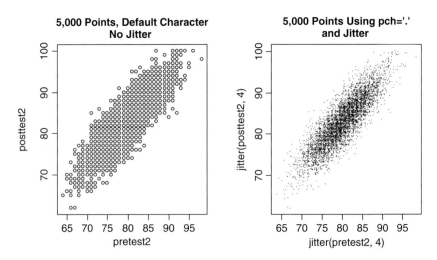

Fig. 21.29 Scatterplots showing how to handle large datasets. The plot on the left using default settings, leaving many points obscured. The one on the right uses much smaller points and jitters them, allowing us to see many more points

Fig. 21.30 A hexbin plot that divides large amounts of data into hexagonal bins to show structure in large datasets. Each bin represents up to 96 original points

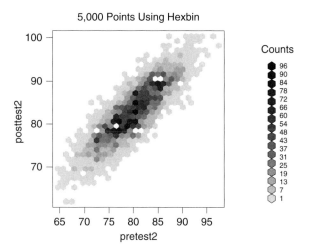

Another way of plotting large amounts of data is a hexbin plot (Fig. 21.30). This type of plot bins the points into little hexagons before plotting them. More points result in darker hexagons. This plot is provided via the `hexbin` package, written by Dan Carr and ported by Nicholas Lewin-Koh and Martin Maechler [34]. Note that hexbin uses the `lattice` package, which in turn uses the grid graphics system. That means that you cannot put multiple graphs on a page, nor set any other parameters using the `par` function.

```
> library("hexbin")

Loading required package: grid
Loading required package: lattice

> plot( hexbin(pretest2,posttest2),
+    main="5,000 Points Using Hexbin")

> detach("package:hexbin")
```

21.9.3 Scatterplots with Lines

You can add straight lines to your plots with the `abline` function (Fig. 21.31). It has several different types of arguments. Let us start with a scatterplot with only points on it.

```
> plot(posttest ~ pretest )
```

Now let us add a horizontal line and a vertical line at the value 75. You might do this if there were a cutoff below which students were not allowed to take the next workshop.

```
> abline( h=75, v=75  )
```

Fig. 21.31 Scatterplot
demonstrating how to add
various types of lines as well
as a legend and title

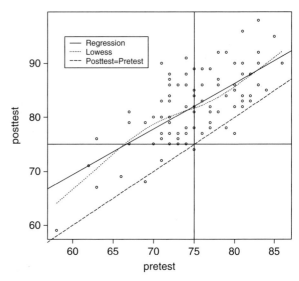

The `abline` function exists to add straight lines to the last plot you did, so there is no `add = TRUE` argument. Next, let us draw a diagonal line that has pretest equal to posttest. If the workshop training had no effect, the scatter would lie on this line. This line would have a *y*-intercept of 0 and a slope of 1. The `abline` function does formulas in the form $y = a + bx$, so we want to specify $a = 0$ and $b = 1$. We will also set the linetype to dashed with `lty = 5`.

```
> abline( a=0, b=1, lty=5 )
```

Next, let us add a regression fit using the `lm` function within the call to `abline`. The `abline` function draws straight lines so let us use the `lines` function along with the `lowess` function to draw smoothly fitting lowess curve. The legend allows you to choose the order of the labels; so I have listed them as they appear from top to bottom in the upper right corner of the plot.

```
> abline( lm( posttest ~ pretest ),      lty=1 )
> lines( lowess( posttest ~ pretest ), lty=3 )
> legend( 60, 95,
+   c( "Regression", "Lowess", "Posttest=Pretest"  ),
+   lty=c (1,3,5 )  )
>
```

21.9.4 Scatterplots with Linear Fit by Group

As we saw in the last section, it is easy to add regression lines to plots using R's traditional graphics. Let us now turn our attention to fitting a regression line separately for each group (Fig. 21.32). First, we will use the `plot` function to

Fig. 21.32 A plot that
displays different points and
regression lines for each
gender

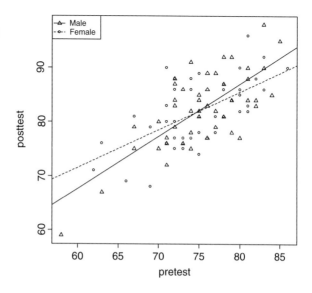

display the scatter, using the pch argument to set the *point characters* based
on gender. Gender is a factor, which cannot be used directly to set the point
characters. Therefore, we are using the as.numeric function to convert it on
the fly. That will result in two symbols used. If we used the as.character
function instead, it would plot the actual characters M and F.

```
> plot (posttest ~pretest,
+    pch=as.numeric (gender)  )
```

Next, we simply use the abline function as we did before but basing our
regression on the males and females separately.

```
> abline ( lm( posttest[ which(gender=="Male")]
+             ~ pretest[ which(gender=="Male")]  ),
+         lty=1  )
>
> abline ( lm( posttest[ which(gender=="Female")]
+             ~ pretest[ which(gender=="Female")]  ),
+         lty=2  )
>
> legend ( "topleft", c("Male", "Female"),
+         lty=c(1,2), pch=c(2,1)   )
```

21.9.5 Scatterplots by Group or Level (Coplots)

Coplots are scatterplots conditioned on the levels of a third variable. For
example, a scatterplot for each type of workshop. In Fig. 21.33, the box

Given : workshop

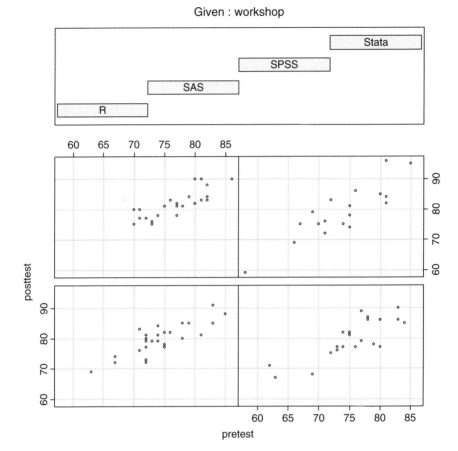

Fig. 21.33 A set of scatterplots for each workshop

above the scatterplots indicates which plot is which. The bottom left is for R, the bottom right is for SAS, the top left is for SPSS, and the top right is for Stata. This is a rather odd layout for what could have been a simple 2 × 2 table of labels, but it makes more sense when the third variable is continuous. Both the lattice and ggplot2 packages do a better job of labeling such plots. The coplot function is easy to use. Simply specify a formula in the form of y~x, and list your conditioning variable after a vertical bar "l".

```
> coplot( posttest ~ pretest | workshop )
```

The next plot, Fig. 21.34, is conditioned on the levels of q1. Now the length of the bars *and their overlap* do an excellent job of showing the values of q1 that apply to each plot. The values of q1 in each plot overlap to prevent you from missing an important change that occurs at a single value of q1.

```
> coplot( posttest ~ pretest | q1 )
```

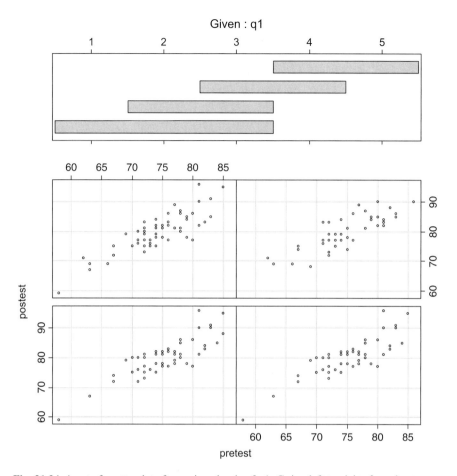

Fig. 21.34 A set of scatterplots for various levels of q1. Going left to right, from bottom to top, the values of q1 increase. The bars in the top frame show the values of q1 for each plot and how they overlap

The functions that modify plots, apply to plots you create one at a time. That is true even when you build a multiframe plot one plot at a time. However, when you use a single R function that creates its own multiframe plot, such as `coplot`, you can no longer use those functions. See the help file for ways to modify coplots.

21.9.6 Scatterplots with Confidence Ellipse

Confidence ellipses help visualize the strength of a correlation as well as provide a guide for identifying outliers. The `data.ellipse` function in the `car` package makes these quite easy to plot (Fig. 21.35). It works much like the `plot` function with the first two arguments being your x and y variables,

Fig. 21.35 Scatterplot with
95% confidence ellipse

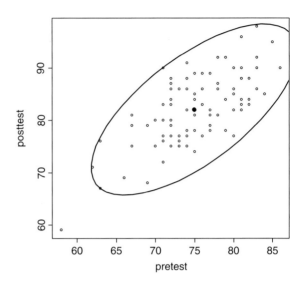

respectively. The `levels` argument lets you specify one confidence limit as
shown below, or you could specify a set in the usual way, for example
`levels=c (.25, .50, .75)`. If you leave the `col="black"` argument
off, it will display in its default color of red.

```
> library("car")
> data.ellipse(pretest, posttest,
+    levels =.95,
+    col="black")
> detach("package:car")
```

21.9.7 Scatterplots with Confidence and Prediction Intervals

As we saw previously, adding a regression line to a scatterplot is easy. However,
adding confidence intervals is somewhat complicated. The `ggplot2` package
covered in the next chapter makes getting a line and 95% confidence band
about the line easy, but getting confidence limits about the predicted points is
complicated even with it.

Let us start with a simple example. We will create a vector x and three vectors
y1, y2 and y3. The three y's will represent a lower confidence limit, the prediction line, and the upper confidence limit.

```
> x   <- c(1,2,3,4)
> y1  <- c(1,2,3,4)
> y2  <- c(2,3,4,5)
> y3  <- c(3,4,5,6)
```

```
> yMatrix  <- cbind(y1,y2,y3)

> yMatrix

     y1 y2 y3
[1,]  1  2  3
[2,]  2  3  4
[3,]  3  4  5
[4,]  4  5  6
```

Now we will use the plot function to plot x against y2 (Fig. 21.36). We will specify the xlim and ylim arguments to ensure the axes will be big enough to hold the other y variables. The result is rather dull! I have used the cex=1.5 argument to do plot *character expansion* of 50% to make it easier to see in this small size.

```
> plot(x,y2, xlim=c(1,4 ), ylim=c(1,6 ), cex=1.5 )
```

Next, we will use the matlines function (Fig. 21.37). It can plot a vector against every column of a matrix. We will use the lty argument to specify *line types* of dashed, solid, dashed for y1, y2, and y3, respectively. Finally, we will specify the line color as col = "black" to prevent it from providing a different color for each line as it does by default.

```
> matlines( x, yMatrix, lty=c (2,1,2), col="black")

> rm( x, y1, y2, y3, yMatrix )
```

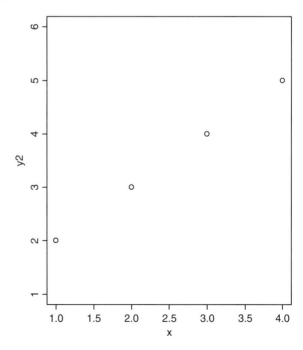

Fig. 21.36 Scatterplot of just four points on a straight line, a simple foundation to build upon

Fig. 21.37 A scatterplot that demonstrates the basic idea of a regression line with confidence intervals. With a more realistic example, the confidence bands would be curved

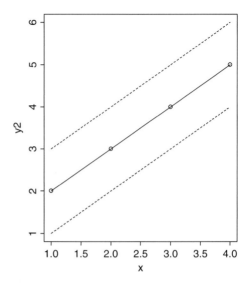

This plot represents the essence of our goal. Now let us fill in the details. First, we need to create a new data frame that will hold a "well-designed" version of the pretest score. We want one that covers the range of possible values evenly. That will not be important for linear regression, since the spacing of points will not change the line, but we want to use a method that would work even if we were fitting a polynomial regression.

```
> myIntervals  <-
  data.frame( pretest=seq(from=60, to=100, by=5)  )
> myIntervals
  pretest
1      60
2      65
3      70
4      75
5      80
6      85
7      90
8      95
9     100
```

Now let us create a regression model using the `lm` function, and store it in myModel. Note that we have attached the "real" dataset, mydata100, so the model is based on its pretest and posttest scores.

```
> myModel  <- lm( posttest ~ pretest  )
```

Next, we will use the `predict` function to apply myModel to the myInter-vals data frame we just created. There are two types of intervals you might wish to plot around a regression line. The 95% *prediction* interval (also called tolerance interval) is the wider of the two and is for predicted values of Y for new values of X (`interval ="prediction"`). The 95% *confidence* interval is for the mean of the Y values at a given value of X (`interval="confi-dence"`). We will run the `predict` function twice to get both types of intervals, and then look at the data. Notice that the `newdata` argument tells the predict function which dataset to use.

```
> myIntervals$pp  <- predict( myModel,
+  interval="prediction", newdata=myIntervals)

> myIntervals$pc  <- predict( myModel,
+  interval="confidence", newdata=myIntervals)

> myIntervals
```

```
  pretest  pp.fit  pp.lwr  pp.upr  pc.fit   pc.lwr   pc.upr
1      60  69.401  59.330  79.472  69.401   66.497   72.305
2      65  73.629  63.768  83.491  73.629   71.566   75.693
3      70  77.857  68.124  87.591  77.857   76.532   79.183
4      75  82.085  72.394  91.776  82.085   81.121   83.050
5      80  86.313  76.579  96.048  86.313   84.980   87.646
6      85  90.541  80.678 100.405  90.541   88.468   92.615
7      90  94.770  84.696 104.843  94.770   91.855   97.684
8      95  98.998  88.636 109.359  98.998   95.207  102.788
9     100 103.226  92.507 113.945 103.226   98.545  107.906
```

Now we have all the data we need. Look at the names pp.fit, pp.lwr, pp.upr. They are the fit and lower/upper prediction confidence intervals. The three variables whose names begin with pc are the same variables for line's narrower confidence interval. But why the funny names? Let us check the class of just "pp."

```
> class( myIntervals$pp )
  [1]  "matrix"

> myIntervals$pp
```

```
       fit     lwr      upr
1   69.401  59.330   79.472
2   73.629  63.768   83.491
3   77.857  68.124   87.591
4   82.085  72.394   91.776
5   86.313  76.579   96.048
6   90.541  80.678  100.405
7   94.770  84.696  104.843
8   98.998  88.636  109.359
9  103.226  92.507  113.945
```

Fig. 21.38 Here we finally
see the complete plot with
both types of confidence
intervals using our practice
dataset

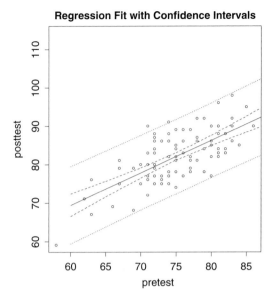

The `predict` function has added two matrices to the myIntervals data frame. Since the `matlines` function can plot all columns of a matrix at once, this is particularly helpful. Now let us use this information to complete our plot with both types of confidence intervals. The only argument that is new is setting the limits of the *y*-axis using the `range` function. I did that to ensure the *y*-axis was wide enough to hold both the pretest scores and the wider prediction interval, pp. Finally, we see the plot in Fig. 21.38.

```
> plot( pretest, posttest,
+    ylim =range( myIntervals$pretest, myIntervals$pp,
        na.rm=TRUE ),
+    main="Regression Fit with Confidence Intervals" )

> matlines(myIntervals$pretest, myIntervals$pc,
+    lty=c(1,2,2 ), col="black"  )

> matlines(myIntervals$pretest, myIntervals$pp,
+    lty=c(1,3,3 ), col="black"  )
```

21.9.8 Plotting Labels Instead of Points

When you do not have too much data, it is often useful to plot labels instead of symbols (Fig. 21.39). If your label is a single character, the pch argument set point characters. If you have a character variable, pch will accept it directly. In our case, gender is a factor, so we can use the as.character function to covert it on the fly.

Fig. 21.39 Scatterplot using gender as its point character. This method works with a single character though

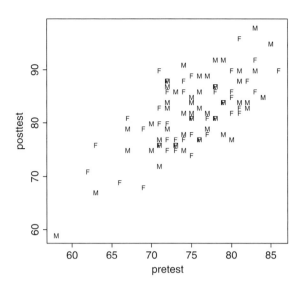

```
> plot(pretest, posttest,
+     pch=as.character(gender)   )
```

Unfortunately, the pch argument can handle only a single character. If you provide it a longer label, it will plot only the first character of each. To see the whole label, you must first plot an empty graph with type="n", for no points, and then add to it with the text function. The text function works just like the plot function but it plots labels instead of points. In Fig. 21.40, we are using the row.names function to provide labels. If we had wanted to plot the workshop value instead, we could have used label=as.character(workshop).

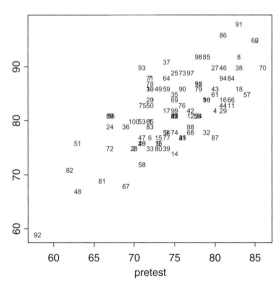

Fig. 21.40 Scatterplot with row.names as labels. The text function is required for labels longer than one character

```
> plot(pretest, posttest, type ="n" )
> text(pretest, posttest,
+    label=row.names(mydata100)   )
```

21.9.9 Scatterplot Matrices

When you have many variables to study it can be helpful to get pairwise
scatterplots of them all (Fig. 21.41). This is easy to do with the `plot`
function. The first one below will suffice, but it inserts gaps between each
pair of graphs. The second one removes those gaps and shrinks the size of the
labels by 10%. The second plot is the only one shown.

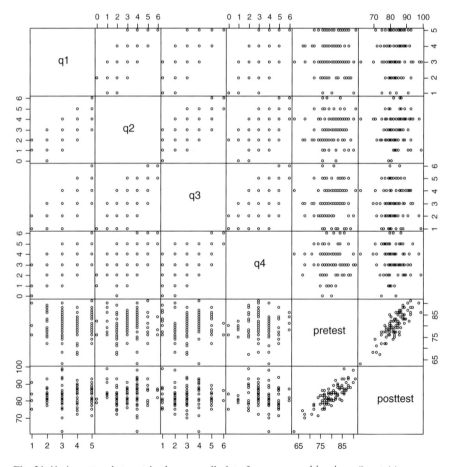

Fig. 21.41 A scatterplot matrix shows small plots for every combination of variables

```
> plot( mydata100[3:8])   #Not shown.

> plot( mydata100[3:8], gap=0, cex.labels=0.9)
```

You can use the entire data frame in this type of plot, and it will convert factors to numeric and give you the following warning. The plots involving factors would appear as strip plots.

```
Warning message:
In data.matrix(x) : class information lost from one or
   more columns
```

As with any generic function, you can see what other functions plot will call given different classes of data. The methods function will show you.

```
> methods(plot)
 [1] plot.acf*          plot.data.frame* plot.Date*
 [4] plot.decomposed.ts* plot.default    plot.dendrogram*
 [7] plot.density       plot.ecdf        plot.factor*
[10] plot.formula*      plot.hclust*     plot.histogram*
[13] plot.HoltWinters*  plot.isoreg*     plot.lm
[16] plot.medpolish*    plot.mlm         plot.POSIXct*
[19] plot.POSIXlt*      plot.ppr*        plot.prcomp*
[22] plot.princomp*     plot.profile.nls* plot.spec
[25] plot.spec.coherency plot.spec.phase  plot.stepfun
[28] plot.stl*          plot.table*      plot.ts
[31] plot.tskernel*     plot.TukeyHSD
```

When you use the plot function on a data frame, it passes the data onto the plot.data.frame function. When you read the help file on that, you find that it then calls the pairs function! So to find out the options to use, you can finally enter help(pairs). That will show you some interesting options, including adding histograms on the main diagonal and Pearson correlations on the upper right panels. In the next example, I have added the panel.smooth values to draw lowess smoothing. Although I call the pairs function directly here, I could have used the plot function to achieve the same result. Since the pairs function is creating a multiframe plot by itself, you must use its own options to modify the plot. In this case we cannot add a smoothed fit with the lines function; we must use panel.smooth instead (Fig. 21.42).

```
> pairs(mydata100[3:8], gap=0,
+    lower.panel=panel.smooth,
+    upper.panel=panel.smooth)
```

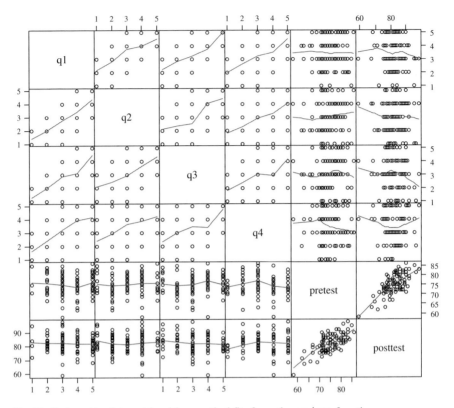

Fig. 21.42 A scatterplot matrix with smoothed fits from the `pairs` function.

21.10 Dual Axes Plots

The usual plot has a *y*-axis on only the left side. However, to enhance readability you may wish to place it on the right as well. If you have the same *Y* variable measured in two different units (Dollars & Euros, Fahrenheit & Celsius...) you may wish a second axis in those units. Plotting two different variables on the y axes is usually not a very good idea. We will simply place the same axis on both sides, which can enhance interpretation. We will plot the same graph, once without axes, then adding one on the right, then the left.

First, we will need to add space in the margins, especially the right side where the new axis will need labeling. That is done with the following command, which changes the margins from their default values of $(5,2,2,4) + 0.1$ to 5,5,4,5 as it applies to the bottom, left, top, and right sides, respectively.

```
> par( mar=c(5,5,4,5)   )
```

Next we will plot the points but without the axes using the argument, axes = FALSE. We will also fix the limits of the *x*-axis with xlim=c (55,100). The limits of the *Y*-axis might be different of course but we need to keep the *x* values in each overlaid plot in the exact same locations.

```
> plot( pretest, posttest, axes=FALSE, xlim=c(55,90),
+    main="Scatterplot with Dual Axes"  )
```

The points and labels will appear now but without axes. The next call to the axis function asks R to place the current axis on the right side (side 4).

```
> axis(4)
```

Next we can add text to the margin around that axis with the mtext function. It is placing its text, in a bold font (font = 2) on line 3 of side 4 (the right).

```
> mtext("Axis label on right side",
+    font=2, side=4, line=3)
```

Now comes the "secret sauce." If we were to use the plot function again, it would completely replace what we have done so far. But if we set the graphics parameter new = TRUE, it does not erase the screen, allowing us to complete our plot (Fig. 21.43).

```
> par( new = TRUE )
> plot( pretest, posttest, xlim=c(60,90)  )
```

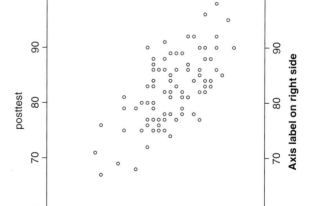

Fig. 21.43 A scatterplot demonstrating an additional axis on the right side.

21.11 Boxplots

Boxplots put the middle 50% of the data in a box with a median line in the middle and lines called "whiskers" extending to $+/-1.5$ times the height of the box (i.e., the 75th percentile minus the 25th). Points that lie outside the whiskers are considered outliers. You can create boxplots using the `plot` function when the first variable you provide is a factor (Fig. 21.44).

```
> plot( workshop, posttest,
+ main="Boxplot using plot function")
```

There are several variations we can do. First, let us use the `mfrow` argument of the `par` function to create a 2×2 multiframe plot.

```
> par( mfrow=c (2,2)  )
```

Next we will do a boxplot of a single variable, posttest. It will appear in the upper left of Fig. 21.45. We will use the `boxplot` function since it gives us more flexibility.

```
> boxplot(posttest)
```

Then we will put pretest and posttest side-by-side in the upper right of Fig. 21.45. The `notch` argument tells it to create notches that, when they do not overlap, provide "strong evidence" that the medians differ. It appears in the upper right.

```
> boxplot( pretest, posttest, notch=TRUE )
```

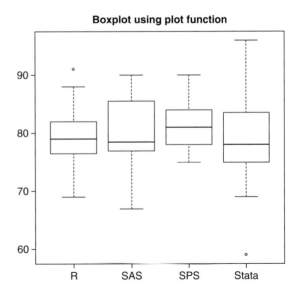

Fig. 21.44 A boxplot of posttest scores for each workshop

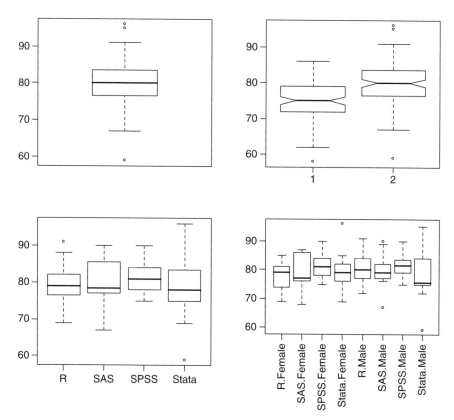

Fig. 21.45 Various boxplots. The upper left is the posttest score. The upper right shows pretest and posttest, with notches indicating possible median differences. The lower left shows posttest scores for each workshop. The lower right shows posttest for the gender and workshop combinations, as well as labels perpendicular to the *x*-axis

Next we will use a formula to get boxplots for each workshop, side-by-side. It appears in the bottom left of Fig. 21.45.

```
> boxplot( posttest ~ workshop )
```

Finally, we will create a boxplot for each workshop:gender combination. However, this will create long labels so we will need to change their angle to be perpendicular to the *x*-axis using las = 2 and increase the bottom margin with mar = c (8, 4, 4, 2) + 0.1. We will set those parameters back to their defaults immediately after using the boxplot function. See Table 21.2 for more on the mar parameter. The plot appears in the bottom right of Fig. 21.45.

```
> par( las=2, mar=c (8,4,4,2)+0.1  )
> boxplot(posttest  ~ workshop:gender)
> par( las=0, mar=c(5,4,4,2)+0.1  )
```

21.12 Error Bar and Interaction Plots

The gplots package by Gregory R. Warnes et al., has a plotmeans function that plots means with 95% confidence bars around each [35]. It is very easy to use. The confidence intervals assume the data comes from a normal distribution (Fig. 21.46).

```
> library("gplots")

> plotmeans( posttest ~ workshop,
+    main="Plotmeans from gplots Package")

> detach("package:gplots")
```

R also has the built-in interaction.plot function that plots the means for a two-way interaction (Fig. 21.47). For a three-way, you can use the by function to repeat the interaction plot for each level of the third variable. In Fig. 21.47, the males seem to be doing better with R and the females with Stata. Since the plot does not display variability, we do not have any test of significance for this.

```
> interaction.plot( workshop, gender, posttest,
+    main="Means Using interaction.plot function")
```

21.13 Adding Equations and Symbols to Graphs

Any of the functions that add text to a graph, such as main, sub, xlab, and ylab can display mathematical equations. For example, a well-known formula for multiple regression is

$$\hat{\beta} = (X^t X)^{-1} X^t Y$$

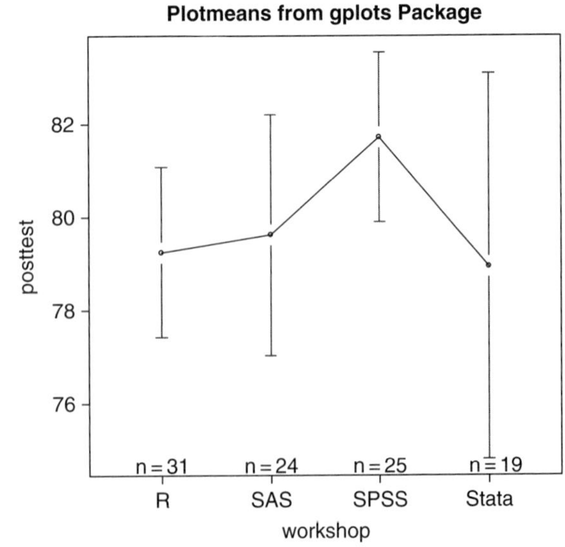

Fig. 21.46 An error barplot of posttest scores for each workshop

Fig. 21.47 An interaction plot of posttest score by the workshop and gender combinations

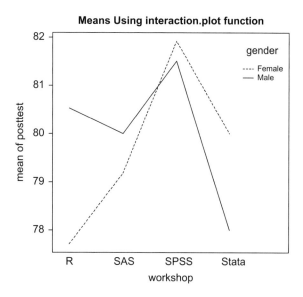

You can add this to any existing graph using the following call to the `text` function.

```
text(66, 88,   "My Example Formula")
text(65, 85,
   expression( hat(beta)   ==
   (X^t   *  X)^{-1}  *  X^t   *  Y)   )
```

The `text` function adds any text at the *x, y* position you specify, in this case 66 and 88. So the use of it above adds, "My Example Formula" to an existing graph at that position. In the second call to the `text` function, we also call the `expression` function. When used on any of the text annotation functions, the `expression` function tells R to interpret its arguments in a special way that allows it to display a wide variety of symbols. In this example, beta will cause the Greek letter β to appear. Two equal signs in a row (x= =y) result in the display of one (x=y). Functions like `hat` and `bar` will cause those symbols to appear over their arguments. So `hat(beta)` will display a ˆ symbol over the Greek beta. This example formula appears on the plot in Fig. 21.48. The help files contain several tables of symbols that you can display with `? plotmath`.

21.14 Summary of Graphics Elements and Parameters

As we have seen in the examples above, R's traditional graphics have a range of functions and arguments that you can use to embellish your plots. Tables 21.1 through 21.4 summarize them.

Table 21.1 Graphics arguments for use with traditional high-level graphics functions

main	Supplies the text to the main title.
	Example: `plot (pretest,posttest,main="My Scatterplot")`
sub	Supplies the text to the sub-title.
	Example: `plot (pretest,posttest...sub="My Scatterplot")`
xlab	An argument that supplies the *x*-axis label (variable labels from `Hmisc` package are ignored).
	Example: `plot (pretest,posttest...xlab="Score Before Training")`
xlim	An argument that specifies the lower and upper limits of the *x*-axis.
	Example: `plot (pretest, posttest, xlim=c (50,100) )`
ylab	An argument that supplies the *y*-axis label (variable labels from `Hmisc` package are ignored).
	Example: `plot (pretest,posttest...ylab="Score After Training")`
xlim	An argument that specifies the lower and upper limits of the *y*-axis.
	Example: `plot (pretest, posttest, ylim=c (50,100) )`

Table 21.2 Graphics parameters to set or query using only par()

ask	par(ask = TRUE) causes R to prompt you before showing a new graphic. The default setting of FALSE causes it to automatically replace any existing plot with a new one. If you ran a program in batch mode, you do *not* want to set this to TRUE!
family	Sets font family for text. The default setting of "sans" provides Helvetica or Arial. "sans" = sans serif like Helvetica or Arial; "serif" = serif like Times Roman; "mono" = monospaced like Courier; "symbol" = math and Greek symbols.
mar	Sets margin size in number of lines. The default setting is: `par ( mar (5,4,4,2)+0.1 )` which sets number of margin lines in order: (bottom, left, top, right). An example that sets label angles to perpendicular and provides eight lines on the left side is: `par ( las = 2, mar=c(4,8,4,1)+0.1 )`.
mfrow	An argument to the `par` function, it sets up a *multi*/*frame* plot to contain several other plots. R will plot to them left to right, top to bottom. Example: `par ( mfrow=c (3,2) )` yields three rows of two plots. Example: `par ( mfrow=c (1,1) )` returns setting to single plot.
mfcol	Works like `mfrow` but writes plots in *co*l*umns* from top to bottom, left to right.
new	Setting `par (new = TRUE)` tells R that a new plot has already been started so that it will not erase what is there before adding to it.
par()	Will display all traditional graphics parameters.
ps	Sets the *point size* of text. For example, `par (ps=12)` selects 12-point text.
usr	Shows the user coordinates of a plot in the form xstart, xstop, ystart, ystop.
xlog	Setting `par (xlog=TRUE)`, log transformed *x*-axis, including tick marks.
ylog	Setting `par (ylog=TRUE)`, log transformed *y*-axis, including tick marks.

Table 21.3 Graphics parameters for both par and graphics functions

adj	Sets justification of text with 0 = left, 0.5 = center, 1 = right.
cex	Sets size of text as a multiplication factor. For example, cex = 1.5 would make all the text in a plot 50% larger. You can control individual elements via `cex.axis`, `cex.lab` (for axis labels), `cex.main`, and `cex.sub` (latter two for main and subtitles.)
col	Sets *col*or of bars, lines, and points. You can control individual elements via `col.axis`, `col.lab` (for axis labels), `col.main`, and `col.sub` (main and subtitles). Enter `colors()` to see them all. Example: `barplot( table(gender), col=c("gray90","gray60"))`
font	Sets font for text elements. You can control individual elements with font.axis, font.lab, font.main, font.sub. Levels are: 1 = plain, 2 = bold, 3 = italic, 4 = bold and italic, 5 = symbol. Example: `plot (x,y...font.main=3,main="My Italic Title")`
las	Sets label angles in margins. Levels are: 0 = text parallel to its axis; 1 = horizontal; 2 = text perpendicular to its axis; 3 = vertical. Example: `par ( las=2)`
lty	The line type argument applies to functions that create lines such as `abline` and `lines`, as well as the `par` and `legend` functions. The types of lines you can request are: 0 = blank; 1 = solid; 2 = dashed; 3 = dotted; 4 = dot-dash; 5 = long dash; 6 = two dashes. Example: `lines ( lowess (posttest ~ pretest), lty=3 )`
lwd	Determines line width. The default value is 1 and the resulting width varies by graphics device. Example: `lines ( lowess (posttest ~ pretest), lwd=2)`
mtext	A function that adds text to the *m*argins of a plot. Its arguments are the line of text from 0 to N, the side of the margin (1 = bottom, 2 = left, 3 = top, 4 = right) and the "at" value, which is in terms of the axis scale itself. Example: `mtext ("line=2", side=2, line=2, at=65 )`
pch	An argument to function that plot *p*oint *ch*aracters like a scatterplot. "." = tiny dot, good for large datasets; 1 = circle; 2 = triangle up; 3 = +; 4 = X; 5 = diamond; 6 = triangle pointing down see? points for more. Example: `plot (pretest,posttest,pch=2)`
srt	Sets rotation of text in plot region. Example text at 45-degree angle: text (65,85,"My Tilted Text", srt = 45)

Table 21.4 Graphics functions to add elements to *existing* plots

abline	A function that adds straight line(s) to an existing plot. Example of intercept zero, slope 1: `abline (a=0, b=1, lty=5)` Example of linear model line: `abline ( lm(posttest ~ pretest), lty=1)`
arrows	A function that draws an arrow with the arguments (fromX, fromY, toX, toY). The optional length argument sets the length of the arrowhead lines. Example: `arrows (65,85,58.5,60.5, length=0.1)`
axis	A function that adds an axis to an existing plot. Example: `axis (4)` adds it to the right side (1 = bottom, 2 = left, 3 = top, 4 = right).
box	A function that adds a box around an existing plot. Example: `box ()`
grid	A function that adds a set of vertical and horizontal lines to an existing plot. Example: `grid()`

	Table 21.4 (continued)
lines	A function that applies line(s) to an existing plot. Example of lowess fit: `lines( lowess(posttest ~ pretest), lty=3)`
text	A function that adds text to an existing plot. Its first two arguments are the *x, y* position of the plot, the text and finally the position of the text. Its pos argument positions text relative to *x, y* values with 1 = bottom, 2 = left, 3 = top, 4 = right. Example: `text (65,85,"Fit is linear", pos=3)`

21.15 Plot Demonstrating Many Modifications

Below is a program that creates a rather horrible looking plot (Fig. 21.48), but it does demonstrate many of the options you are likely to need. The very repetitive `mtext` functions that place labels all over the margins could be done with a loop, making it much more compact but less clear to beginners unfamiliar with loops.

```
par( mfrow=c (1,1)  )
par(family="serif")

plot(  (pretest,posttest,
  main="My Main Title" ,
  xlab="My xlab text" ,
  ylab="My ylab text",
  sub="My subtitle",
  pch=2)

text(66, 88,  "My Example Formula")
text(65, 85,
  expression( hat (beta)  ==
  (X^t  * X)^{-1}  * X^t * Y)  )

text  (85,65, "My label with arrow", pos=3)
arrows(85,65,64,62, length=0.1)
abline(h=75,v=75)
abline(a=0, b=1, lty=5)
abline( lm(posttest~pretest),     lty=1 )
lines( lowess(posttest~pretest), lty=3 )
legend( 64, 99,
  c( "Regression", "Lowess", "Posttest =Pretest" ),
  lty=c(1,3,5)  )

mtext("line=0", side=1, line=0, at=57 )
mtext("line=1", side=1, line=1, at=57 )
mtext("line=2", side=1, line=2, at=57 )
mtext("line-3", side-1, line=3, at=57 )
mtext("line=4", side=1, line=4, at=57 )
```

```
mtext("line=0", side=2, line=0, at=65  )
mtext("line=1", side=2, line=1, at=65  )
mtext("line=2", side=2, line=2, at=65  )
mtext("line=3", side=2, line=3, at=65  )

mtext("line=0", side=3, line=0, at=65  )
mtext("line=1", side=3, line=1, at=65  )
mtext("line=2", side=3, line=2, at=65  )
mtext("line=3", side=3, line=3, at=65  )

mtext("line=0", side=4, line=0, at=65  )
mtext("line=1", side=4, line=1, at=65  )
```

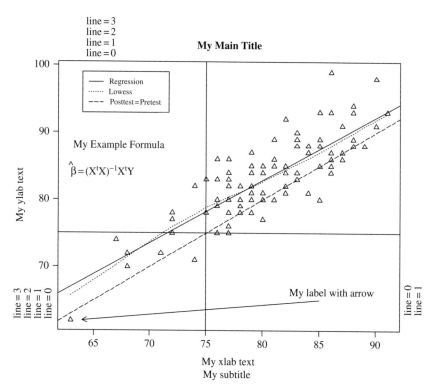

Fig. 21.48 A plot demonstrating many types of text and line annotations. The "line = n" labels around the margins display how the line numbers start at zero right next to each axis and move outward as the line numbers increase

21.16 Example Traditional Graphics Programs

The SAS and SPSS examples in Table 21.5 below are particularly sparse compared to those for R. This is due to space constraints rather than due to lack of capability. The SPSS examples below are done only using their legacy graphics. I present a parallel set of examples using SPSS GPL in the next chapter.

Table 21.5 Example traditional graphics programs

SAS programming statements

```
* SAS Program for Basic Graphics;
* GraphicsTraditional.sas ;

LIBNAME myLib 'C:\myRfolder';
OPTIONS _LAST_=myLib.mydata100;

* Histogram of q4;
PROC GCHART; VBAR q4; RUN;

* Bar charts of workshop & gender;
PROC GCHART; VBAR workshop gender; RUN;

* Scatterplot of pretest by posttest;
PROG GPLOT; PLOT posttest*pretest; RUN;

* Scatterplot matrix of all vars but gender;
* Gender would need to be recoded as numeric;

PROC INSIGHT;
SCATTER workshop q1-q4 * workshop q1-q4;
RUN;
```

SPSS programming statements

```
* SPSS Program for Basic Graphics.
* GraphicsTraditional.sps .

CD 'C:\myRfolder'.
GET FILE='mydata100.sav'.

* Legacy SPSS commands for histogram of q4.
GRAPH  /HISTOGRAM=q4 .

* Legacy SPSS commands for bar chart of gender.
GRAPH  /BAR(SIMPLE)=COUNT BY gender .

* Legacy syntax for scatterplot of q1 by q2.
GRAPH /SCATTERPLOT(BIVAR)=pretest WITH posttest.
```

Table 21.5 (continued)

```
* Legacy SPSS commands for scatterplot matrix
* of all but gender. * Gender cannot be used until

* it is recoded numerically.
GRAPH  /SCATTERPLOT(MATRIX)=workshop q1 q2 q3 q4.

execute.
```

R programming statements

```
#Example Programs for Traditional Graphics in R.
#GraphicsTraditional.R

setwd("/myRfolder")
load(file="mydata100.Rdata")
attach(mydata100)
options(width=64)

# Request it to ask you to click for new graph.
par(ask=FALSE, mfrow=c(1,1) )

#—Barplots—

# Barplots of counts via table

barplot( c(60,40) )

barplot(q4)
table(q4)
barplot( table(q4) )
barplot( workshop )
barplot( table(workshop) )
barplot(gender)
barplot( table(gender) )

barplot( table(workshop), horiz=TRUE)

barplot( as.matrix( table(workshop) ),
  beside = FALSE)

# Grouped barplots & mosaic plots

barplot( table(gender,workshop) )

plot( workshop,gender )

mosaicplot( table(workshop,gender) )

mosaicplot(~ Sex + Age + Survived,
  data = Titanic, color = TRUE)
```

Table 21.5 (continued)

```
barplot( table(gender,workshop), beside=TRUE )

# Barplots of means via tapply

myMeans <- tapply(q1, gender, mean, na.rm=TRUE)
barplot(myMeans)

myMeans <- tapply(q1, list(workshop,gender), mean,na.rm=TRUE)
barplot(myMeans, beside=TRUE)

#—Adding main title, color and legend—

barplot( table(gender,workshop),
  beside=TRUE,
  col=c("gray90","gray60"),
  main="Number of males and females \nin each workshop" )
legend( "topright",
  c("Female","Male"),
  fill=c("gray90","gray60") )

# A manually positioned legend at 10,15.
legend( 10,15,
  c("Female","Male"),
  fill=c("gray90","gray60") )

#—Mulitple Graphs on a Page—

par()
head( par() )

par( mar=c(3,3,3,1)+0.1 )
par( mfrow=c(2,2) )
barplot( table(gender,workshop) )
barplot( table(workshop,gender) )
barplot( table(gender,workshop), beside=TRUE )
barplot( table(workshop,gender), beside=TRUE )

par( mfrow=c(1,1) ) #Sets back to 1 plot per page.
par( mar=c(5,4,4,2)+0.1 )

#—Dotcharts—

dotchart( table(workshop,gender),
  main="Dotchart of Workshop Attendance",
  cex=1.5)

#—Piecharts—
pie( table(workshop),
  col=c("white","gray90","gray60","black" ),
  main="Piechart of Workshop Attendance" )
```

Table 21.5 (continued)

```
# —Histograms—

hist(posttest)

# More bins plus density and ticks at values.
hist(posttest, breaks=20, probability=TRUE,
  main="Histogram with Density and Points")
lines( density(posttest) )
myZeros <- rep(0, each=length(posttest) )
myZeros
points( posttest, myZeros, pch="|" )

# Histogram of males only.
hist( posttest[ which(gender=="Male") ],
  col="gray60")

# Plotting above two on one page,
# matching breakpoints.
par(mfrow=c(2,1) )
hist(posttest, col="gray90",
  breaks=c(50,55,60,65,70,75,80,85,90,95,100) )
hist(posttest[ which(gender=="Male") ],
  col="gray60",
  breaks=c(50,55,60,65,70,75,80,85,90,95,100) )
par(mfrow=c(1,1) )

# Could have used either of these:
# breaks=seq(from=50, to=100, by=5) )
# breaks=seq(50,100,5) )

# Histograms overlaid

hist( posttest, col="gray90",
  breaks=seq(from=50, to=100, by=5) )
hist(posttest[ which(gender=="Male") ],
  col="gray60",
  breaks=seq(from=50, to=100, by=5),
  add=TRUE )
legend( "topright", c("Female","Male"),
  fill=c("gray90","gray60") )

# Same plot but extracting $breaks
# from previous graph.

myHistogram <- hist(posttest, col="gray90")
names(myHistogram)
myHistogram$breaks
myHistogram$xlim
hist(posttest[ which(gender=="Male") ],
  col=' gray60' ,
  add=TRUE, breaks=myHistogram$breaks)
```

Table 21.5 (continued)

```
legend( "topright", c("Female","Male"),
  fill=c("gray90","gray60") )

# What else does myHistogram hold?
class(myHistogram)
myHistogram

#—Q-Q plots—

library("car")
qq.plot(posttest,
  labels=row.names(mydata100),
  col="black" )
detach("package:car")
myQQ <- qqnorm(posttest) #Not shown in text.
identify(myQQ)

#—Stripcharts—

par( mar=c(4,3,3,1)+0.1 )
par(mfrow=c(2,1) )
stripchart(posttest, method="jitter",
  main="Stripchart with Jitter")
stripchart(posttest, method="stack",
  main="Stripchart with Stacking")
par( mfrow=c(1,1) )
par( mar=c(5,4,4,2)+0.1 )

par( las=2, mar=c(4,8,4,1)+0.1 )
stripchart(posttest~workshop, method="jitter")
par( las=0, mar=c(5,4,4,2)+0.1 )

# — Scatterplots —

plot(pretest,posttest)

# Find low score interactively.
# Click 2nd mouse button to choose stop.
identify(pretest,posttest)

# Check it manually.
mydata100[ pretest<60, ]

# Different types of plots.
par( mar=c(5,4,4,2)+0.1 )
par( mfrow=c(2,2) )
plot( pretest, posttest, type="p", main="type=p" )
plot( pretest, posttest, type="l", main="type=l" )
plot( pretest, posttest, type="b", main="type=b" )
plot( pretest, posttest, type="h", main="type=h" )
par( mfrow=c(1,1) )
```

Table 21.5 (continued)

```
# Scatterplots with Jitter

par ( mar=c (5,4,4,2)+0.1 )
par ( mfrow=c (1,2) )
plot ( q1, q4,
  main="Likert Scale Without Jitter")
plot ( jitter (q1,3), jitter (q4,3),
  main="Likert Scale With Jitter")

# Scatterplot of large data sets.

# Example with pch="." and jitter.
par (mfrow=c (1,2) )
pretest2  <- round ( rnorm ( n=5000, mean=80, sd=5) )
posttest2 <-
  round ( pretest2 + rnorm ( n=5000, mean=3, sd=3) )
pretest2[ pretest2>100] <- 100
posttest2[ posttest2>100] <- 100
plot ( pretest2, posttest2,
  main="5,000 Points, Default Character \nNo Jitter")
plot ( jitter (pretest2,4), jitter (posttest2,4), pch=".",
  main="5,000 Points Using pch=' .' \nand Jitter")
par (mfrow=c (1,1) )

# Hexbins (resets mfrow automatically).
library ("hexbin")
plot ( hexbin (pretest2,posttest2),
  main="5,000 Points Using Hexbin")
detach ("package:hexbin")

rm (pretest2,posttest2) # Cleaning up.
# Scatterplot with different lines adde d.
plot (posttest~pretest)
abline (h=75,v=75)
abline (a=0, b=1, lty=5)
abline ( lm (posttest~pretest),   lty=1 )
lines ( lowess (posttest~pretest), lty=3 )
legend ( 60, 95,
  c ( "Regression", "Lowess", "Posttest=Pretest" ),
  lty=c (1,3,5) )

# Scatterplot of q1 by q2 separately by gender.
plot (posttest~pretest,
 pch=as.numeric (gender) )

abline ( lm ( posttest[ which (gender=="Male") ]
       ~ pretest[ which (gender=="Male") ] ),
       lty=1 )

abline ( lm ( posttest[ which (gender=="Female") ]
```

Table 21.5 (continued)

```
            ~ pretest[ which(gender=="Female") ] ),
         lty=2 )

legend( "topleft", c("Male","Female"),
        lty=c(1,2), pch=c(2,1) )

# Coplots: conditioned scatterplots
coplot( posttest~pretest | workshop)
coplot( posttest~pretest | q1)

# Scatterplot plotting text labels.

plot(pretest, posttest,
   pch=as.character(gender) )

plot(pretest, posttest, type="n" )
text(pretest, posttest,
   label=row.names(mydata100) )

# Scatterplot matrix of whole data frame.
plot(mydata100[ 3:8] ) #Not shown with text.
plot(mydata100[ 3:8] , gap=0, cex.labels=0.9)

pairs(mydata100[ 3:8] , gap=0,
   lower.panel=panel.smooth,
   upper.panel=panel.smooth)

# Dual axes
#Adds room for label on right margin.
par( mar=c(5,5,4,5) )
plot( pretest, posttest, axes=FALSE, xlim=c(55,90),
   main="Scatterplot with Dual Axes" )
axis(4)
mtext("Axis label on right side",
   font=2, side=4, line=3)
par(new=TRUE)
plot(pretest, posttest, xlim=c(55,90) )

# Scatterplot with Confidence Ellipse.
library("car")
data.ellipse(pretest, posttest,
   levels=.95,
   col="black")
detach("package:car")

# Confidence Intervals: A small example
x    <-   c  (1,2,3,4)
y1   <-   c  (1,2,3,4)
y2   <-   c  (2,3,4,5)
y3   <-   c  (3,4,5,6)
yMatrix  <-   cbind  (y1,y2,y3)
```

Table 21.5 (continued)

```
yMatrix
plot( x, y2, xlim=c(1,4), ylim=c(1,6), cex=1.5 )
matlines( x, yMatrix, lty=c(2,1,2), col="black" )
rm( x, y1, y2, y3, yMatrix)

# Confidence Intervals: A realistic example
myIntervals <-
  data.frame(pretest=seq(from=60, to=100, by=5))
myIntervals
myModel <- lm( posttest~pretest )
myIntervals$pp <- predict( myModel,
  interval="prediction", newdata=myIntervals)
myIntervals$pc <- predict( myModel,
  interval="confidence", newdata=myIntervals)
myIntervals
class( myIntervals$pp )
myIntervals$pp
plot( pretest, posttest,
  ylim=range( myIntervals$pretest,
   myIntervals$pp, na.rm=TRUE),
 main="Regression Fit with Confidence Intervals" )
matlines(myIntervals$pretest, myIntervals$pc,
  lty=c(1,2,2), col="black" )
matlines(myIntervals$pretest, myIntervals$pp,
  lty=c(1,3,3), col="black" )

#—Boxplots—
plot(workshop, posttest,
  main="Boxplot using plot function")

par( mfrow=c(2,2) )
boxplot(posttest)
boxplot(pretest,posttest,notch=TRUE)
boxplot(posttest~workshop)
par( las=2, mar=c(8,4,4,2)+0.1 )
boxplot(posttest~workshop:gender)
par( las=1, mar=c(5,4,4,2)+0.1 )

#—Error bar plots—
library("gplots")
par( mfrow=c(1,1) )
plotmeans( posttest~workshop,
  main="Plotmeans from gplots Package")
detach("package:gplots")

interaction.plot( workshop, gender, posttest,
  main="Means Using interaction.plot function")

# —Adding Labels—

# Many annotations at once.
```

Table 21.5 (continued)

```
par( mar=c(5,4,4,2)+0.1 )
par( mfrow=c(1,1) )
par(family="serif")
plot(pretest,posttest,
   main="My Main Title" ,
   xlab="My xlab text" ,
   ylab="My ylab text",
   sub="My subtitle ",
   pch=2)

text(66, 88, "My Example Formula")
text(65, 85,
 expression( hat(beta) ==
   (X^t * X)^{ -1} * X^t * Y) )

text (85,65,"My label with arrow", pos=3)
arrows(85,65,64,62, length=0.1)
abline(h=75,v=75)
abline(a=0, b=1, lty=5)
abline( lm(posttest~pretest),   lty=1 )
lines( lowess(posttest~pretest), lty=3 )
legend( 64, 99,
 c( "Regression", "Lowess", "Posttest=Pretest" ),
 lty=c(1,3,5) )

mtext("line=0", side=1, line=0, at=57 )
mtext("line=1", side=1, line=1, at=57 )
mtext("line=2", side=1, line=2, at=57 )
mtext("line=3", side=1, line=3, at=57 )
mtext("line=4", side=1, line=4, at=57 )

mtext("line=0", side=2, line=0, at=65 )
mtext("line=1", side=2, line=1, at=65 )
mtext("line=2", side=2, line=2, at=65 )
mtext("line=3", side=2, line=3, at=65 )

mtext("line=0", side=3, line=0, at=65 )
mtext("line=1", side=3, line=1, at=65 )
mtext("line=2", side=3, line=2, at=65 )
mtext("line=3", side=3, line=3, at=65 )

mtext("line=0", side=4, line=0, at=65 )
mtext("line=1", side=4, line=1, at=65 )

#—Scatterplot with bells & whistles—
plot(pretest,posttest,pch=19,
   main="Scatterplot of Pretest and Postest",
   xlab="Test score before taking workshop",
   ylab="Test score after taking workshop" )
myModel <- lm(posttest~pretest)
abline(myModel)
```

Table 21.5 (continued)

```
arrows(60,82,65,71, length=0.1)
text(60,82,"Linear Fit", pos=3)
arrows(70,62,58.5,59, length=0.1)
text(70,62,"Double check this value", pos=4)
# Use locator() or:
# predict(myModel,data.frame(pretest=75) )
```

Chapter 22
Graphics with `ggplot2` (GPL)

As we discussed in Chap. 20, the `ggplot2` package is an implementation of Wilkinson's Grammar of Graphics. SPSS implements these concepts in its Graphics Production Language (GPL). Chapter 21 focused on R's traditional graphics functions. Many plots were easy, but other plots seemed to be quite a bit of work compared to SAS or SPSS. In particular, adding things like legends and confidence intervals were rather complicated. The `ggplot2` package makes those things easier as you will now see as we replicate many of the same graphs. Since `ggplot2` has a shorter `qplot` function (also called `quickplot`) and a more powerful `ggplot` function, we will use both so you can learn the difference. The built-in `lattice` package is also capable of doing these examples well.

While traditional graphics come with R, you will need to install the `ggplot2` package. For details, see Chap. 5. Once installed, we need to load the package using the `library` function.

```
> library("ggplot2")

Loading required package: grid
Loading required package: reshape
Loading required package: proto
Loading required package: splines
Loading required package: MASS
Loading required package: RColorBrewer
Loading required package: colorspace
```

Notice that it requires the grid package. That is a completely different graphics system than the traditional graphics system. That means that the `par` function used to set graphics parameters like fonts in the last chapter does not work with `ggplot2`. Nor do any of the base functions that we have covered, including abline, arrows, axis, box, grid, lines, and text.

22.1 Overview qplot and ggplot

With the ggplot2 package, you create your graphs by specifying the elements below.

Aesthetics – the aesthetics map your data to the graph telling it what role each variable will play. Some variables will map to an axis, some will determine the color, shape, or size of a point in a scatterplot. Different groups might have differently shaped or colored points. The size or color of a point might reflect the magnitude of a third variable. Other variables might determine how to fill the bars of a bar chart with colors or patterns so, for example, you can see the number of males and females within each bar.

Geoms – short for *geom*etric objects; geoms determine the objects that will represent the data values. Possible geoms include: bar, boxplot, errorbar, histogram, jitter, line, path, point, smooth, and text.

Statistics – provide functions for features like adding regression lines to a scatterplot, or dividing a variable up into bins to form a histogram.

Scales – match your data to the aesthetic features, for example in a legend that tells us that triangles represent males and circles represent females.

Coordinate system – For most plots this is the usual rectangular Cartesian coordinate system. However, for pie charts it is the circular polar coordinate system.

Facets – these describe how to repeat your plot for each subgroup, perhaps creating a separate scatterplot for males and females. A helpful feature with facets is that they standardize the axes on each plot, making comparisons across groups much easier.

The qplot function tries to simplify graph creation by (a) allowing you to skip specifying as many of the items above as possible and (b) looking a lot like the traditional plot function. As usual, you do not have to name the x and y arguments with the prefix "x=" or "y=" so long as you supply them in that order. Nor do you have to specify the data frame if it is attached. Finally, as you would expect, elements are specified by an *argument*. For example, geom= "bar".

The ggplot function sacrifices those R standards in exchange for a more complete implementation of the grammar of graphics. It requires you to specify the data frame, since you can use different data frames in different layers of the graph. Its options are specified by additional *functions* rather than the usual arguments. Rather than the geom= "bar" format of qplot, they follow the form +geom_bar(options). The form is quite consistent so if you know there is a geom named "smooth" you can readily guess how to specify it in either qplot or ggplot.

Although SPSS' GPL is also an implementation of the grammar of graphics, the ggplot2 package differs in several important ways. It depends upon R's ability to transform data, so you can use `log(x)` or any other function within `qplot` or `ggplot`. It also uses R's ability to reshape or aggregate data, so `ggplot2` does not include its own algebra for these steps as GPL does. Also, `ggplot2` displays axes and legends automatically so there is no "guide" function. (Table 22.1) summarizes the major differences between `qplot` and

Table 22.1 Comparison of the `qplot` and `ggplot` functions

	The qplot function	The ggplot function
Goal	Designed to be as quick and standard R as possible.	Designed to get the full flexibility of the grammar of graphics concept.
Specifying variable aesthetics	Like most R functions: `qplot( myVar...` `qplot( x=myVar...` `qplot(...fill=myVar...` `qplot(...color=myVar...` `qplot(...shape=myVar...`	You must specify the mapping between each graphical element, even *x* and *y* axes, and the variable(s): `ggplot(data=mydata, aes(` `x=myVar,` `y=myVar,` `fill=myVar,` `color=myVar,` `shape=myVar) )`
Specifying data frame	Optional as with most R functions. That is, it finds attached variables.	You *must* specify `data=` argument. `ggplot(data=mydata, aes(...`
Geom abline	`,geom="abline",` `intercept=a,` `slope=b`	`geom_abline(intercept=a,` `slope=b)`
Geom bars	`...geom="bar",` `position="stack"` or dodge.	`+geom_bar(position="stack")` or dodge.
Geom points	`...geom="point"` That is the default.	`+geom_point(size=2)` There is no default geom for ggplot. The default size is 2.
Geom histogram	`...geom="histogram"` `binwidth=1`	`+geom_bar(binwidth=1)`
Geom density	`...geom="density"`	`+geom_density()`
Geom jitter	`...geom="jitter"`	`+geom_jitter()`
Geom line	`geom="line"`	`+geom_line()`
Geom vline	`geom="vline",` `intercept=?`	`geom_vline(intercept=?)`
Geom hline	`geom="hline",` `intercept=?`	`geom_hline(intercept=?)`
Geom Smooth	`geom="smooth",` `method="lm"` Lowess is default method.	`geom_smooth(method="lm")` Lowess is default method.
Geom Smooth w/o Confidence	`geom="smooth",` `method="lm",` `se=FALSE`	`geom_smooth(method="lm",` `se=FALSE)`

Table 22.1 (continued)

	The qplot function	The ggplot function
Titles	Just like plot function: ...main="My Title"	+ opts(title="My Title")
Axis labels	Just like plot function: ...xlab="My Text"	+scale_x_discrete("My Text") +scale_y_discrete("My Text") +scale_x_continuous("My...") +scale_y_continuous("My...")
Creating plots for subgroups	..., facets=gender ~.	+ facet_grid(gender ~ .)
Filling bars	...posttest, fill=gender	aes(x=posttest, fill=gender))
Axis flipping	+coord_flip()	+coord_flip()
Axis logarithmic	+scale_x_log10() +scale_x_log2() +scale_x_log()	+scale_x_log10() +scale_x_log2() +scale_x_log()
Pie (polar coordinates)	+coord_polar (theta="y")	+coord_polar(theta="y")
Fill grey	+scale_fill_grey (start=0, end=1)	+scale_fill_grey(start=0,end=1)
Stat QQ	stat="qq"	stat_qq=()
Aspect ratio	Leave out for interactive adjustment. + coord_equal(ratio= height/width) +coord_equal() is square	Leave out for interactive adjustment. + coord_equal(ratio= height/width) +coord_equal() is square

ggplot. For a more detailed comparison, see the book, ggplot, by Hadley Wickham [28].

Now let us look at some examples. Each is done using both qplot and ggplot. You can decide which you prefer.

22.2 Bar Charts

Let us do a simple bar chart of counts for our workshop variable (Fig 22.1.). Both of the following commands will do it. First, let us look at the qplot approach.

```
> attach(mydata100)
> qplot(workshop, geom="bar")
```

Here is the same plot done using the ggplot function. Notice that the first line ends with a "+" sign that you have to enter. Since R prompts you with "+"

Fig. 22.1 A barplot of
workshop attendance

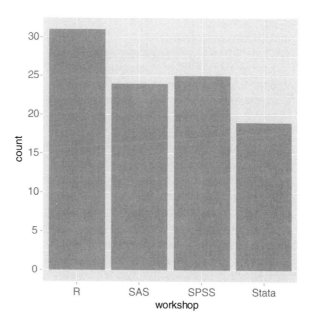

at the beginning of a continued line, that looks a bit confusing at first. The plus
on the left R typed; the plus on the right, you type.

```
> ggplot(mydata100, aes(workshop) ) +
+   geom_bar()
```

Both methods look pretty easy.

The qplot function call above has only two arguments.

1. The variable workshop. It appears in order so we do not need to say
   x = workshop. The qplot function finds attached data, so we need not
   add data = mydata100, since we have attached it.
2. The argument geom = "bar" tells it to make this a bar chart.

   The ggplot function call above uses three arguments.

1. Unlike most other R functions, it requires that you specify the data frame.
   As we will see later, that is because ggplot can plot multiple layers, and
   each layer can use a different data frame.
2. The aes function maps the variable workshop to the x-axis.
3. The geom_bar() function tells it to make this a bar chart. This function is
   tied to the first one through the "+" sign.

   We did the same plot using traditional graphics, but they required us to
summarize the data using table(workshop). The ggplot2 package is
more like SAS and SPSS in this regard; it does that type of summarization
for you.

Fig. 22.2 A horizontal
barplot demonstrating the
impact of the
`coord_flip()` function

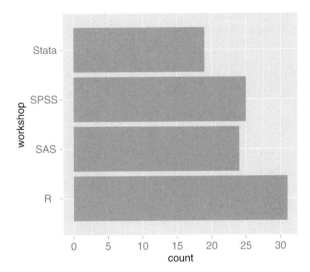

If we want to change to a horizontal bar chart (Fig. 22.2), all we need to do is
flip the coordinates. It is pretty easy to see how it works. With this task, it is
clear we simply added the `cord_flip` function to the end of both `qplot` and
`ggplot`. There is no argument to `qplot` like `coord="flip"`. This brings up
an interesting point. Both methods create the exact same graphics object. Even
if there is a `qplot` equivalent, you can always add a `ggplot` function to the
`qplot` function.

The `qplot` approach:

```
> qplot(workshop, geom="bar") + coord_flip()
```

The `ggplot` approach:

```
> ggplot(mydata100, aes(workshop) ) +
+   geom_bar() + coord_flip()
```

You can create the usual types of grouped barplots. Let us start with a simple
stacked one (Fig. 22.3). You can use either function below. They contain only two
new arguments. Although we are requesting only a single bar, we must still supply
a variable for the x-axis. The function call `factor("")` provides the variable we
need, and it has simply a factor value that is empty. We use the `factor` function
to keep it from labeling the x-axis from 0 to 1, which it would do if the variable
were continuous. The `fill=workshop` aesthetic argument tells the function to
fill the bars with the number of students who took each workshop.

On qplot, we are clearing labels on the x-axis with xlab = "". Otherwise the
word "factor" would occur there from our `factor("")` statement. The
equivalent way to label ggplot is to use the `scale_x_discrete` function,
also providing an empty label for the x-axis.

Fig. 22.3 A stacked barplot of workshop attendance

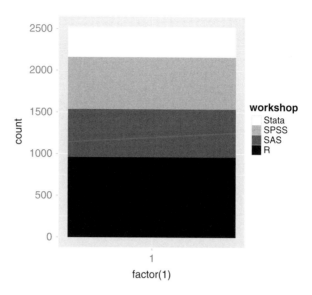

Finally, the scale_fill_grey function tells each function to use shades of grey. You can leave this out of course and both functions will choose the same nice color scheme. The start and end values tell the function to go all the way to black and white, respectively. If you use just scale_fill_grey(), it will use four shades of grey.

The qplot approach:

```
> qplot(factor(""), fill=workshop,
+    geom="bar", xlab="") +
+    scale_fill_grey(start=0, end=1)
```

The ggplot approach:

```
> ggplot(mydata100,
+    aes(factor(""), fill=workshop) ) +
+    geom_bar() +
+    scale_x_discrete("") +
+    scale_fill_grey(start=0, end=1)
```

22.3 Pie Charts

One interesting aspect to the Grammar of Graphics concept is that a pie chart (Fig. 22.4) is just a single-stacked bar chart drawn in polar coordinates. So we can use the same commands as we used for the bar chart above, but convert to polar afterwards using the coord_polar function. This is a plot that only the ggplot function can do correctly. The geom_bar(width=1) function call

Fig. 22.4 A pie chart of
workshop attendance

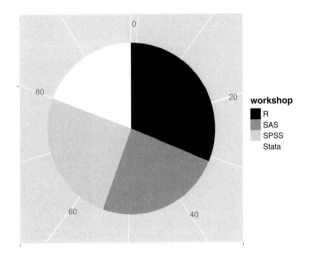

tells it to put the slices right next to each other. If you included that on a
standard bar chart, it would also put the bars right next to each other.

```
> ggplot(mydata100,
+    aes( factor(""), fill=workshop ) ) +
+    geom_bar( width=1 ) +
+    scale_x_discrete("") +
+    coord_polar(theta="y") +
+    scale_fill_grey(start=0, end=1)
```

22.4 Bar Charts with Subgroups

Let us now look at bar charts that handle subgroups. This requires having factors
named for both the x argument and the `fill` argument. By default, the `posi-
tion` argument stacks the fill groups, in this case the workshops. That graph is
displayed in the upper left frame of Fig. 22.5. Changing to `position = "fill"`
makes every bar fill the y-axis, which then converts from displaying number to
proportion. That type of graph is called a spine plot and it is displayed in the upper
right of Fig. 22.5. Finally, if you set `position = "dodge"` the filled segments
appear beside one another, dodging each other. That takes a bit more room on the
x-axis, so it appears across the whole bottom row of Fig. 22.5. We will discuss how
to do multiframe plots in Sect. 22.25.

The `qplot` approach:

```
> qplot(gender, geom="bar",
+    fill=workshop, position="stack") +
+    scale_fill_grey(start=0, end=1)
```

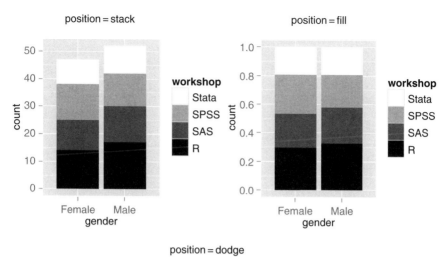

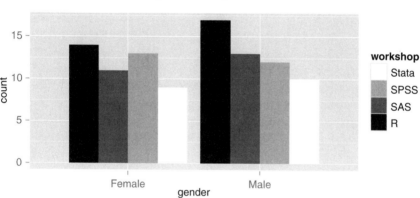

Fig. 22.5 A multiframe plot showing the impact of the various position settings

The ggplot approach:

```
> ggplot(mydata100, aes(gender, fill=workshop) ) +
+    geom_bar(position="stack") +
+    scale_fill_grey( start=0, end=1 )
```

22.5 Plots by Group or Level

One of the nicest features of the ggplot2 package is its ability to easily plot subgroups within a single plot (Fig. 22.6). To fully appreciate all the work it is doing for us, let us consider one way we could do this with traditional graphics functions. We would set up a multiframe plot, say for males and females. Then we might create a bar chart on workshop, selecting gender = = "Female" and

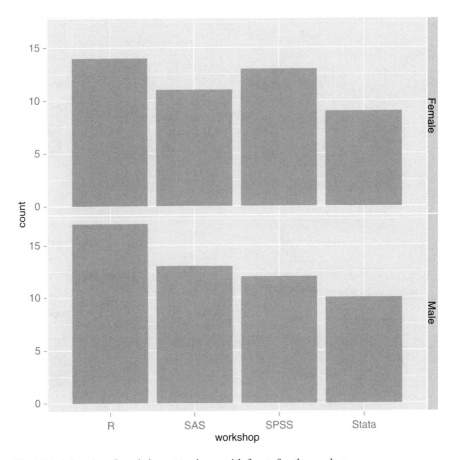

Fig. 22.6 A barplot of workshop attendance with facets for the genders

another for males. We probably also would want to standardize the axes to better enable comparisons and add legend. All those steps become one using either of the following commands:

The qplot approach:

```
> qplot(workshop, geom="bar", facets=gender ~ . )
```

The ggplot approach:

```
> ggplot(mydata100, aes(workshop) ) +
+   geom_bar() + facet_grid( gender ~ . )
```

The new feature is the facets argument in qplot and the facet_grid function in ggplot. The formula it uses is in the form "rows ~ columns." In this case we have gender ~ . so we will get rows of plots for each gender and no columns. The "." represents "1" row or column. If we instead did . ~ gender we would have one row and two columns of plots side-by-side. You can extend this idea with the various rules for formulas described in Sect. 8.5.2. Given the constraints of space, the most you are likely to find useful is the addition of one more variable, such as `facets=workshop ~ gender`. In our current example, that leaves us little to plot, but we will look at a scatterplot example of that later.

22.6 Pre-summarized Data

I mentioned earlier that the `ggplot2` package assumed your data needed summarizing, which is the opposite of the traditional R graphics functions. But what if the data are already summarized? The `qplot` function makes it quite easy to deal with as you can see in the program below and in the resulting Fig. 22.7. However, the `ggplot` function requires the data to be in a data frame. I find it much easier to create a temporary data frame containing the summary data. Trying to nest a data frame creation within the `ggplot` function will work, but you end up with so many parentheses that it can be a challenge.

The `qplot` approach:

```
> qplot( factor(c(1,2)), c(40, 60), geom="bar")
```

Here is the much more complicated approach used by `ggplot`. Either approach result in axis labels that are not very good. See Sec. 22.8 to see how to change those.

```
>   myTemp <- data.frame( myX=factor( c(1,2) ), myY=
      c(40, 60) )
```

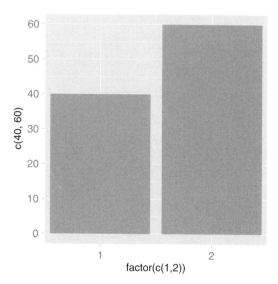

Fig. 22.7 A barplot of pre-summarized data with default axis labels

```
>   myTemp

    myX     myY
1    1       40
2    2       60

>   ggplot(data=myTemp, aes(myX, myY) ) +
+       geom_bar()

>   rm(myTemp) #Cleaning up.
```

22.7 Dotcharts

Dotcharts are bar charts reduced to just points on lines, so you can take any of
the above bar chart examples and turn into dotcharts (Fig. 22.8). Dotcharts are
particularly good at packing in a lot of information on a single plot, so let us
look at the counts for the attendance in each workshop, for both males and
females. This example points out how very different qplot and ggplot can
be. It also shows how flexible ggplot is and that it is sometimes much easier to
understand than qplot.

First, let us look at how qplot does it. The variable workshop is in the x
position, so this is the same as saying x = workshop. If you look at Figure 22.8,

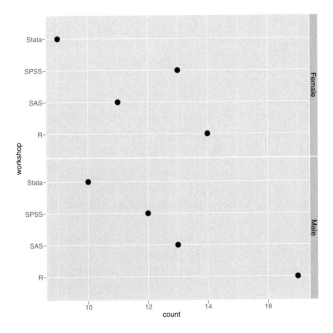

Fig. 22.8 A dotchart of workshop attendance with facets for the genders

workshop is on the *y*-axis. However, qplot requires an x variable so we cannot simply say y=workshop and not specify an x variable. Next it specifies stat = "bin", which applies to the default geom = "point". Not having to list that default saves time but also obscures the relationship between the two arguments. Finally, we have to add the coord_flip function to get it in the direction we desire.

```
> qplot(workshop, stat="bin",
+          facets=gender ~ ., size=4 ) +
>    coord_flip()
```

Now let us look at two ways ggplot can do the same plot. The aes function supplies the x variable, and the y variable uses the special ..count.. computed variable. Then you have a choice. You can either plot the point geoms with the stat = "bin" setting as shown in the code below, or you can use the stat_bin function with the geom = "point" argument. Those are two ways of saying the same thing. The coord_flip function then reverses the axes. Finally, the facet_grid function specifies the same formula used above in qplot.

The ggplot approach:

```
> ggplot(mydata100,
+    aes(workshop, ..count..) ) +
+    geom_point(stat="bin", size=4) + coord_flip()+
+    facet_grid( gender~. )
```

22.8 Adding Titles and Labels

Sprucing up your graphs with titles and labels is easy to do (Fig. 22.9). The qplot function adds them exactly like the traditional plot functions do. You supply the main title with the main argument, and the x and y labels with xlab and ylab, respectively. There is no subtitle argument. As with all labels in R, the characters, "\n" causes it to go to a new line, so "\nWorkshops" below will put just the word "Workshops" at the beginning of a new line.

The qplot approach:

```
> qplot(workshop, geom="bar",
+    main="Workshop Attendance",
+    xlab="Statistics Package \nWorkshops")
```

The ggplot approach:

```
> ggplot(mydata100, aes(workshop, ..count..)) +
+    geom_bar() +
```

Fig. 22.9 A barplot
demonstrating titles and
x-axis labels

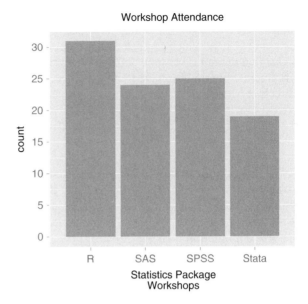

```
+    opts( title="Workshop Attendance" ) +
+    scale_x_discrete("Statistics Package \nWorkshops")
```

Adding titles and labels in ggplot is slightly more verbose. The opts function sets various options, one of which is title. The axis labels are attributes of the axes themselves. They are controlled by the functions, scale_x_discrete, scale_y_discrete, scale_x_continuous, and scale_y_continuous, which are clearly named according to their function. I find this a bit odd since you use different functions for labeling axes if they are discrete or continuous but it is one of the trade-offs you make when getting all of the flexibility that ggplot offers.

22.9 Histograms

Many statistical methods make assumptions about the distribution of your data, or at least of your model residuals. As long as you have 30 or more observations, histograms (Fig. 22.10) are a good way to examine those assumptions. You can use either of the following examples to create one. As you can see from the message, 30 bins is the default.

The qplot approach:

```
> qplot(posttest, geom="histogram")

stat_bin: bin width unspecified, using 30 bins as default.
```

Fig. 22.10 Histogram of
posttest

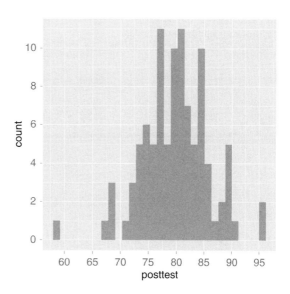

The ggplot approach:

```
> ggplot(mydata100, aes(posttest)) +
+     geom_histogram()
```

stat_bin: bin width unspecified, using 30 bins as default.

If you narrow the width of the bins, you will get more bars, showing more structure in the data (Fig. 22.11). If you prefer qplot, simply add the binwidth argument. If you prefer ggplot, add the geom_bar function with its binwidth argument. Smaller numbers result in more bars.

The qplot approach:

```
> qplot(posttest, geom="histogram", binwidth=0.5)
```

The ggplot approach:

```
> ggplot(mydata100, aes(posttest) ) +
+   geom_bar( binwidth=0.5 )
```

If you prefer to see a density curve, just change the geom argument or function to density (Fig. 22.12).

The qplot approach:

```
> qplot(posttest, geom="density" )
```

The ggplot approach:

```
> ggplot(mydata100, aes(posttest)) +
+   geom_density()
```

Fig. 22.11 Histogram of
posttest with smaller bin
widths

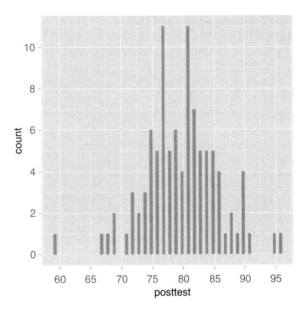

Overlaying the density on the histogram, as in Fig. 22.13, is only slightly
more complicated. The variable that qplot or ggplot computes in the back-
ground for the *y*-axis is named "..density..". To ask for both a histogram
and the density, you must explicitly list ..density.. as the *y* variable. Then
for qplot, you provide both histogram and density to the geom argument. For
ggplot, you simply call both functions.

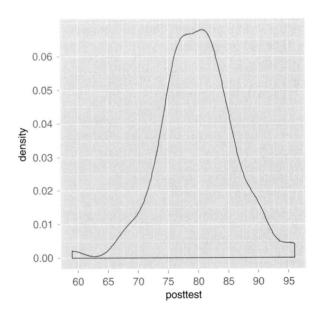

Fig. 22.12 A density plot

Fig. 22.13 A density plot
overlaid on a histogram

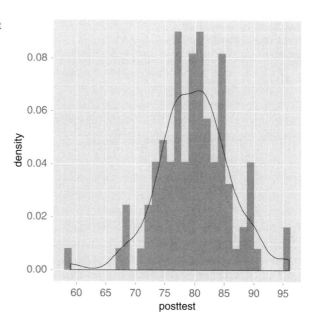

The `qplot` approach:

```
> qplot(posttest, y=..density..,
+    geom=c("histogram","density") )
```

The `ggplot` approach:

```
> ggplot(mydata100, aes(posttest, ..density..)) +
+    geom_histogram() + geom_density()
```

```
stat_bin: bin width unspecified, using 30 bins as default.
```

What if we want to compare the histograms for males and females
(Fig. 22.14)? Using base graphics, we had to set up a multiframe plot and
learn how to control breakpoints for the bars so that they would be comparable.
Using ggplot2, the `facet` feature makes the job trivial. By default, missing
values for factors in your facets argument will create a whole new plot. That
might be fine in your initial analyses but you are unlikely to want that in a plot
for publication. See Sect. 14.5 for ways to remove those observations.

The `qplot` approach:

```
> qplot(posttest, geom="histogram", facets=gender ~ .)
```

The `ggplot` approach:

```
> ggplot(mydata100, aes(posttest) ) +
+    geom_histogram() + facet_grid( gender ~ . )
```

Fig. 22.14 Histograms of
posttest with facets for the
genders

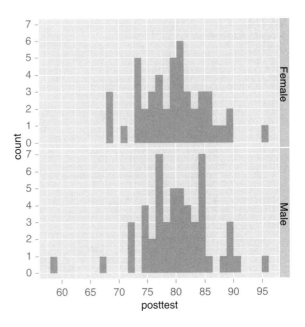

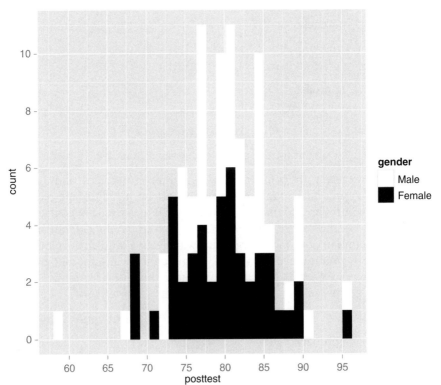

Fig. 22.15 Histogram with bars filled by gender

We can also compare males and females by filling the bars by gender (Fig. 22.15). As before, if you leave off the `scale_fill_grey` function, the bars will come out in two colors rather than black and white.

The `qplot` approach:

```
> qplot( posttest, geom="histogram", fill=gender ) +
+    scale_fill_grey( start=0, end=1 )
```

The `ggplot` approach:

```
> ggplot(mydata100, aes(posttest, fill=gender) ) +
+    geom_bar() + scale_fill_grey( start=0, end=1 )
```

22.10 Normal QQ Plots

We discussed QQ plots in the previous chapter on traditional graphics. Creating them in the `ggplot2` package is straightforward (Fig. 22.16). If you prefer the `qplot` function, use the sample=variable and stat=qq argument. In ggplot, the similar `stat_qq` function will do the trick.

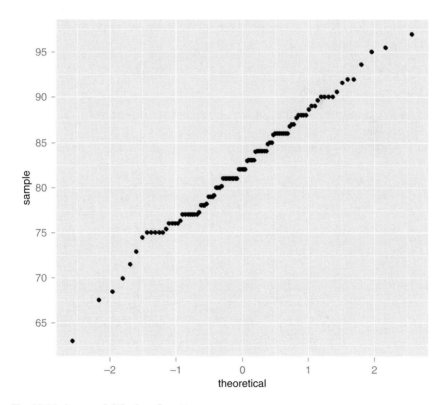

Fig. 22.16 A normal QQ plot of posttest

The qplot approach:

```
> qplot( sample=posttest, stat="qq" )
```

The ggplot approach:

```
> ggplot( mydata100, aes(sample=posttest) ) +
+    stat_qq()
```

22.11 Strip Plots

Strip plots are scatterplots of single continuous variables, or a continuous variable displayed at each level of a factor like workshop. As with the single-stacked bar chart, the case of a single strip plot still requires a variable on the *x*-axis (Fig. 22.17). As you see below, factor("") will suffice. The variable to actually plot is the y argument. Reversing the *x* and *y* variables will turn the plot on its side, the default way the traditional graphics function stripchart does it. I prefer the vertical approach as it matches the style of boxplots and error barplots when you use them to compare groups. The geom = "jitter" adds some noise to separate points that would otherwise obscure other points by plotting on top of them. Finally, the xlab="" and scale_x_discrete("") labels erase what would have been a meaningless label about factor("") for qplot and ggplot, respectively.

The qplot approach:

```
> qplot( factor(""), posttest, geom="jitter", xlab="" )
```

The ggplot approach:

```
> ggplot(mydata100, aes(factor(""), posttest) ) +
+    geom_jitter() +
+    scale_x_discrete("")
```

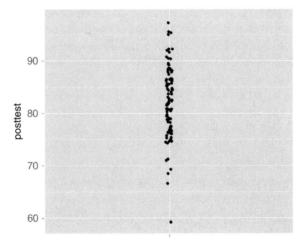

Fig. 22.17 A strip chart done using the jitter goem

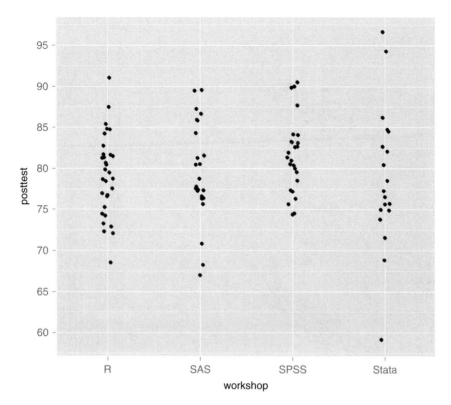

Fig. 22.18 A strip chart with facets for the workshops

Placing a factor like workshop on the *x*-axis will result in a strip chart for each level of the factor (Fig. 22.18).

The qplot approach:

```
> qplot(workshop, posttest, geom="jitter")
```

The ggplot approach:

```
> ggplot(mydata100, aes(workshop, posttest ) ) +
+   geom_jitter()
```

22.12 Scatterplots

The simplest scatterplot hardly takes any effort in qplot. Just list x and y in that order. You could add the geom = "point" argument, but it is the default. The ggplot function is a bit more complicated, but the geom_point function is the only function below that we have not seen before. The resulting graph appears in the upper left of the multiframe plot shown in Fig. 22.19.

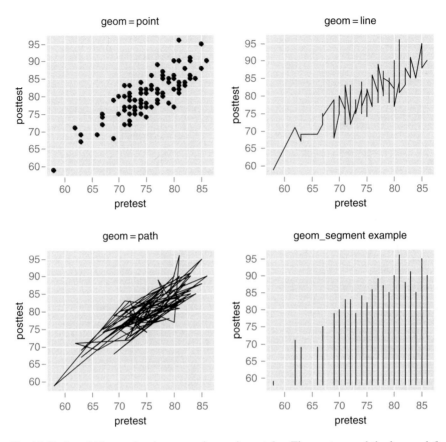

Fig. 22.19 A multiframe plot demonstrating various styles. The *top* two and the *bottom left* show different geoms. The *bottom right* is done a very different way, by drawing line segments from each point to the *x*-axis

The `qplot` approach:

```
> qplot(pretest, posttest)
```

The `ggplot` approach:

```
> ggplot(mydata100, aes(pretest, posttest) ) +
+    geom_point()
```

We can connect the points using the line geom, as you see below. However, the result is different from what you get in traditional R graphics, or in SAS or SPSS for that matter. The line connects the points in the order that they appear on the *x*-axis. That almost makes our data appear as a time series, when it is not. The resulting graph is on the upper right of Fig. 22.19.

The qplot approach:

```
qplot( pretest, posttest, geom="line")
```

The ggplot approach:

```
> ggplot(mydata100, aes(pretest, posttest) ) +
+    geom_line()
```

Although the line geom ignored the order of the points in the data frame, the path geom will connect them in that order. You can see the result in the lower left quadrant of Fig. 22.19. The order of the points in our dataset has no meaning, so it is just a mess!

The qplot approach:

```
> qplot( pretest, posttest, geom="path")
```

And here is the ggplot approach:

```
> ggplot(mydata100, aes(pretest, posttest) ) +
+    geom_path()
```

Now let us run a vertical line to each point. When we did that using traditional graphics, it was a very minor variation. In ggplot2, it is quite different but an interesting example. It is a plot that is much more clear in ggplot form, so we will skip qplot for this one.

In the ggplot code below, the first line is the same as the above examples. Where it gets interesting is the geom_segment function. It has its own aes function, repeating the x and y arguments, but in this case, they are the beginning points for drawing line segments! It also has the arguments xend and yend which tell it where to end the line segments.

This may look overly complicated compared to the simple "type=h" argument from the plot function, but you could use this approach to draw all kinds of line segments. You could easily draw them coming from the top or either side, or among sets of points. The "type=h" approach is a one trick pony. With that approach, adding features to a function leads to a very large number of options, and the developer is still unlikely to think of all the interesting variations in advance.

Here is the code, and the result is in the lower right panel of Fig. 22.19.

```
> ggplot(mydata100, aes(pretest, posttest) ) +
+    geom_segment( aes(  pretest, posttest,
+                     xend=pretest, yend=50) )
```

22.13 Scatterplots with Jitter

We discussed the benefits of jitter in the previous chapter. To get a non-jittered plot of q1 and q4, we will just use qplot (Fig. 22.20, *left*).

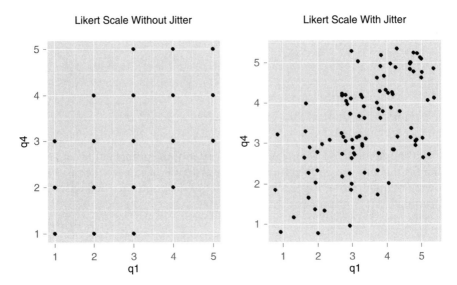

Fig. 22.20 A multiframe plot showing the impact of jitter on five-point Likert-scale data. The plot on the *left* is not jittered, so many points are obscured. The plot on the *right* is jittered, randomly moving points out from behind one another

```
> qplot(q1,q4)
```

To add jitter (Fig. 22.20, *right*), below are both the `qplot` and `gglot` approaches. Note that the `geom="point"` argument is the default in `qplot`. Since that default is not shown, the fact that the `position` argument applies to it is not obvious. That relationship is more clear in the `ggplot` code, where the `position` argument is clearly part of the `geom_point` function.

The `qplot` approach:

```
> qplot( q1, q4, position=position_jitter(x=5,y=5) )
```

The `ggplot` approach:

```
> ggplot( mydata100, aes(q1, q2) ) +
+    geom_point( position = position_jitter(x=5,y=5) )
```

22.13.1 Scatterplots with Large Datasets

Unlike SPSS' GPL, the `ggplot2` package does not yet have a hexbin function. But it does have some other features to help you deal with large amounts of data. The jitter capability we discussed above comes in handy. Another approach is to shrink the plot symbol size down as small as possible. You can do that using the `size` argument in `qplot` or `ggplot`. Finally, you can set the color to have alpha transparency. By fiddling around with point size, amount of

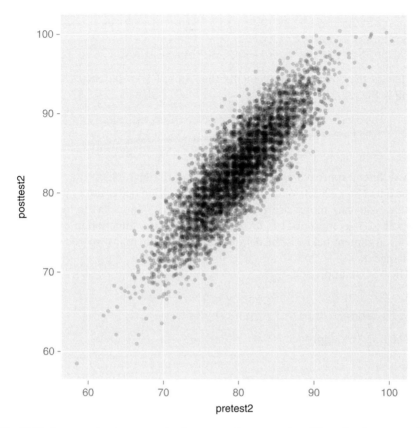

Fig. 22.21 A scatterplot demonstrating how to see the most points. The point sizes are small, jittered and have some transparency, allowing you to see through them. This is a lower resolution bitmap image due to the use of transparency

jitter, and the amount of transparency, you can find a good combination that lets you see through points into the heart of a dense scatterplot. Using all of these options leads to complexity that is best left to ggplot.

Unfortunately, transparency is not yet supported in Windows metafiles. So if you are a Windows user, choose "Copy as bitmap" when cutting and pasting graphs into your word processor. It leads to a low-quality image but it works. You can also write the image to Portable Document Format, using the pdf funtion. That is one of the few high-resolution formats that supports transparency. See section 20.7 for details.

To get 5000 points to work with, I generated it with the following:

```
pretest2  <- round( rnorm( n=5000, mean=80, sd=5) )
posttest2 <- round( pretest2 + rnorm( n=5000, mean=3,
    sd=3) )
```

```
pretest2[pretest2>100] <- 100
posttest2[posttest2>100] <- 100
temp=data.frame(pretest2,posttest2)
```

Here is the code that implements the "alpha" transparency. The resulting graph is Fig. 22.21.

```
> ggplot(temp, aes( pretest2, posttest2) +
+    size=2, position=position_jitter(x=2,Y=2) ) +
+    geom_jitter(colour=alpha("black",0.15) )
```

A different approach is to draw density contours graphics, ggplot2 on top of the data (Fig. 22.22). With this approach it is often better not to jitter the data so that you can more clearly see the contours. You can do this with the density2d geom in qplot, or the geom_density function in ggplot. The size = 1 argument below is the default. I included it here just so you can see how to change it.

The qplot approach:

```
> qplot(pretest2, posttest2,
+    geom=c("point","density2d") )
```

The ggplot approach:

```
> ggplot(temp, aes( pretest2, posttest2) ) +
+    geom_point( size=1 ) + geom_density_2d()
```

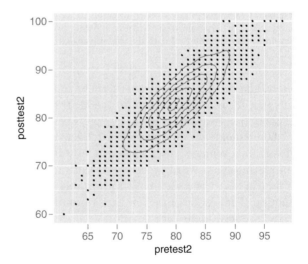

Fig. 22.22 This scatterplot shows an alternate way to see the structure in a large dataset. These points are small, but not jittered, making more space for us to see the density contour lines

22.14 Scatterplots with Fit Lines

While the traditional graphics plot function took quite a bit of extra effort to add
confidence lines around a regression fit (Fig. 22.23), the ggplot2 package makes
that automatic. "Transparency used to create the confidence band. That is not
supported when you cut and paste the image as a metafile in Windows. You can
copy it as a lower resolution bitmap, or write it to a PDF file. See section 20.7 for
details." To get a regression line in qplot, simply specify geom = "smooth".
However, that alone will replace the default of geom = "point", so if you want
both, you need to specify geom = c("point", "smooth"). In ggplot, you use
both the geom_point and geom_smooth functions. The default smoothing
method is a lowess function so if you prefer a linear model, include the method = lm argument.

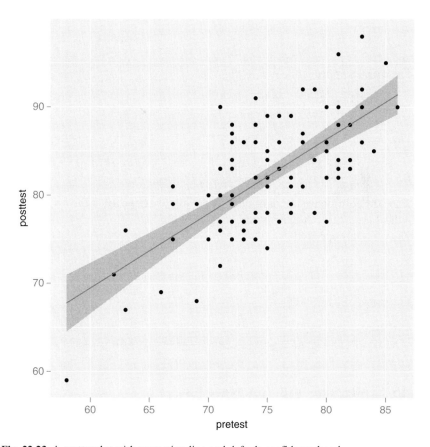

Fig. 22.23 A scatterplot with regression line and default confidence band.

Fig. 22.24 Scatterplot with
default confidence band
removed. No longer
relevant.

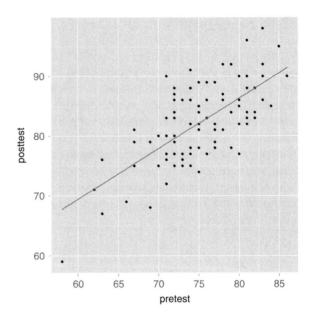

The qplot approach:

```
> qplot(pretest, posttest,
+   geom=c("point","smooth"), method=lm )
```

The ggplot approach:

```
> ggplot(mydata100, aes(pretest, posttest) ) +
+   geom_point() + geom_smooth(method=lm)
```

Since the confidence bands appear by default, we have to set the se argument (*standard error*) to FALSE to turn it off (Fig. 22.24).

The qplot approach:

```
> qplot(pretest, posttest,
+   geom=c("point","smooth"), method=lm, se=FALSE )
```

The ggplot approach:

```
> ggplot(mydata100, aes(pretest, posttest) ) +
+   geom_point() + geom_smooth(method=lm, se=FALSE)
```

22.15 Scatterplots with Reference Lines

To place an arbitrary straight line on the plot, use the abline geom in qplot. You specify your slope and intercept using clearly named arguments. Here we are using intercept = 0 and slope = 1 since this is the line where posttest = pretest. If the students did not learn anything in the workshops, the

Fig. 22.25 A scatterplot with a line added where pretest = posttest. Most of the points lying above this line show the students did learn from the workshop

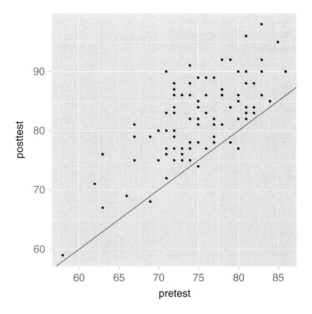

data would fall on this line (assuming a reliable test). The ggplot function adds the abline function with arguments for intercept and slope (Fig. 22.25).

The qplot approach:

```
> qplot(pretest, posttest,
+    geom=c("point","abline"),
+    intercept=0, slope=1 )
```

The ggplot approach:

```
> ggplot(mydata100, aes(pretest, posttest) ) +
+    geom_point()+ geom_abline( intercept=0, slope=1 )
```

Vertical or horizontal reference lines can help emphasize points or cutoffs. For example, if our students are required to get a score greater than 75 before moving on, we might want to display those cutoffs on our plot (Fig. 22.26). This is an area where ggplot can do something that qplot cannot. As you can see in the qplot example below, adding the vline and hline geoms, along with the intercept = 75 argument will meet our current goal. For ggplot, the geom_vline and goem_hline functions do the same task. The big difference between the two is that ggplot can accept a different value for each axis. So if you had dual criteria, say 75 was good enough on the pretest but after taking the workshop an 80 was required, then you would have to use ggplot to display those two values (not shown).

The qplot approach:

```
> qplot(pretest, posttest,
+    geom=c("point","vline","hline"),
+    intercept=75)
```

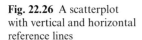
Fig. 22.26 A scatterplot
with vertical and horizontal
reference lines

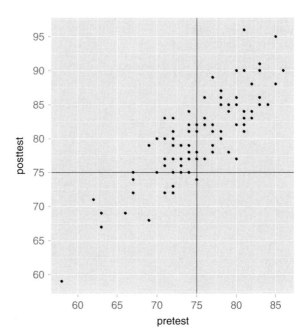

The ggplot approach:

```
> ggplot(mydata100, aes(pretest, posttest)) +
+    geom_point() +
+    geom_vline( intercept=75 ) +
+    geom_hline( intercept=75 )
```

To add a series of reference lines, we need to use the geom_vline or geom_hline functions (Fig. 22.27). The qplot example does not do much with qplot itself since it cannot accept multiple reference lines. Note that in the geom_vline function below both qplot and ggplot examples are identical. It includes the seq function, without which we could have used intercept=c(70,72,74,...80).

The qplot approach:

```
> qplot(pretest, posttest, type="point") +
+    geom_vline( intercept=seq(from=70,to=80,by=2) )
```

The ggplot approach:

```
> ggplot(mydata100, aes(x = pretest, y=posttest)) +
+    geom_point() +
+    geom_vline( intercept=seq(from=70,to=80,by=2) )
```

Fig. 22.27 Scatterplot with multiple vertical reference lines

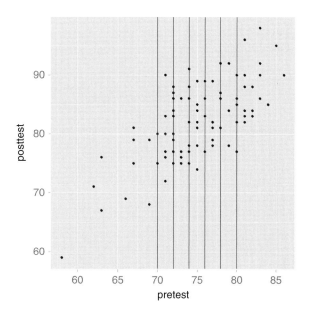

22.16 Plotting Labels Instead of Points

If you do not have much data, or you are only interested in points around the edges, you can plot labels instead of plots symbols (Fig. 22.28). The labels can be identifiers such as ID numbers, people's names or row names, or they could be values of other variables of interest to add a third dimension to the plot. You do this using the geom = "text" argument in qplot or the geom_text function in ggplot. In either case, the label argument points to the values to use. Recall that in R, row.names(mydata) gives you the stored row names, even if these are just the sequential characters, "1," "2," etc. We will store that in a variable named mydata$id and then use it with the label argument. The reason we do not use the form label = row.names (mydata100) is that the ggplot2 package puts all the variables it uses into a separate temporary data frame before running.

```
> mydata100$id <- row.names(mydata100)
```

The qplot approach:

```
> qplot(pretest, posttest, geom="text",
+    label=mydata100$id )
```

The ggplot approach:

```
> ggplot(mydata100, aes(pretest, posttest,
+    label=mydata100$id ) ) + geom_text()
```

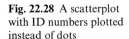
Fig. 22.28 A scatterplot
with ID numbers plotted
instead of dots

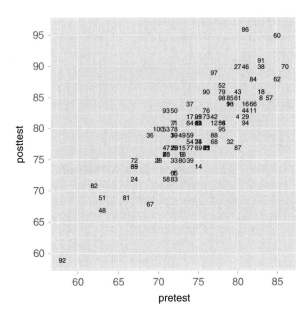

22.17 Changing Plot Symbols by Group

Using different points to indicate group membership is easy (Fig. 22.29). The
`qplot` function adds the `shape` argument, while `ggplot` uses the `aes` function with the `shape=gender` argument. You can also set color and size by
substituting either of those words for shape.

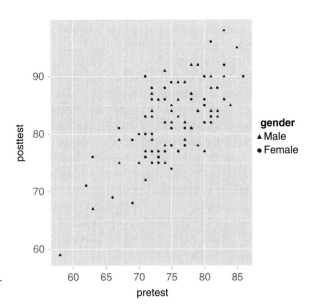

Fig. 22.29 Scatterplot with
differently shaped points for
each gender

The `qplot` approach:

```
> qplot(pretest, posttest, shape=gender)
```

The `ggplot` approach:

```
> ggplot(mydata100, aes(pretest, posttest) ) +
+    geom_point( aes(shape=gender ) )
```

22.18 Adding Linear Fits by Group

We can fit a different linear regression line to each gender (Fig 22.30). All you need to add is the `shape` argument. The `method` argument will then apply a smoothing method to each level of your group.

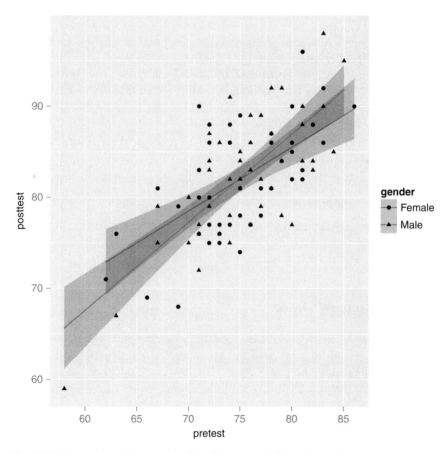

Fig. 22.30 Scatterplot with regression lines fit separately for each gender

The qplot approach:

```
> qplot(pretest, posttest, geom=c("smooth","point"),
+    method="lm", shape=gender)
```

The ggplot approach:

```
> ggplot(mydata100,
+    aes(pretest, posttest, shape=gender) ) +
+    geom_smooth( method="lm" ) + geom_point()
```

22.19 Scatterplots Faceted by Groups

Another way to compare groups on scatter with or without lines of fit is through facets (Fig. 22.31). As we have seen several times before, simply adding the facets argument to the qplot function allows you to specify rows ~ columns of categorical variables. So facets = workshop~gender is requesting a grid of plots for each workshop–gender combination, with workshop determining the rows and gender determining the columns.

The ggplot function works similarly, using the facet_grid function to do the same. If you have a continuous variable to condition on, you can use the chop function from the ggplot2 package or the cut function that is built in R to break the variable into groups.

The qplot approach:

```
> qplot(pretest, posttest, geom=c("smooth","point"),
+    method="lm", shape=gender, facets=workshop ~ gender)
```

The ggplot approach:

```
> ggplot(mydata100, aes( pretest, posttest) ) +
+    geom_smooth( method="lm" ) + geom_point() +
+    facet_grid( workshop ~ gender )
```

22.20 Scatterplot Matrix

When you have many variables to plot, a scatterplot matrix is helpful (Fig. 22.32). You lose a lot of detail compared to a full-sized plot, but you usually get the gist of the relationship. The ggplot2 package has a separate plotmatrix function for this type of plot. Simply entering the following will plot variables 3 through 8 against one another (not shown).

```
> plotmatrix( mydata100[3:8] )
```

You can embellish the plots with many of the options we have covered above. You could assign different symbol shape and linear fits per group with

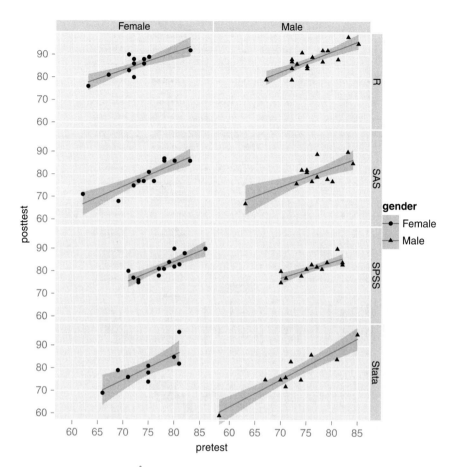

Fig. 22.31 Scatterplots with facets showing linear fits for each workshop and gender combination

the following (plot not shown.)

```
> plotmatrix( mydata100[3:8],
+    aes( shape=gender ) ) +
+    geom_smooth(method=lm)
```

Figure 22.32 is a scatterplot matrix with smoothed lowess fits for the entire dataset (i.e., not by group). The lowess fit generated some warnings but that is not a problem. The density plots on the diagonals appear by default.

```
> plotmatrix( mydata100[3:8] ) +
+    geom_smooth()
There were 50 or more warnings (use warnings() to see
the first 50)
```

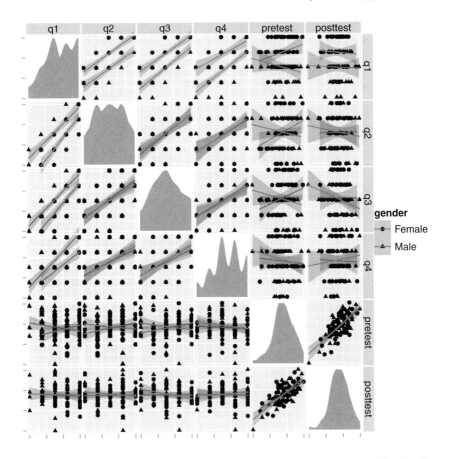

Fig. 22.32 Scatterplot matrix with lowess curve fits on the off diagonal plots and density plots on the diagonal. This was copied as a lower-resolution bitmap image

22.21 Boxplots

The boxplots we produced using traditional graphics offered "notches" to indicate possible group differences. The ggplot2 package does everything with boxplots except the notches.

The simplest type of boxplot is for a single variable (Fig. 22.33). The qplot function uses the simple form of factor("") to act as its x-axis value. The y value is the variable to plot, in this case, posttest. The geom of boxplot specifies the main display type. The xlab = "" argument blanks out the label on the x-axis, which would have been a meaningless factor("").

The equivalent ggplot approach is almost identical with its ever-present aes arguments for x and y and the geom_boxplot function to

Fig. 22.33 A boxplot of posttest

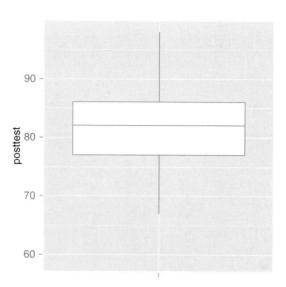

draw the box. The `scale_x_discrete` function simply blanks out the *x*-axis label.

The `qplot` approach:

```
> qplot(factor(""), posttest,
+    geom="boxplot", xlab="")
```

The `ggplot` approach:

```
> ggplot(mydata100,
+    aes(factor(""), posttest) ) +
+    geom_boxplot() +
+    scale_x_discrete("")
```

Adding a grouping variable like workshop makes boxplots much more informative (Fig. 22.34, ignore the overlaid strip plot points for now). These are the same commands as above but with the *x* variable specified as workshop. We will skip showing this one in favor of the next.

The `qplot` approach:

```
> qplot(workshop, posttest, geom="boxplot" )
```

The `ggplot` approach:

```
> ggplot(mydata100,
+    aes( workshop, posttest) ) +
+    geom_boxplot()
```

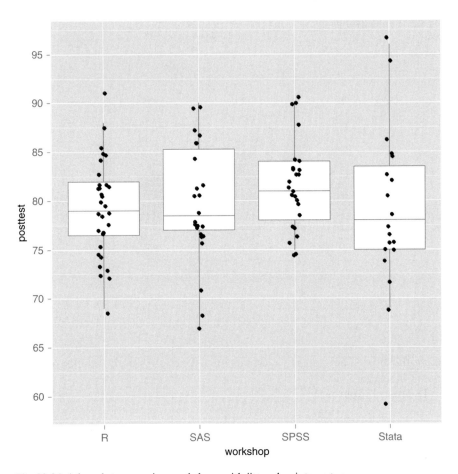

Fig. 22.34 A boxplot comparing workshops with jittered points on top

Now we will do the same plot but display a jittered strip plot on top of it (Fig. 22.34). This way we get the boxplot information about the median and quartiles plus we get to see any interesting structure in the points that would otherwise have been lost. As you can see, the qplot now has jitter added to its geom argument, and ggplot has an additional geom_jitter function. The plot is shown below.

The qplot approach:

```
> qplot(workshop, posttest,
+    geom=c("boxplot","jitter") )
```

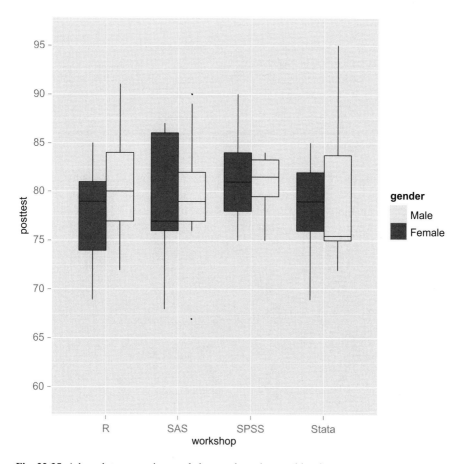

Fig. 22.35 A boxplot comparing workshop and gender combinations

The `ggplot` approach:

```
> ggplot(mydata100,
+   aes(workshop, posttest) ) +
+   geom_boxplot() + geom_jitter()
```

To add another grouping variable, as in Fig. 22.35, you only need to add the fill argument to either `qplot` or `ggplot`.

The `qplot` approach:

```
> qplot(workshop, posttest,
+   geom="boxplot", fill=gender ) +
+   scale_fill_grey( start=0, end=1 )
```

The ggplot approach:

```
> ggplot(mydata100,
+    aes(workshop, posttest) ) +
+    geom_boxplot( aes(fill=gender), color="black") +
+    scale_fill_grey( start=0, end=1 )
```

22.22 Error Barplots

Plotting means and 95% confidence intervals, as in Fig. 22.36, is a task that stretches what qplot was designed to do. As you can see from the two examples below, there is very little typing saved by using qplot over ggplot. In both cases we are adding a jittered strip plot of points, as we did before in the

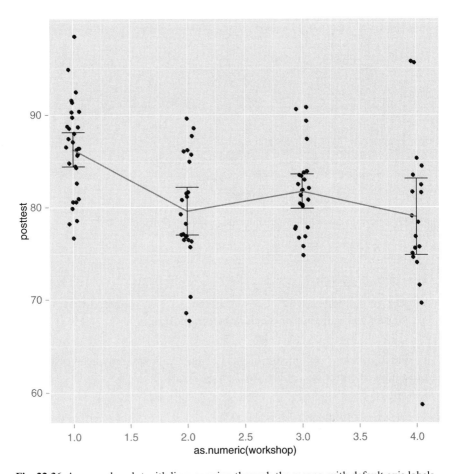

Fig. 22.36 An error barplot with lines running through the means, with default axis labels.

sections on strip plots and boxplots. Notice that we had to use the as. numeric function for our *x* variable, workshop. Since workshop is a factor, the software would not connect the means across the levels of *x*. Workshop is not a continuous variable, so that makes sense! However, the errorbar geoms lack a center line for the mean, so the only way to display means is to add the line. Hopefully a future version of the ggplot2 package will add means to its errorbar geom.

The key function for this plot is stat_summary, which we use twice. First, we use the argument *mean* to connect the means with a smoothed line. Secondly, on the following line, we use it to get error bars and set their widths.

The qplot approach:

```
> qplot ( as.numeric(workshop),
+            posttest, geom="jitter") +
+    stat_summary(fun="mean",
+      geom="smooth", se=FALSE) +
+    stat_summary(fun="mean_cl_normal",
+      geom="errorbar", width=.2)
```

The ggplot approach:

```
> ggplot(mydata100,
+    aes(as.numeric(workshop), posttest) ) +
+    geom_jitter() +
+    stat_summary(fun="mean",
+      geom="smooth", se=FALSE) +
+    stat_summary(fun="mean_cl_normal",
+      geom="errorbar", width=.2)
```

22.23 Logarithmic Axes

If your data has a very wide range of values, working in a logarithmic scale is often helpful. In ggplot2 you can approach this in three different ways. First, you can take the logarithm of the data before plotting.

```
qplot ( log(pretest), log(posttest) )
```

Another approach is to use evenly placed tick marks on the plot, but have the axis values use logarithmic values such as 10^1, 10^2, etc. This is what the scale_x_log10 function does (similarly for the *y*-axis of course). There are similar functions for natural logarithms, scale_x_log and base 2 logarithms, scale_x_log2.

```
qplot(pretest, posttest, data=mydata100) +
  scale_x_log10() + scale_y_log10()
```

Finally, you can have the tick marks spaced unevenly and use values on your original scale. The `coord_trans` function does that. Its arguments for the various bases of logarithms are `log10`, `log`, and `log2`.

```
qplot(pretest, posttest, data=mydata100) +
   coord_trans("log10", "log10")
```

With our dataset, the range of values is too small to use for good examples. Therefore I am not showing any of these plots.

22.24 Aspect Ratio

Changing the aspect ratio of a graph can be far more important than you might first think. Research has shown that when most of the lines or scatter on a plot are angled at 45°, people make more accurate comparisons to those parts that are not [30].

Unless you specify an aspect ratio for your graph, `qplot` and `ggplot` will match the dimensions of your output window and allow you to change those dimensions using your mouse, as you would any other window. If you are routing your output to a file however, it is helpful to be able to set it using code.

You set the aspect ratio using the `coord_equal` function. If you leave it empty, as in `coord_equal()`, it will make the x and y axes of equal lengths. If you are working interactively, you can still reshape your window but the graph within it will remain square. Specifying a ratio parameter follows the form "height/width." For a mnemonic, think of how R specifies [rows,columns]. The example below would result in a graph that is four times wider than it is high (plot not shown).

```
qplot(pretest, posttest) + coord_equal(ratio=1/4)
```

22.25 Multiple Plots on a Page

In the last chapter on traditional graphics, we discussed how to put multiple plots on a page. However, `ggplot2` uses the grid graphics system, so that method does not work. We saw the multiframe plot Figure 22.37 in the section on barplots. Let us now look at how I created it. We will skip the barplot details here and focus on how we combined them.

We first clear the page with the `grid.newpage` function. This is an important step as otherwise plots printed using the following methods will appear on top of any previous plots.

```
> grid.newpage()
```

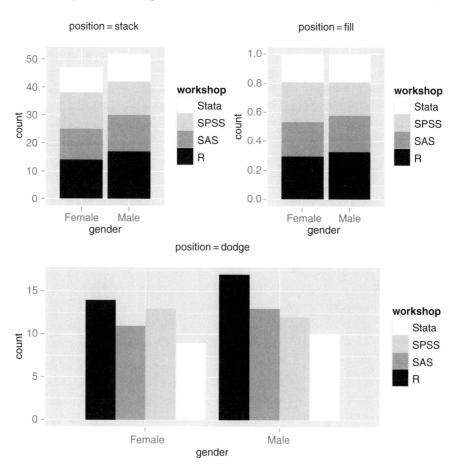

Fig. 22.37 A multiframe barplot, shown previously, is repeated here to discuss how to combine plots rather than how to do barplots

Next, we use the `pushViewport` function to define the various frames called *viewports* in the grid graphics system. The `grid.layout` argument uses R's common format of (rows, columns). The example below sets up a 2×2 grid for us to use.

```
> pushViewport( viewport(layout=grid.layout(2,2) ) )
```

In traditional graphics, you would just do the graphs in order and they would find their place. However, in grid we must instead save the plot to an object[1] and

---

[1] You could nest the plot within the print function but it would be messy and error-prone.

then use the print function to print it into the viewport we desire. The object name of "p" is commonly used as an object name for the *plot*. Since there are many ways to add to this object, it is helpful to keep it short. However, to clarify this is something I get to choose, I will use "myPlot".

The `print` function has a vp argument that lets you specify the *viewport's* position in row(s) and column(s). In the example below, we will print the graph to row 1 and column 1.

```
> myPlot <- ggplot(mydata100,
+       aes(gender, fill=workshop) ) +
+    geom_bar(position="stack") +
+    scale_fill_grey(start = 0, end = 1) +
+    opts( title="position=stack " )
> print(myPlot, vp=viewport(
+    layout.pos.row=1,
+    layout.pos.col=1) )
```

The next plot prints to row 1 and column 2.

```
> myPlot <- ggplot(mydata100,
+       aes(gender, fill=workshop) ) +
+    geom_bar(position="fill") +
+    scale_fill_grey(start = 0, end = 1) +
+    opts( title="position=fill" )
> print(myPlot, vp=viewport(
+    layout.pos.row=1,
+    layout.pos.col=2) )
```

The third and final plot is much wider than the first two. So we will print it to row 2 in both columns 1 and 2. Since we did not set the aspect ratio explicitly, the graph will resize to fit the double-wide viewport.

```
> myPlot <- ggplot(mydata100,
+       aes(gender, fill=workshop) ) +
+    geom_bar(position="dodge")   +
+    scale_fill_grey(start = 0, end = 1) +
+    opts( title="position=dodge" )
> print(myPlot, vp=viewport(
+    layout.pos.row=2,
+    layout.pos.col=1:2) )
```

The next time you print a plot without specifying a viewport, the screen resets back to its previous full-window display.

The code for the other multiframe plots is in the examples program see Table 22.2.

22.26 Saving **ggplot2** Graphs to a File

The ggplot2 package has its own command to save graphs to a file. To save the last graph you created, with either qplot or ggplot, use the ggsave function. It will choose the proper graphics device from the file extension. See the help files for additional options.

```
ggsave(file = "mygraph.pdf")
```

22.27 An Example Specifying All Defaults

Now that you have seen some examples of both qplot and ggplot, let us take a brief look at the full power of ggplot by revisiting the scatterplot with a regression line (Fig. 22.38). First, done with qplot, it is quite easy.

```
> qplot(pretest, posttest,
+    geom=c("point","smooth"), method="lm" )
```

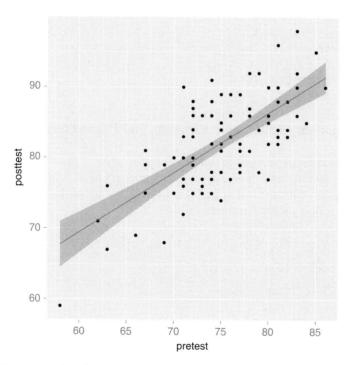

Fig. 22.38 A scatterplot, shown previously, repeated here and created using three different sets of code

Next, let us do it using `ggplot` with as many default settings as possible. It is not too much more typing.

```
> ggplot(mydata100, aes(pretest, posttest) ) +
+     geom_point() +
+     geom_smooth(method="lm")
```

Finally, here it is again in ggplot but with no default settings. We see that the plot is actually two layers, each of which could use different data frames, variables, geometric objects, statistics, and so on. If you need graphics flexibility, ggplot2 is the package for you!

```
> ggplot() +
+ layer(
+     data=mydata100,
+     mapping=aes(pretest, posttest),
+     geom="point",
+     stat="identity"
+ ) +
+ layer(
+     data=mydata100,
+     mapping=aes(pretest, posttest),
+     geom="smooth",
+     stat="smooth",
+     method="lm"
+ ) +
+ coord_cartesian()
```

22.28 Summary of Graphic Elements and Parameters

We have seen many ways to modify plots in the ggplot2 package. The ggopt function is another way. You can set the parameters of all future graphs in the current session with the following function call. See the help file on the ggopt function for many more parameters (Table 22.2).

```
ggopt(
background.fill = "black",
background.color ="white",
axis.color = "black"   #default axis fonts are grey.
)
```

The example SPSS and R program for these topics are in Table 22.2. SAS does not offer this type of graphics. See previous chapter for SAS/Graph examples. The SPSS examples are sparse compared to those in R. That is due to space constraints, not lack of capability. For examples using SPSS legacy graphics, see previous chapter

Table 22.2 Example programs for ggplot2

SPSS programming statements

```
        * SPSS Programs for Grammar of Graphics.
        * GraphicsGG.sps .

        CD 'C:\myRfolder'.
        GET FILE='mydata.sav'.

        * GPL statements for histogram of q1.
        GGRAPH
          /GRAPHDATASET NAME="graphdataset" VARIABLES=q1
        MISSING=LISTWISE
          REPORTMISSING=NO
          /GRAPHSPEC SOURCE=INLINE.
        BEGIN GPL
         SOURCE: s=userSource(id("graphdataset"))
         DATA: q1=col(source(s), name("q1"))
         GUIDE: axis(dim(1), label("q1"))
         GUIDE: axis(dim(2), label("Frequency"))
         ELEMENT: interval(position(summary.count(bin.rect(q1))),
        shape.interior(
          shape.square))
        END GPL.

        * GPL statements for bar chart of gender.
        GGRAPH
          /GRAPHDATASET NAME="graphdataset" VARIABLES=gender
        COUNT()[name=
          "COUNT"] MISSING=LISTWISE REPORTMISSING=NO
          /GRAPHSPEC SOURCE=INLINE.
        BEGIN GPL
         SOURCE: s=userSource(id("graphdataset"))
         DATA: gender=col(source(s), name("gender"),
        unit.category())
         DATA: COUNT=col(source(s), name("COUNT"))
         GUIDE: axis(dim(1), label("gender"))
         GUIDE: axis(dim(2), label("Count"))
         SCALE: cat(dim(1))
         SCALE: linear(dim(2), include(0))
         ELEMENT: interval(position(gender*COUNT),
        shape.interior(shape.square))
        END GPL.

        * GPL syntax for scatterplot of q1 by q2.
        GGRAPH
          /GRAPHDATASET NAME="graphdataset" VARIABLES=q1 q2
        MISSING=LISTWISE
          REPORTMISSING=NO
          /GRAPHSPEC SOURCE=INLINE.
        BEGIN GPL
         SOURCE: s=userSource(id("graphdataset"))
         DATA: q1=col(source(s), name("q1"))
```

Table 22.2 (continued)

```
  DATA: q2=col(source(s), name("q2"))
  GUIDE: axis(dim(1), label("q1"))
  GUIDE: axis(dim(2), label("q2"))
  ELEMENT: point(position(q1*q2))
END GPL.
* Chart Builder.
GGRAPH
  /GRAPHDATASET NAME="graphdataset" VARIABLES=workshop q1
q2 q3 q4
  MISSING=LISTWISE REPORTMISSING=NO
  /GRAPHSPEC SOURCE=INLINE.
BEGIN GPL
  SOURCE: s=userSource(id("graphdataset"))
  DATA: workshop=col(source(s), name("workshop"))
  DATA: q1=col(source(s), name("q1"))
  DATA: q2=col(source(s), name("q2"))
  DATA: q3=col(source(s), name("q3"))
  DATA: q4=col(source(s), name("q4"))
  TRANS: workshop_label = eval("workshop")
  TRANS: q1_label = eval("q1")
  TRANS: q2_label = eval("q2")
  TRANS: q3_label = eval("q3")
  TRANS: q4_label = eval("q4")
  GUIDE: axis(dim(1.1), ticks(null()))
  GUIDE: axis(dim(2.1), ticks(null()))
  GUIDE: axis(dim(1), gap(0px))
  GUIDE: axis(dim(2), gap(0px))
  ELEMENT: point(position((
workshop/workshop_label+q1/q1_label+q2/q2_label+q3/
q3_label+q4/q4_label)*(

workshop/workshop_label+q1/q1_label+q2/q2_label+q3/
q3_label+q4/q4_label)))
END GPL.

* GPL statements for scatterplot matrix of workshop to q4
excluding gender.
* Gender cannot be used in this context.
GGRAPH
  /GRAPHDATASET NAME="graphdataset" VARIABLES=workshop q1
q2 q3 q4
  MISSING=LISTWISE REPORTMISSING=NO
  /GRAPHSPEC SOURCE=INLINE.
```

Table 22.2 (continued)

```
BEGIN GPL
 SOURCE: s=userSource(id("graphdataset"))
 DATA: workshop=col(source(s), name("workshop"))
 DATA: q1=col(source(s), name("q1"))
 DATA: q2=col(source(s), name("q2"))
 DATA: q3=col(source(s), name("q3"))
 DATA: q4=col(source(s), name("q4"))
 TRANS: workshop_label = eval("workshop")
 TRANS: q1_label = eval("q1")
 TRANS: q2_label = eval("q2")
 TRANS: q3_label = eval("q3")
 TRANS: q4_label = eval("q4")
 GUIDE: axis(dim(1.1), ticks(null()))
 GUIDE: axis(dim(2.1), ticks(null()))
 GUIDE: axis(dim(1), gap(0px))
 GUIDE: axis(dim(2), gap(0px))
 ELEMENT: point(position((

workshop/workshop_label+q1/q1_label+q2/q2_label+q3/
q3_label+q4/q4_label)*(

workshop/workshop_label+q1/q1_label+q2/q2_label+q3/
q3_label+q4/q4_label)))
END GPL.
```

R programming statements

```
#Grammar of Graphics in R using ggplot2.
#Graphics GG.R
setwd("/myRfolder")
load(file="mydata100.Rdata")
detach(mydata100) #In case I'm running repeatedly.
#Get rid of missing values for facets
mydata100 <- na.omit(mydata100)
attach(mydata100)
library(ggplot2)

#Barplot - Vertical

qplot(workshop, geom="bar")

ggplot(mydata100, aes( workshop ) ) +
   geom_bar()

# ---Barplot - Horizontal

qplot(workshop, geom="bar") + coord_flip()

ggplot(mydata100, aes( workshop ) ) +
```

Table 22.2 (continued)

```
geom_bar() + coord_flip()

# ---Barplot - Single Bar Stacked

qplot(factor(""), fill=workshop,
  geom="bar", xlab="") +
  scale_fill_grey(start=0, end=1)

ggplot(mydata100,
  aes(factor(""), fill=workshop) ) +
  geom_bar() +
  scale_x_discrete("") +
  scale_fill_grey(start=0, end=1)

# ---Pie charts, same as stacked bar but polar coordinates

# This is almost it. See book for details.
qplot(factor(""), fill=workshop,
  geom="bar", xlab="") +
  coord_polar(theta="y") +
  scale_fill_grey(start=0, end=1)

ggplot(mydata100,
  aes( factor(""), fill=workshop ) ) +
  geom_bar( width=1 ) +
  scale_x_discrete("") +
  coord_polar(theta="y") +
  scale_fill_grey()

# --- Barplots - Grouped

qplot(gender, geom="bar",
    fill=workshop, position="stack") +
  scale_fill_grey(start = 0, end = 1)
qplot(gender, geom="bar",
    fill=workshop, position="fill")   +
  scale_fill_grey(start = 0, end = 1)
qplot(gender, geom="bar",
    fill=workshop, position="dodge") +
  scale_fill_grey(start = 0, end = 1)

ggplot(mydata100, aes(gender, fill=workshop) ) +
  geom_bar(position="stack") +
```

Table 22.2 (continued)

```
    scale_fill_grey(start = 0, end = 1)
ggplot(mydata100, aes(gender, fill=workshop) ) +
  geom_bar(position="fill") +
  scale_fill_grey(start = 0, end = 1)
ggplot(mydata100, aes(gender, fill=workshop ) ) +
  geom_bar(position="dodge") +
  scale_fill_grey(start = 0, end = 1)


# --- Barplot - Faceted

qplot(workshop, geom="bar", facets=gender~.)

ggplot(mydata100, aes(workshop) ) +
  geom_bar() + facet_grid( gender~. )

# --- Barplot - Pre-summarized data

qplot( factor(c(1,2)), c(40, 60), geom="bar")

myTemp <- data.frame( myX=factor( c(1,2) ), myY=c(40, 60)
)
myTemp
ggplot(data=myTemp, aes(myX, myY) ) +
  geom_bar()

# --- Adding Titles and Labels

qplot(workshop, geom="bar",
  main="Workshop Attendance",
  xlab="Statistics Package \nWorkshops")

ggplot(mydata100, aes(workshop, ..count..)) +
  geom_bar() +
  opts( title="Workshop Attendance" ) +
  scale_x_discrete("Statistics Package \nWorkshops")
ggplot(mydata100, aes(pretest,posttest ) ) +
  geom_point() +
  scale_x_continuous("Test Score Before Training") +
  scale_y_continuous("Test Score After Training")   +
  opts( title="The Relationship is Linear" )

# ---Dotcharts
```

Table 22.2 (continued)

```
qplot(workshop, stat="bin",
      facets=gender~., size=4 ) +
  coord_flip()

ggplot(mydata100,
  aes(workshop, ..count.. ) ) +
  geom_point(stat="bin", size=4) + coord_flip()+
  facet_grid( gender~. )

# or with stat_bin:
ggplot(mydata100,
  aes(workshop, ..count.. ) ) +
  stat_bin(geom="point", size=4) + coord_flip()+
  facet_grid( gender~. )

# --- Histograms

# Simle Histogram
qplot(posttest, geom="histogram")

ggplot(mydata100, aes(posttest) ) +
  geom_histogram()

# Histogram with more bars.
qplot(posttest, geom="histogram", binwidth=0.5)
ggplot(mydata100, aes(posttest) ) +
  geom_histogram( binwidth=0.5 )

# Density plot
qplot(posttest, geom="density")

ggplot(mydata100, aes(posttest)) +
  geom_density()

# Histogram with Density

qplot(data=mydata100,posttest, ..density..,
  geom=c("histogram","density") )

ggplot(mydata100, aes(posttest, ..density.. ) ) +
  geom_histogram() + geom_density()
```

Table 22.2 (continued)

```
# Histogram - Separate plots by group

qplot(posttest, geom="histogram", facets=gender~.)

ggplot(mydata100, aes(posttest) ) +
  geom_histogram() + facet_grid(gender~. )

# Histogram with Stacked Bars

qplot( posttest, geom="histogram", fill=gender ) +
  scale_fill_grey(start = 0, end = 1)

ggplot( mydata100, aes(posttest, fill=gender) ) +
  geom_bar() +
  scale_fill_grey(start = 0, end = 1)

# - - - QQ plot
qplot(sample=posttest, stat="qq")

ggplot( mydata100, aes(sample=posttest) ) +
  stat_qq()

# - - - Strip plot
qplot( factor(""), posttest, geom="jitter", xlab="")

ggplot( mydata100, aes(factor(""), posttest) ) +
  geom_jitter() +
  scale_x_discrete("")

# Strip plot by group.
qplot(workshop, posttest, geom="jitter")

ggplot(mydata100, aes(workshop, posttest) ) +
  geom_jitter()

# - - - Scatterplots

# Simple scatterplot

qplot(pretest, posttest)
qplot(pretest, posttest, geom="point")
```

Table 22.2 (continued)

```
ggplot (mydata100, aes (pretest, posttest)) +
  geom_point ()

# Scatterplot connecting points sorted on x.
qplot ( pretest, posttest, geom="line")

ggplot (mydata100, aes (pretest, posttest) ) +
  geom_line ()

# Scatterplot connecting points in data set order.

qplot ( pretest, posttest, geom="path")

ggplot (mydata100, aes (pretest, posttest) ) +
  geom_path ()

# Scatterplot with skinny histogram-like bars to X axis.

#qplot does not do this in 0.5.7 but will in future release:
qplot (pretest,posttest,
  xend=pretest, yend=50,
  geom="segment")

ggplot (mydata100, aes (pretest,      posttest) ) +
geom_segment ( aes (   pretest,      posttest,
               xend=pretest,      yend=50) )

# Scatterplot with jitter
qplot (q1, q4) #First without

qplot (q1, q4, position=position_jitter (x=5, y=5) )

# or
ggplot (mydata100, aes (x=q1, y=q2) ) +
  geom_point (position = position_jitter (x=5, y=5) )

# Scatterplot on large data sets

pretest2   <- round ( rnorm ( n=5000, mean=80, sd=5) )
posttest2  <- round ( pretest2 + rnorm ( n=5000, mean=3, sd=3) )
pretest2[ pretest2>100] <- 100
```

Table 22.2 (continued)

```
posttest2[ posttest2>100] <- 100
temp=data.frame(pretest2,posttest2)

# Small, jittered, transparent points.
# No qplot example for this one.

ggplot(temp, aes(pretest2, posttest2),
  size=2, position = position_jitter(x=2,y=2) ) +
  geom_jitter(colour=alpha("black",0.15) )

# Using density contours and small points.
  qplot( pretest2, posttest2,
  geom=c("point", "density2d") )

ggplot(temp, aes( x=pretest2, y=posttest2) ) +
  ggplot(temp, aes(pretest2, posttest2) )+
  geom_point( size=1 ) + geom_density_2d()

rm(pretest2,posttest2,temp)

# Scatterplot with regression line, 95% confidence intervals.

qplot(pretest, posttest,
  geom=c("point","smooth"), method=lm )

ggplot(mydata100, aes(pretest, posttest) ) +
  geom_point() + geom_smooth(method=lm)

# Scatterplot with regression line but NO confidence intervals.

qplot(pretest, posttest,
  geom=c("point","smooth"),
  method=lm, se=FALSE )
ggplot(mydata100, aes(pretest, posttest) ) +
  geom_point() +
  geom_smooth(method=lm, se=FALSE)

# Scatter with x=y line

qplot(pretest, posttest,
  geom=c("point","abline"),
  intercept=0, slope=1 )

  ggplot(mydata100, aes(pretest, posttest) ) +
  geom_point()+
```

Table 22.2 (continued)

```
geom_abline(intercept=0, slope=1)


# Scatterplot with different point shapes for each group.


qplot(pretest, posttest, shape=gender)

ggplot(mydata100, aes(pretest, posttest) ) +
  geom_point( aes(shape=gender ) )


# Scatterplot with regressions fit for each group.

qplot(pretest, posttest,
  geom=c("smooth","point"),
  method="lm", shape=gender)

ggplot(mydata100,
  aes(pretest, posttest, shape=gender) ) +
  geom_smooth( method="lm" ) + geom_point()


# Scatterplot faceted for groups

qplot(pretest, posttest,
  geom=c("smooth", "point"),
  method="lm", shape=gender,
  facets=workshop~gender)

ggplot(mydata100, aes(pretest, posttest) ) +
  geom_smooth( method="lm" ) +
  geom_point() +
  facet_grid( workshop~gender )
# Scatter with vertical or horizontal lines


#qplot requires the values be equal
!qplot(pretest, posttest,
  geom=c("point", "vline", #x0022;hline"),
  intercept=75)


# ggplot allows them to differ.
ggplot(mydata100, aes(pretest, posttest)) +
  geom_point() +
  geom_vline( intercept=75 ) +
  geom_hline( intercept=75 )


# Scatterplot with a set of vertical lines
```

Table 22.2 (continued)

```
qplot(pretest, posttest, type="point") +
  geom_vline( intercept=seq(from=70,to=80,by=2) )

ggplot(mydata100, aes(pretest, posttest)) +
  geom_point() +
  geom_vline( intercept=seq(from=70,to=80,by=2) )

ggplot(mydata100, aes(pretest, posttest)) +
  geom_point() +
  geom_vline( intercept=70:80 )

# Scatter plotting text labels

qplot(pretest, posttest, geom="text",
label=rownames(mydata100) )

ggplot(mydata100,
  aes(pretest, posttest,
  label=rownames(mydata100) ) ) +
  geom_text()

# Scatterplot matrix

plotmatrix( mydata100[ 3:8] )

# Small points & lowess fit.
plotmatrix( mydata100[ 3:8] , aes( size=1 ) ) +
  geom_smooth()

# Shape and gender fits.
plotmatrix( mydata100[ 3:8] ,
  aes( shape=gender ) ) +
  geom_smooth(method=lm) #

# - - - Boxplot

# Boxplot of one variable

qplot(factor(""), posttest,
  geom="boxplot", xlab="")

ggplot(mydata100,
```

Table 22.2 (continued)

```
  aes(factor(""), posttest) ) +
  geom_boxplot() +
  scale_x_discrete("")


# Boxplot by group

qplot(workshop, posttest, geom="boxplot" )

ggplot(mydata100,
  aes(workshop, posttest) ) +
  geom_boxplot()


# Boxplot by group with jitter

qplot(workshop, posttest,
  geom=c("boxplot","jitter") )

ggplot(mydata100,
  aes(workshop, posttest )) +
  geom_boxplot() + geom_jitter()


# Boxplot for two-way interaction.

qplot(workshop, posttest,
  geom="boxplot", fill=gender ) +
  scale_fill_grey(start = 0, end = 1)

ggplot(mydata100,
  aes(workshop, posttest) ) +

  geom_boxplot( aes(fill=gender), color="grey50") +
  scale_fill_grey(start = 0, end = 1)


# Error bar plot

qplot( as.numeric(workshop),
      posttest, geom="jitter") +
  stat_summary(fun="mean",
    geom="smooth", se=FALSE) +
  stat_summary(fun="mean_cl_normal",
    geom="errorbar", width=.2)

ggplot(mydata100,
  aes( as.numeric(workshop), posttest) ) +
```

Table 22.2 (continued)

```
  geom_jitter() +
stat_summary(fun="mean",
geom="smooth", se=FALSE) +
stat_summary(fun="mean_cl_normal",
geom="errorbar", width=.2)#

#---Logarithmic Axes

# Change the variables
qplot( log(pretest), log(posttest) )

ggplot(mydata100,
  aes( log(pretest), log(posttest) ) ) +
  geom_point()

# Change axis labels

qplot(pretest, posttest, log="xy")

ggplot(mydata100, aes( x=pretest, y=posttest) ) +
  geom_point() + scale_x_log10() + scale_y_log10()

# Change axis scaling

# Tickmarks remain uniformly spaced,
# because scale of data is too limited.

qplot(pretest, posttest, data=mydata100)   +
  coord_trans(x="log10", y="log10")

ggplot(mydata100, aes( x=pretest, y=posttest) ) +
  geom_point() + coord_trans(x="log10", y="log10")

#---Aspect ratio

# This forces x and y to be equal.
qplot(pretest, posttest) + coord_equal()

# This sets aspect ratio to height/width.
qplot(pretest, posttest) + coord_equal(ratio=1/4)

#---Multiframe Plots: Barchart Example---
```

Table 22.2 (continued)

```
# Clears the page, otherwise new plots
# will appear on top of old.

grid.newpage()

# Sets up a 2 by 2 grid to plot into.
pushViewport( viewport(layout=grid.layout(2,2) ) )

# Barchart dodged in row 1, column 1.
myPlot <- ggplot(mydata100,
    aes(gender, fill=workshop) ) +
  geom_bar(position="stack") +
  scale_fill_grey(start = 0, end = 1) +
  opts( title="position=stack " )
print(myPlot, vp=viewport(
  layout.pos.row=1,
  layout.pos.col=1) )

# Barchart stacked, in row 1, column 2.
myPlot <- ggplot(mydata100,
    aes(gender, fill=workshop) ) +
  geom_bar(position="fill") +
  scale_fill_grey(start = 0, end = 1) +
  opts( title="position=fill" )
print(myPlot, vp=viewport(
  layout.pos.row=1,
  layout.pos.col=2) )

# Barchart dodged, given frames,
# in row 2, columns 1 and 2.
myPlot <- ggplot(mydata100,
    aes(gender, fill=workshop) ) +
  geom_bar(position="dodge")   +
  scale_fill_grey(start = 0, end = 1) +
  opts( title="position=dodge" )
print(myPlot, vp=viewport(
  layout.pos.row=2,
  layout.pos.col=1:2) )

#---Multiframe Scatterplot
```

Table 22.2 (continued)

```
# Clears the page, otherwise new plots will appear on top of old.
grid.newpage ()

# Sets up a 2 by 2 grid to plot into.
pushViewport ( viewport (layout=grid.layout (2,2) ) )

# Scatterplot of points
myPlot <- qplot (pretest, posttest, main="geom=point")
print (myPlot, vp=viewport (
  layout.pos.row=1,
  layout.pos.col=1) )

myPlot <- qplot ( pretest, posttest,
        geom="line", main="geom=line" )
print (myPlot, vp=viewport (
  layout.pos.row=1,
  layout.pos.col=2) )

myPlot <- qplot ( pretest, posttest,
        geom="path", main="geom=path" )
print (myPlot, vp=viewport (
layout.pos.row=2,
layout.pos.col=1) )

myPlot <- ggplot ( mydata100, aes (pretest, posttest) ) +
        geom_segment ( aes (x=pretest, y=posttest,
                        xend=pretest, yend=58) ) +
        opts ( title="geom_segment example" )
print (myPlot,
vp=viewport (layout.pos.row=2, layout.pos.col=2) )

#---Multiframe Scatterplot for jitter
grid.newpage ()
pushViewport ( viewport (layout=grid.layout (1,2) ) )
# Scatterplot without
myPlot <- qplot ( q1, q4,
        main="Likert Scale Without Jitter")
print (myPlot, vp=viewport (
  layout.pos.row=1,
  layout.pos.col=1) )
```

Table 22.2 (continued)

```
myPlot <- qplot ( q1, q4,
        position=position_jitter(x=5,y=5),
        main="Likert Scale With Jitter" )
print(myPlot, vp=viewport(
  layout.pos.row=1,
  layout.pos.col=2) )


#---Detailed Comparison of qplot and ggplot

qplot(pretest, posttest,
  geom=c("point","smooth"), method="lm" )

# Or ggplot with default settings:

ggplot(mydata100, aes(x=pretest, y=posttest) ) +
  geom_point() +
  geom_smooth(method="lm")

# Or with all the defaults displayed:
ggplot() +
layer(
  data=mydata100,
  mapping=aes(x=pretest, y=posttest),
  geom="point",
  stat="identity"
) +
layer(
  data=mydata100,
  mapping=aes(x=pretest, y=posttest),
  geom="smooth",
  stat="smooth",
  method="lm"
) +
coord_cartesian()
```

Chapter 23
Statistics

This Chapter demonstrates some basic statistical methods. More importantly, it shows how even in the realm of fairly standard analyses, R differs sharply from the approach used by SAS and SPSS. Since this book is aimed at people who already know SAS or SPSS, I assume you are already familiar with most of these methods. I will briefly list each test's goal and assumptions, and how to get R to perform them. For more statistical coverage, see Dalgaard's *Introductory Statistics with R* [36], or Venable and Ripley's much more advanced *Modern Applied Statistics in S* [18].

The examples in this chapter will use the mydata100 dataset described in Chap. 20.

Many of examples in this chapter assume that we have attached this data frame with the command attach(mydata100). This enables us to use short variable names like "gender" rather than a longer form like "mydata["gender"]". To get things to fit well on these pages, I have set:

```
options (linesize=64)
```

You can use that if you want your output to match perfectly, but that is not necessary.

23.1 Scientific Notation

While SAS and SPSS tend to print their small probability values as 0.000, R often uses scientific notation. An example is 7.447e–5 which means 7.447×10^{-5} or 0.00007447. When the number after the "e" is negative, you move the decimal place that many places to the left.

You may also see p-values of just "0." That value is controlled by the digits option, which is set to be seven significant digits by default. If you wanted to increase the number of digits to 10, you could do so with the following command.

```
options(digits=10)
```

SAS has a similar option, OPTIONS PROBSIG = 10, but it applies only to the p-values it prints. SPSS has a command that works more like R, SET SMALL 0.0000000001, which affects the printing of all values that contain decimals. In all

R.A. Muenchen, *R for SAS and SPSS Users*, DOI: 10.1007/978-0-387-09418-2_23,   403
© Springer Science+Business Media, LLC 2009

three packages, setting the number of digits affects only their display. The full precision of *p*-values is always available where the numbers are stored.

Supplying a positive number to R's *sci*entific *pen*alty option `scipen` biases the printing away from scientific notation and more towards fixed notation. A negative number does the reverse. So if you want to completely block scientific notation, you can do so with the following command. This is equivalent to "SET SMALL 0." in SPSS.

```
options (scipen=999)
```

23.2 Descriptive Statistics

Let us start with functions that are most like SAS or SPSS and then move onto functions that are more fundamental to R.

The `Hmisc` package [6], written by Frank Harrell, offers a wide selection of functions that have more comprehensive output than the standard R functions. One of these is the `describe` function, which we use below. It is similar to the `summary` function we have used throughout this book.

Before using the package, you must install it. For details, see chap. 5. Then you must load it with either the *Packages> Load Packages* menu item, or the command:

```
library ("Hmisc")
```

You can select variables in many ways. See Chap. 10 for details. One of the nicest features of the `describe` function is that it works on non-factors as well as factors. With our dataset that makes sense, so we can get basic statistics on all variables with one simple command:

```
> describe (mydata100)

mydata100L

8  Variables       100  Observations
------------------------------------------------------
gender
        n missing  unique
       99       1       2

Female (47, 47%), Male (52, 53%)
------------------------------------------------------
workshop
        n missing  unique
       99       1       4

R (31, 31%), SAS (24, 24%), SPSS (25, 25%), Stata (19, 19%)
------------------------------------------------------
q1 : The instructor was well prepared.
```

```
        n missing    unique    Mean
      100        0         5    3.45
              1  2  3  4  5
Frequency  4 14 36 25 21
%            4 14 36 25 21
```
---
q2 : The instructor communicated well.
```
        n missing    unique    Mean
      100        0         5    3.06

              1  2  3  4  5
Frequency 10 28 21 28 13
%           10 28 21 28 13
```
---
q3 : The course materials were helpful.
```
        n missing    unique    Mean
       99        1         5    3.081

              1  2  3  4  5
Frequency 10 20 34 22 13
%           10 20 34 22 13
```
---
q4 : Overall, I found this workshop useful.
```
        n missing    unique    Mean
      100        0         5    3.4

              1  2  3  4  5
Frequency  6 14 34 26 20
%            6 14 34 26 20
```
---
pretest
```
        n  missing unique    Mean    .05    .10    .25    .50
      100        0     23   74.97  66.95  69.00  72.00  75.00
      .75      .90    .95
    79.00    82.00  83.00
```
lowest : 58 62 63 66 67, highest: 82 83 84 85 86
---
posttest
```
        n missing unique    Mean    .05    .10    .25    .50
      100       0     28   82.06  71.95  75.00  77.00  82.00
      .75     .90    .95
    86.00   90.00  92.00
```
lowest : 59 67 68 69 71, highest: 91 92 95 96 98

Unlike SAS and SPSS, the describe function does not provide percents that include missing values. You can change that by setting the exclude. missing argument to FALSE. The describe function will automatically provide a table of frequencies whenever a variable has no more than 20 unique values[1]. Beyond that, it will print the five largest and five smallest values, just like the SAS UNIVARIATE procedure and the SPSS EXPLORE procedure.

Notice that the survey questions themselves appear in the output. That is because this version of the data frame was prepared as described in Chap. 16, using the Hmisc package's approach.

R's built-in function for univariate statistics is summary. We have used the summary function extensively throughout this book, but I repeat its output here for comparison.

```
> summary(mydata100)
    gender      workshop          q1                  q2
 Female:47    R    :31    Min.   :1.00     Min.   :1.00
 Male  :52    SAS  :24    1st Qu.:3.00     1st Qu.:2.00
 NA's  : 1    SPSS :25    Median :3.00     Median :3.00
              Stata:19    Mean   :3.45     Mean   :3.06
              NA's : 1    3rd Qu.:4.00     3rd Qu.:4.00
                          Max.   :5.00     Max.   :5.00


        q3                 q4              pretest
 Min.   :1.000    Min.   :1.0     Min.   :58.00
 1st Qu.:2.000    1st Qu.:3.0     1st Qu.:72.00
 Median :3.000    Median :3.0     Median :75.00
 Mean   :3.081    Mean   :3.4     Mean   :74.97
 3rd Qu.:4.000    3rd Qu.:4.0     3rd Qu.:79.00
 Max.   :5.000    Max.   :5.0     Max.   :86.00
 NA's   :1.000


      posttest
 Min.   :59.00
 1st Qu.:77.00
 Median :82.00
 Mean   :82.06
 3rd Qu.:86.00
 Max.   :98.00
```

As you can see, it is much more sparse lacking percents for factors, frequencies and percents for numeric variables (even if they have a small number of values), number of non-missing values, and so on. However, it is much more compact. The numbers labeled "1st Qu." and "3rd Qu." are the first and third quartiles, or the 25th and 75th percentiles, respectively. Notice that the variable

---

[1] In SPSS, this is what the /FORMAT = LIMIT(20) option does in the FREQUENCIES procedure.

labels are now ignored (at the time of this writing, only Hmisc functions display them) and adding value labels to the q variables would convert them to factors, so we would no longer get means and quartiles.

The built-in summary function is applicable to a wide range of R objects as we will see. Hmisc's describe function works only with data frames, vectors, matrices, or formulas. For data frames, choose the function that meets your needs.

Now let us review R's built-in functions for frequencies and proportions. We have covered those in earlier sections also, but I repeat them here for ease of comparison and elaboration.

R's built-in function for frequency counts provide output that is much more sparse than those of the describe function.

```
> table(workshop)

workshop
    R   SAS  SPSS  Stata
   31    24    25    19
> table(gender)

gender
Female    Male
    47      52
```

The above output is quite minimal, displaying only the frequencies. As minimal as this is, it is very easy to use data in this form in other functions such as barplot, see Chap. 21 for examples.

We can get proportions by using the prop.table function.

```
> prop.table( table(workshop) )

workshop
        R        SAS       SPSS       Stata
0.3131313 0.2424242 0.2525253 0.1919192
>
> prop.table( table(gender) )

gender
   Female       Male
0.4747475 0.5252525
```

You can round off the proportions using the round function. The only arguments you need are the object to round and the number of decimals you would like to keep.

```
> round( prop.table( table(gender) ), 2 )
gender
Female    Male
  0.47    0.53
```

Converting that to percents is of course just a matter of multiplying by 100. If you multiply before rounding, you will not even need to specify the number of decimals to keep since the default is to round to whole numbers.

```
> round( 100* ( prop.table( table(gender) ) ) )
gender
Female    Male
    47      53
```

A word of caution about the `table` function. Unlike the `summary` function, if you use it on a whole data frame, it will not give you all one-way frequency tables. Instead, it will cross-tabulate all the variables at once. You can use it on a surprising number of factors at once and, when convert its output to a data frame, have a concise listing of how many observations you have of all possible combinations. For an example, see Sect. 14.12.4 in Chap. 14.

R's built-in functions offer similarly sparse output for univariate statistics. To get just the means of variables q1 through posttest (variables 3 through 8), we can use:

```
> sapply( mydata100[3:8], mean,   na.rm=TRUE)

    q1       q2       q3       q4   pretest  posttest
3.4500   3.0600   3.0808   3.4000   74.9700   82.0600
```

Similarly, for the standard deviations:

```
> sapply( mydata100[3:8], sd,     na.rm=TRUE)

    q1       q2       q3       q4   pretest  posttest
1.0952   1.2212   1.1665   1.1371    5.2962    6.5902
```

You can also substitute the `var` function for variance or the `median` function for that statistic. You can apply several of these functions at once by combining them into your own single function. For an example of that, see Sect. 8.5.6 in Chap. 8. For details about the `sapply` function, see Sect. 14.2.

23.3 Cross-Tabulation

You can compare groups on categorical measures with the Chi-squared test. For example, testing to see if males and females attended the various workshops in the same proportions.

Assumptions:

- No more than 20% of the cells in your cross-tabulation have counts less than 5. If you have sparse tables, the exact `fisher.test` function is more appropriate.
- Observations are independent. For example, if you measured the same subjects repeatedly, it would be important to take that into account in a more complex model.
- The variables are not the same thing measured at two times. If that is the case, the `mcnemar.test` function may be what you need.

To get output most like that from SAS and SPSS, we will first use the functions from Gregory Warnes' gmodels package [37]. Then we will cover the cross-tabulation functions that are built into R.

```
> library("gmodels")

> CrossTable(workshop, gender,

+   chisq=TRUE, format="SAS")
```

```
   Cell Contents
|--------------------|
|                 N  |
| Chi-square contribution  |
|         N / Row Total  |
|         N / Col Total  |
|       N / Table Total  |
|--------------------|
```

Total Observations in Table: 99

workshop	gender Female	Male	Row Total
R	14	17	31
	0.035	0.032	
	0.452	0.548	0.313
	0.298	0.327	
	0.141	0.172	
SAS	11	13	24
	0.014	0.012	
	-0.458	0.542	0.242
	0.234	0.250	
	0.111	0.131	
SPSS	13	12	25
	0.108	0.097	
	0.520	0.480	0.253
	0.277	0.231	
	0.131	0.121	
Stata	9	10	19
	0.000	0.000	
	0.474	0.526	0.192
	0.191	0.192	
	0.091	0.101	

```
Column Total |         47 |         52 |         99 |
             |      0.475 |      0.525 |            |
-----------|--------|--------|--------|
```

Statistics for All Table Factors

Pearson's Chi-squared test

```
-------------------------------------------------------
Chi^2 =  0.2978553     d.f. =  3       p =  0.9604313
```

The CrossTable function call above has three arguments.

1. The variable that determines the table rows.
2. The variable that determines the table columns.
3. The chisq = TRUE argument tells R to perform that test. As with SAS and SPSS, if you leave this argument out, it will perform the cross-tabulation but not the Chi-squared test.
4. The format="SAS" argument tells it to create a table as SAS would. That is the default so you do not need to list it if that is what you want. Placing "SPSS" here would result in a table in the style of SPSS.

R also has built-in functions to do cross-tabulation and the Chi-squared test. As usual, the built-in functions present sparse results. We will first use the table function. To simplify the coding and to demonstrate a new type of data structure, we will save the table and name it myWG for *workshop* and *gender*.

```
> myWG <- table(workshop, gender)
```

Printing myWGtable will show us that it contains the form of counts to which SAS and SPSS users are accustomed.

```
> myWG
         gender
workshop Female Male
   R         14   17
   SAS       11   13
   SPSS      13   12
   Stata      9   10
```

You may recall from our discussion of factors that you can create factor levels (and their labels) that do not exist in your data. That will help if you were to enter more data later that is likely to contain those values, or if you were to merge your data frame with others that have a full set of values. However, when performing a cross-tabulation, the levels with zero values will become part of the table. These empty cells will affect the resulting Chi-squared statistic which will drastically change its value. To get rid of the unused levels, append [,drop = TRUE] to the variable reference. For example:

```
myWG <- table( workshop[ ,drop=TRUE], gender)
```

Some R functions work better with this type of data in a data frame. You probably associate this style of tabular data with output from the SAS SUM-MARY procedure or SPSS AGGREGATE procedure. The as.data.frame function can provide it.

```
> myWGdata <- as.data.frame(myWG)

> myWGdata

  workshop gender Freq
1        R Female   14
2      SAS Female   11
3     SPSS Female   13
4    Stata Female    9
5        R   Male   17
6      SAS   Male   13
7     SPSS   Male   12
8    Stata   Male   10
```

The functions we discuss now work well on table objects, so we will use myWG. We can use the chisq.test function to perform the Chi-squared test.

```
> chisq.test(myWG)

        Pearson's Chi-squared test

data:  myWG
X-squared = 0.2979, df = 3, p-value = 0.9604
```

The table function does not calculate any percents or proportions. To get row or column proportions, we can use the prop.table function. The arguments in the example below are the table and the margin to analyze where 1 = row and 2 = column. So this will calculate row proportions.

```
> prop.table(myWG, 1)

          gender
workshop   Female      Male
     R    0.45161   0.54839
   SAS    0.45833   0.54167
  SPSS    0.52000   0.48000
 Stata    0.47368   0.52632
```

Similarly, changing the 1 to 2 requests column proportions.

```
> prop.table(myWG, 2)

          gender
workshop   Female      Male
     R    0.29787   0.32692
   SAS    0.23404   0.25000
```

```
    SPSS     0.27660   0.23077
   Stata     0.19149   0.19231
```

If you do not provide the margin argument, the function will calculate total proportions.

```
> prop.table (myWG)
         gender
workshop   Female      Male
    R      0.14141   0.17172
   SAS     0.11111   0.13131
   SPSS    0.13131   0.12121
   Stata   0.09091   0.10101
```

The round function will round off unneeded digits by telling it how many decimal places you want, in this case 2. These are the row proportions.

```
> round ( prop.table(myWG, 1 ), 2  )
         gender
workshop  Female  Male
    R        0.45  0.55
   SAS       0.46  0.54
   SPSS      0.52  0.48
   Stata     0.47  0.53
```

To convert proportions to percents, multiply by 100 and round off. If you want to round to the nearest whole percent, multiply by 100 before rounding off. That way you do not even have to tell the round function how many decimal places to keep as its default is to round off to whole numbers.

```
> round ( 100 *    ( prop.table(myWG, 1 )  )  )
         gender
workshop  Female  Male
    R          45    55
   SAS         46    54
   SPSS        52    48
   Stata       47    53
```

If you wish to add marginal totals, the addmargins function will do so. It works much like the prop.table function; in that its second argument is 1 to add a row with totals or 2 to add a column.

```
> addmargins ( myWG, 1 )
         gender
workshop  Female  Male
    R          14    17
   SAS         11    13
```

```
SPSS           13      12
Stata           9      10
Sum            47      52
```

If you do not specify a preference for row or column totals, you will get both.

```
> addmargins ( myWG )
         gender
workshop  Female  Male  Sum
   R          14    17   31
   SAS        11    13   24
   SPSS       13    12   25
   Stata       9    10   19
   Sum        47    52   99
```

23.4 Correlation

Correlations measure the strength of linear association between two continuous variables.

Assumptions:

- Scatterplots of the variables shows essentially a straight line. The function plot(x, y) would do the scatterplots. If you have a curve, transformations such as square roots or logarithms often help.
- The spread in the data is the same at low, medium, and high values. Transformations often help with this assumption also.
- For a Pearson correlation, the data should be at least interval-level and normally distributed. As discussed in Chap. 21, hist(myvar) or qqnorm(myvar) is a quick way to examine the data. If your data are not normally distributed, or are just ordinal measures (e.g., low, medium, high), you can use the non-parametric Spearman correlation or the (less popular) Kendall correlation.

The Hmisc package has a good function for doing correlations. We will use it first, and then cover the functions that are built into R. Before using the Hmisc package, you must install it. See Section 5.1 for details. Then you must load it with either the *Packages> Load Packages* menu item, or the command:

```
library ("Hmisc")
```

To correlate two variables, simply use their names as the arguments to the rcorr function.

```
> rcorr ( q1,q4 )

      x     y
x  1.00  0.58
y  0.58  1.00
```

```
n = 100

P
   x  y
x     0
y  0
```

The first piece of output is the Pearson correlation of 0.58. That is a strong positive relationship, at least for the social sciences. The more a student likes the instructor (q1) the more he or she likes the workshop overall. Next it reports the number of valid observations, 100. Since the rcorr function uses pairwise deletion of missing values by default, each correlation is done on the maximum number of cases. Finally, it reports the p-value of 0. So we can reject the null hypothesis that the correlation is zero; a significant result. When the p-value gets very small, it prints just "0" but the full accuracy is available if you save the output (example below).

To get correlations on a set of variables, you must first put them into a matrix. That is easily done with the cbind function that binds the columns. Below, I place the survey questions into a matrix called myQs and then use it as the first argument for the rcorr function call. I also specify the type of correlation as Pearson. That is the default, so I am just doing it here for educational purposes. Note that although Pearson in a proper noun, as an argument it does *not* begin with a capital "P".

```
> myQs  <- cbind (q1,q2,q3,q4)

> rcorr ( myQs, type="pearson"  )

      q1   q2   q3   q4
q1 1.00 0.67 0.69 0.58
q2 0.67 1.00 0.60 0.49
q3 0.69 0.60 1.00 0.47
q4 0.58 0.49 0.47 1.00

n
     q1  q2 q3  q4
q1 100 100 99 100
q2 100 100 99 100
q3  99  99 99  99
q4 100 100 99 100

P
    q1  q2 q3 q4
q1      0  0  0
q2  0      0  0
q3  0  0      0
q4  0  0  0
```

You can change that value to "spearman" to get correlations on variables that are not normally distributed, or are only ordinal in scale (e.g., low, medium, high).

```
> rcorr (myQs, type ="spearman")

       q1   q2   q3   q4
q1 1.00 0.67 0.67 0.55
q2 0.67 1.00 0.61 0.49
q3 0.67 0.61 1.00 0.43
q4 0.55 0.49 0.43 1.00

n
     q1  q2 q3  q4
q1 100 100 99 100
q2 100 100 99 100
q3  99  99 99  99
q4 100 100 99 100

P
    q1 q2 q3 q4
q1      0  0  0
q2 0       0  0
q3 0  0        0
q4 0  0  0
```

If you want to see the very small p-values, you can save the output to a list object like myCorrs. We will use the names function to see what components it contains.

```
> myCorrs  <- rcorr (myQs, type ="pearson")

> names ( myCorrs )

[1 ]   "r"   "n"   "P"
```

Notice that the correlations are stored in a component named "r" even though that name did not appear in our output above. The print function determines how output prints, and when we installed the Hmisc package, it provided the methods for the print function to use on its objects. The method that printed the output from rcorr included the instruction not to print the "r", as it is obvious what the correlations are.

We can guess that the "P" object contains the p-values, so let us print just myCorrs$P. By changing the option digits = 7 to various values (7 is the default) you can adjust the precision of the output. As you can see, all the p-values are so tiny that it is not worth the effort. Getting at p-values like this can be helpful though when you want to do something like a Bonferroni correction for multiple testing. You could also set options (scipen = 999) to block scientific notation, but I do not show this as the numbers are *very* small.

```
> options (digits =7 )

> myCorrs$P

              q1              q2              q3              q4
q1            NA 3.597123e -14 1.554312e -15 3.646581e -10
q2 3.597123e-14             NA 3.602896e-11 2.040895e-07
q3 1.554312e-15 3.602896e-11             NA 7.624040e-07
q4 3.646581e-10 2.040895e-07 7.624040e-07             NA
```

Now let us take a look at R's built-in functions. As usual, they provide more sparse output.

```
> cor ( data.frame (q1, q2, q3, q4 ),
+    method ="pearson", use ="pairwise")

          q1        q2        q3        q4
q1 1.0000000 0.6668711 0.6948419 0.5758860
q2 0.6668711 1.0000000 0.6040746 0.4917447
q3 0.6948419 0.6040746 1.0000000 0.4730732
q4 0.5758860 0.4917447 0.4730732 1.0000000
```

The cor function call above uses three arguments.

1. The variables to correlate. This can be a vector, matrix, or data frame. If it is a vector, then the next argument must be another vector with which to correlate. You can label them x= and y=, or just put them in the first two positions.
2. The method argument can be pearson, spearman, or kendall for those types of correlations. Be careful not to capitalize these names when used as arguments in R.
3. The use value determines how the function will deal with missing data. The value pairwise.complete, abbreviated pairwise above, uses as much data as possible. That is the default value in SAS and SPSS.

The value complete.obs is the SAS/SPSS equivalent of listwise deletion of missing values. This tosses out cases that have *any* missing values for the variables analyzed. If each variable has just a few missing values, but each are missing on different cases, you can lose a very large percent of your data with this option. However, it does ensure that every correlation is done on the exact same cases. That can be important if you plan to use the correlation matrix in additional computations.

As usual, by default R provides no results if it finds missing values. That is the use = all.obs setting. So if you omit the use argument and have missing values, cor will print only an error message telling you it has found missing values.

Unlike Hmisc's rcorr function, the built-in cor function provides only a correlation matrix. Another built-in function, cor.test, provides comprehensive output but only for two variables at a time.

```
> cor.test (q1, q2, use ="pairwise ")

          Pearson 's product -moment correlation

data:  q1 and q2

t  = 8.8593, df  = 98, p -value  = 3.597e -14
alternative hypothesis: true correlation is not equal to 0

95 percent confidence interval:
 0.5413638 0.7633070

sample estimates:
      cor
0.6668711
```

23.5 Linear Regression

Linear regression models the linear association between one continuous dependent variable and a set of continuous independent or predictor variables.
   Assumptions:

- Scatterplots of the dependent variable with each independent variable shows essentially a straight line. The function plot(x, y) would do the scatterplots. If you have a curve, transformations such as square roots or logarithms often help straighten it out.
- The spread in the data is the same at low, medium, and high values. Transformations often help with this requirement also.
- The model residuals (difference between the predicted values and the actual values) are normally distributed. We will use the plot function to generate a normal QQ plot to test this assumption and others.
- The model residuals are independent. If they contain a relationship, such as the same subjects measured through time, or classes of subjects sharing the same teacher, you would want to use a more complex model to take that into account.

When performing a single type of analysis in SAS or SPSS, you prepare your commands and then submit them to get all your results at once. You could save some of the output using the SAS Output Delivery System (ODS), SAS or SPSS Output Management System (OMS), SPSS, but it is not a routine part of a typical analysis (other than ODS graphics on/off). R on the other hand shows you very little output with each command and everyone uses its integrated output management capabilities.

Let us look at a simple example. First, we will use the lm function to do a *l*inear *m*odel predicting the values of q4 from the other survey questions. Although this type of data is viewed as ordinal scale by many, social scientists often view it as interval-level. With a more realistic dataset, we would be working with a scale for each measure that consisted of the means of several questions, resulting in far more than just five values.

```
> lm ( q4 ~ q1+q2+q3, data =mydata100 )

Call:
lm (formula  = q4 ~ q1 + q2 + q3, data  = mydata100 )

Coefficients:
 (Intercept )              q1              q2              q3
    1.20940          0.41134         0.15791         0.09372
```

We see that the results provide only the coefficients to the linear regression model. So the model is:

Predicted q4 = 1.20940 + 0.41134*q1 + 0.15791*q2 + 0.09372*q3

The other results that SAS or SPSS would provide, such as R-squared or tests of significance are not displayed. Now we will run the model again and save its results in an object called myModel.

```
> myModel <- lm (q4 ~ q1 + q2 + q3, data =mydata100 )
```

This time, no printed results appear. We can print the contents of myModel by entering its name just like any other R object.

```
> myModel

Call:
lm (formula  = q4  ~ q1  + q2  + q3, data  = mydata100 )

Coefficients:
 (Intercept )              q1              q2              q3
    1.20940          0.41134         0.15791         0.09372
```

The above results are exactly the same as we saw initially. So what type of object is myModel? The mode and class functions can tell us.

```
> mode (myModel )

[1 ]  "list"

> class (myModel )

[1 ]  "lm"
```

So we see the lm function saved our model as a list with a class of lm. Now that we have the model stored, we can apply a series of *extractor functions* to get much more information. Each of these functions will see the class of lm and will apply the methods that it has available for that class of object. We have used the summary function before. With our data frame object, mydata, the summary function "knew" to get frequency counts on factors and other measures, like means, on continuous variables. Here is what it will do with lm objects.

```
> summary (myModel )

Call:
lm (formula  = q4 ~ q1 + q2 + q3, data  = mydata100 )
```

```
Residuals:
Overall, I found this workshop useful.
    Min      1Q   Median      3Q      Max
-1.9467  -0.6418   0.1175  0.5960   2.0533

Coefficients:
              Estimate Std. Error t value Pr (>|t |)
  (Intercept )  1.20940    0.31787   3.805 0.000251   ***
q1              0.41134    0.13170   3.123 0.002370   **
q2              0.15791    0.10690   1.477 0.142942
q3              0.09372    0.11617   0.807 0.421838
- - - - -
Signif. codes:  0 '***' 0.001 '**' 0.01 '*' 0.05 '.
  ' 0.1 '  ' 1

Residual standard error: 0.9308 on 95 degrees of
  freedom
  (1 observation deleted due to missingness )
Multiple R-squared: 0.3561,  Adjusted R-squared: 0.3358
F-statistic: 17.51 on 3 and 95 DF,  p -value: 3.944e-09
```

That is the much the same output we would get from SAS or SPSS. The main difference is the summary of model residuals. The *t*-tests on the model parameters are partial tests of each parameter, conditional upon the other independent variables being in the model. These are sometimes called type III tests. From the perspective of this table it might appear that neither q2 nor q3 are adding significantly to the model. However, the next table will provide a different perspective. The very small *p*-value of 3.944e–09 or 0.000000003944 is smaller than 0.05, so we would reject the hypothesis that the overall model is worthless. The significant *p*-value of 0.002370 for q1 makes it the only significant predictor, given the other variables in the model. We will test that below.

You can ask for an analysis of variance (ANOVA) table with the anova function.

```
> anova (myModel)

Analysis of Variance Table
Response: q4
           Df Sum Sq Mean Sq F value     Pr (>F )
q1          1 42.306  42.306 48.8278 3.824e -10   ***
q2          1  2.657   2.657  3.0661    0.08317 .
q3          1  0.564   0.564  0.6508    0.42184
Residuals 95 82.312   0.866
- - - - -
Signif. codes:  0 '***' 0.001 '**' 0.01 '*' 0.05 '. '
  0.1 '  ' 1
```

These tests are what SAS and SPSS would call sequential or Type I tests. So the first test is for q1 by itself. The second is for q2 given that q1 is already in the model. The third is for q3 given that q1 and q2 are already in the model. Changing the order of the variables in the `lm` function would change these results. From this perspective, q1 is even more significant.

23.5.1 Plotting Diagnostics

The `plot` function also has methods for lm class objects. The single call to the `plot` function below was sufficient to generate all four of the plots that follow.

```
> plot (myModel)
```

The *Residuals vs. Fitted* plot in Fig. 23.1, upper left, shows the fitted values plotted against the model residuals. If the residuals follow any particular

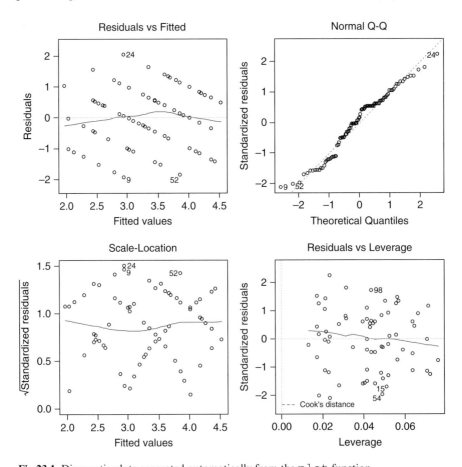

Fig 23.1 Diagnostic plots generated automatically from the `plot` function

pattern, such as a diagonal line, there may be other predictors not yet in the model that could improve it. The fairly flat lowess line looks good.

The *Normal Q-Q Plot* in Fig. 23.1, upper right, shows the quantiles of the standardized residuals plotted against the quantiles you would expect if the data were normally distributed. Since these fall mostly on the straight line, the assumption of normally distributed residuals is met.

The *Scale-Location* plot in Fig. 23.1, lower left, shows the square root of the absolute standardized residuals plotted against the fitted, or predicted, values. Since the lowess line that fits this is fairly flat, it indicates that the spread in the predictions is roughly the same across the prediction line, meeting the assumption of homoscadasticity.

Finally, the *Residuals vs. Leverage* plot in Fig. 23.1, lower right, shows a measure of the influence of each point on the overall equation against the standardized residuals. Since no points stand out far from the pack, we can assume that there are no outliers having undue influence on the fit of the model.

23.5.2 Comparing Models

To do stepwise models, first create the full model and save it, then apply one of the functions step, add1, drop1 or, from the MASS package, stepAIC. Keep in mind that if you start with a large number of variables, stepwise methods make your *p*-values essentially worthless. If you can choose a model on part of your data and see that it still works on the remainder, then the *p*-values obtained from the remainder data should be valid.

The anova function can also compare two models so long as one is a subset of the other. So to see if q2 and q3 combined added significantly to a model over q1, I created the two models and then compared them. One catch with this scenario is that with missing values present, the two models will be calculated on different sample sizes and R will not be able to compare them. It will warn you, "models were not all fitted to the same size of dataset." So we will use the na.omit function discussed in Sect. 14.5 to remove the missing values first.

```
> myNoMissing   <- na.omit (mydata [ , c ("q1", "q2",
  "q3", "q4") ] )
> myFullModel      <- lm ( q4   ~ q1 +q2 +q3,
  data =myNoMissing )
> myReducedModel  <- lm ( q4   ~ q1,
  data =myNoMissing )
> anova ( myReducedModel, myFullModel )
Analysis of Variance Table

Model 1: q4 ~ q1
Model 2: q4 ~ q1  + q2  + q3
```

```
    Res.Df      RSS Df Sum of Sq        F Pr (>F )
1       97 85.532
2       95 82.312  2      3.220 1.8585 0.1615
```

So in this case, the *p*-value of 0.1615 tells us that variables q2 and q3 combined did not add significantly to the model that had only q1. When you compare two models using the `anova` function, you should list the reduced model first.

23.5.3 Making Predictions with New Data

You can apply a saved model to a new set of data using the `predict` function. Of course, the new dataset must have the same variables, with the same names, and same types. Under those conditions, you can apply myModel to myNew-Data using:

```
myPredictions  <- predict (myModel, myNewData )
```

Because the variables must have the exact same names, it is very important *not* to use the $ format in names. The following two statements perform what is statistically the same model, but the one using mydata$q4 will only work on future data frames named "mydata"!

```
myModel  <- lm ( mydata$q4 ~ mydata$q1 )   #Bad for
  predicting.
myModel  <- lm (q4 ~ q1, data =mydata100 )  #Good for
  predicting.
```

23.6 t-Test – Independent Groups

A *t*-test for independent groups compares two groups at a time on the mean of a continuous measure. Its assumptions are:

- The measure is interval-level continuous data. If the data are only ordinal (e.g., low–medium–high), consider the Wilcoxon rank sum test, also known as the Mann–Whitney test.
- The measures are normally distributed. You can examine that with `hist` (myVar) or qqnorm (myvar) as shown in Chap. 21. If they are not, consider the Mann–Whitney–Wilcoxon test. For details see Sect. 23.9.
- Observations are independent. For example, if you measured the same subjects repeatedly, it would be important to take that into account in a more complex model.
- The variances in the two groups should be comparable.

You can perform *t*-tests in R using the t.test function.

```
> t.test ( q1 ~ gender, data =mydata100 )
        Welch Two Sample t -test
data:   q1 by gender
t  = 4.9784, df  = 97.998, p -value  = 2.748e -06
alternative hypothesis: true difference in means
is not equal to 0
95 percent confidence interval:
 0.58789 1.36724
sample estimates:
mean in group Female    mean in group Male
              3.9583                 2.9808
```

The t.test function call above has two arguments, the formula and the data frame to use with it.

The formula q4 $\sim$ gender is in the form dependent $\sim$ independent. For details, see Sect. 8.5.2.

Instead of a formula, you can specify two variables to compare such as t.test(x, y) and they can be selected using any of R's many variable selection approaches. The data argument specifies the data frame to use *only in the case of a formula*. So you might think that this form works:

```
t.test ( q4 [which (gender =='m';),   ],
         q4 [which (gender =='f'),   ] , data = mydata )
```

But unless the data is attached, it does not! You would have to enter attach (mydata) before the command above would work. Alternatively, with an unattached data frame, you could leave off the data argument and use the form:

```
with (mydata,
   t.test ( q4 [ which (gender =='m')   ],
            q4 [ which (gender =='f')   ] )
)
```

Or, you might prefer the subset function:

```
t.test (
   subset (mydata100, gender =="m ", select =q4 ),
   subset (mydata100, gender =="f ", select =q4 )
)
```

For more details see Sect. 10.8, attach *and* with, in Chap. 10.

The results show that the mean for the females is 3.96 and for the males is 2.98. The *p*-value of 2.748e–06, or 0.000002748 is much smaller than 0.05 so we would reject the hypothesis that the means are the same.

Unlike SAS and SPSS, the `t.test` function does not provide a test for homogeneity of variance and two *t*-test calculations for equal and unequal variances. Instead, it provides only the unequal variance test by default. The additional argument `var.equal = TRUE` will perform the other test. See Sect. 23.7 below, for that topic.

23.7 Equality of Variance

SAS and SPSS offer tests for equality (homogeneity) of variance in their *t*-test and ANOVA procedures. R, in keeping with its minimialist perspective, offers such tests in separate functions. The Levene test for equality of variance is the most popular. John Fox's `car` package [10] contains the `levene.test` function. If you have not installed this package, see Sec. 5.1.

```
> library ("car")

> levene.test (posttest, gender)

Levene 's Test for Homogeneity of Variance

        Df F  value   Pr (>F)
group   1    0.4308   0.5131
        97
```

The `levene.test` function has only two arguments of the form (var, group). Its null hypothesis is that your groups have equal variances. If its *p*-value is smaller than 0.05, you reject that hypothesis. So in the case above, we cannot reject the hypothesis that the variances are equal.

Other tests for comparing variances are R's built-in `var.test` and `bartlett.test` functions.

23.8 *t*-Test – Paired or Repeated Measures

The goal of a paired *t*-test is to compare the mean of two correlated measures. These are often the same measure taken on the same subjects at two different times. Its assumptions are:

- The two measures are interval-level continuous data. If the data are only ordinal (e.g., low–medium–high), consider the Wilcoxon signed-rank test. For details see Sect. 23.10 below.
- The *differences* between the measures are normally distributed. You can examine that with `hist (posttest-pretest)` or `qqnorm`

(posttest-pretest) as shown in Chap. 21. If they are not, consider the Wilcoxon signed-rank test. For details, see Sect. 23.10.

- Other than the obvious pairing, observations are independent. For example, if siblings were also in your dataset, it would be important to take that into account in a more complex model.

You can perform the paired t-tests in R using the t.test function. This example assumes the data frame is attached so we can use the short form of variable names.

```
> ttest (posttest, pretest, paired=TRUE)

        Paired t-test

data:  posttest and pretest

t = 144597, df = 99, p-value < 22e-16
alternative hypothesis: true difference in means is
 not equal to 0

95 percent confidence interval:
 6.11708 8.06292
sample estimates:
mean of the differences
              7.09
```

The t.test function call above has three main arguments.

1. The first variable to compare.
2. The second test variable to compare. The function will subtract this from the first to get the mean difference. If these two were reversed, the p-value would be the same but the mean would be negative -7.09 and the confidence interval would be around that.
3. The paired = TRUE argument tells it that the two variables are correlated rather than independent. *It is critical that you set this option for a paired test!* If you forget this, it will perform an independent-groups t-test instead.

The results show that the mean difference of 7.09 is statistically significant with a p-value of 2.2e–16 or 0.00000000000000022. We can reject the null hypothesis that the means are equal.

23.9 Wilcoxon Mann–Whitney Rank Sum Test – Independent Groups

The Wilcoxon rank sum test, also known as the Mann–Whitney U test, compares two groups on the mean rank of a dependent variable that is at least ordinal (e.g., low, medium, high). Its assumptions are:

- The distributions in the two groups must be the same, other than a shift in location. That is, the distributions should have the same variance and skewness. Examining a histogram is a good idea if you have at least 30 subjects. A boxplot is also helpful regardless of the number of subjects in each group.
- Observations are independent. For example, if you measured the same subjects repeatedly, it would be important to take that into account in a more complex model.

The wilcox.test function works very much like the t.test function.

```
> wilcox.test ( q1 ~ gender, data=mydata100 )
        Wilcoxon rank sum test with continuity correction
data:   q1 by gender
W = 18415, p-value = 6.666e-06
alternative hypothesis: true location shift is not
  equal to 0
```

The wilcox.test function call above has two main arguments.

1. The formula q4 $\sim$ gender is in the form dependent $\sim$ independent. For details, see Sect. 8.5.2.
   Instead of a formula, you can specify two variables to compare such as wilcox.test (x, y) and they can be selected using any of R's many variable selection approaches. See the examples in Sect. 23.6.
   For more details see Sect. 10.8 in Chap. 10.
2. The data argument specifies the data frame to use. The wilcox.test function will extract all the variables in the formula from this data frame.

The p-value of 6.666e–06 or 0.000006666 is less than 0.05, so we would reject the hypothesis that the males and females have the same mean ranks on the q4 variable. The median is the more popular measure of location to report on for a non-parametric test and we can apply that function to calculate them using the aggregate function. Recall that aggregate requires its group argument be a list or data frame.

```
> aggregate ( q1, dataframe(gender),
+    median, na.rm=TRUE)

   gender x
1  Female 4
2    Male 3
```

23.10 Wilcoxon Signed-Rank Test – Paired Groups

The goal of a Wilcoxon signed-rank test is to compare the mean ranks of two correlated measures that are at least ordinal (e.g., low, medium, high) in scale. These are often the same measure taken on the same subjects at two different times. Its only assumptions are:

- The distribution of the difference between the two measures has a symmetric distribution.
- Other than the obvious pairing, the observations are independent. For example, if your data also contained siblings, you would want a more complex model to make the most of that information.

This test works very much like the t.test function with the paired argument.

```
> wilcox.test ( posttest, pretest, paired=TRUE)
        Wilcoxon signed rank test with continuity correction

data:   posttest and pretest

V   = 5005, p-value < 2.2e-16
alternative hypothesis: true location shift is not equal to 0
```

The wilcox.test function call above has three main arguments.

1. The first variable to compare.
2. The second variable to compare. The function will subtract this from the first and then convert the difference to ranks.
3. The paired = TRUE argument tells it that the variables are correlated. *Be careful as without this argument, it will perform the Wilcoxon Rank Sum test for independent groups.* That is a completely different test that would be inappropriate for correlated data.

The *p*-value of 2.2e–16 is less than 0.05, so we would conclude that the two tests are not the same. As Dalgaard points out, with a sample size of 6 or fewer, it is impossible to achieve a significant result with this test [36].

The median is the more popular measure of location to report on for a non-parametric test and we can calculate them with the median function.

```
> median (pretest)
      [1] 75
> median (posttest )
      [1] 82
```

23.11 Analysis of Variance

An analysis of variance (ANOVA or AOV) tests for group differences on the mean of a continuous variable divided up by one or more categorical factors. Its assumptions are:

- The measure is interval-level continuous data. If the data are only ordinal (e.g., low–medium–high), consider the Kruskal–Wallis test for a single factor (i.e., one-way ANOVA).
- The measure is normally distributed. You can examine that with hist (myVar) or qqnorm (myvar) as shown in Chap. 21. If they are not, consider the Kruskal–Wallis test for a single factor (i.e., one-way ANOVA).

- The observations are independent. For example, if each group contains subjects that were measured repeatedly over time or who are correlated (e.g., same family), you would want a more complex model to make the most of that information.
- The variance of the measure is the same in each group.

We can get the group means using the aggregate function.

```
> aggregate ( posttest,
+    data.frame (workshop ),
+    mean, na.rm=TRUE)

  workshop      x
1        R 86.258
2      SAS 79.625
3     SPSS 81.720
4    Stata 78.947
```

Similarly, we can get variances by applying the var function.

```
> aggregate ( posttest,
+    dataframe (workshop),
+    var, na.rm=TRUE)

  workshop      x
1        R 24998
2      SAS 37549
3     SPSS 19543
4    Stata 73608
```

You can see the variance for the Stata group is quite a bit higher than the rest. Levene's test will provide a test of significance for that.

```
> levenetest (posttest, workshop)

Levene's Test for Homogeneity of Variance
        Df F value  Pr(>F)
group    3    2.51 0.06337 .
        95

- - - - -
Signif codes:  0 '***' 0001 '**' 001 '*' 005 '.' 01 ' ' 1
```

The Levene test's null hypothesis is that the variances do not differ. Since it calculated a p-value of 0.06337, we can conclude that the differences in variance are not significant.

The aov function calculates the ANOVA.

```
> myModel <- aov( posttest ~ workshop, data=mydata100 )
```

The aov function call above has two arguments.

1. The formula `posttest ~ workshop` is in the form dependent $\sim$ independent. The independent variable must be a factor. See Sect. 8.5.2, for details about models with more factors, interactions, nesting, and so forth.
2. The `data` argument specifies the data frame to use for the formula. Alternatively, you could enter `attach(mydata)` before using the `aov` function if you wanted to use the short form of variable names without specifying the data argument.

We can see some results by printing myModel.

```
> myModel

Call:
    aov(formula = posttest ~ workshop, data = myta100)

Terms:
                  workshop Residuals
    Sum of Squares   875.442   340.7548
    Deg of Freedom         3         95

    Residual standard error: 5.989067
    Estimated effects may be unbalanced
    1 observation deleted due to missingness
```

Those are the same results you will get from `summary(myModel)`. The anova function provides the ANOVA table.

```
> anova(myModel)

Analysis of Variance Table

Response: posttest
             Df Sum Sq Mean Sq F value    Pr (> F )
workshop      3  875.4   291.8  8.1356 7.062e -05 ***
Residuals    95 3407.5    35.9
-----
Signif. codes:   0  '***' 0.001 '**' 0.01 '*' 0.05 '.'
    0.1 ' ' 1
```

Given the p-value of 7.062e–05 or 0.00007062 we would reject the hypothesis that the means are all the same. But which ones differ? The `pairwise.t.test` function provides all possible t-tests, and corrects them for multiple testing using the Holm method by default. In our case, we are doing six t-tests, so the best (smallest) p-value is multiplied by 6, the next best by 5, and so on. This is also called a sequential Bonferroni correction.

```
> pairwise.t.test (posttest, workshop)

Pairwise comparisons using t tests with pooled SD
```

```
data:  posttest and workshop

        R          SAS        SPSS
SAS    0.00048    -          -
SPSS   0.02346    0.44791    -
Stata  0.00038    0.71335    0.39468

P value adjustment method: holm
```

We see that the posttest scores are significantly different for R compared to the other three. The mean scores for the SAS, SPSS, and Stata workshops do not differ significantly among themselves.

An alternate comparison approach is to use Tukey's HSD test. The `TukeyHSD` function call below uses only two arguments, the model and the factor whose means you would like to compare.

```
> TukeyHSD (myModel,  "workshop")

Tukey multiple comparisons of means
  95% family-wise confidence level

Fit: aov (formula = posttest ~ workshop, data = mydata100 )

$workshop
                 diff        lwr        upr      p adj
SAS-R        -6.63306   -10.8914   -2.37472   0.00055
SPSS -R      -4.53806    -8.7481   -0.32799   0.02943
Stata-R      -7.31070   -11.8739   -2.74745   0.00036
SPSS-SAS      2.09500     2.3808    6.57078   0.61321
Stata-SAS    -0.67763    -5.4871    4.13185   0.98281
Stata-SPSS   -2.77263    -7.5394    1.99416   0.42904
```

The *diff* column provides mean differences. The *lwr* and *upr* columns provide lower and upper 95% confidence bounds, respectively. Finally, the *p adj* column provides the *p*-values adjusted for the number of comparisons made. The conclusion is the same, the R group's posttest score differs from the others, but they do not differ among themselves.

We can graph these results using the plot function (Fig. 23.2).

```
> plot( TukeyHSD(myModel, "workshop") )
```

The plot function also provides appropriate diagnostic plots (not shown). These are the same plots shown and discussed in Sect. 23.5.1 above.

```
> plot(myModel)
```

You can perform much more complex ANOVA models in R but they are beyond the scope of this book. See Sect. 8.5.2 for ways to specify interactions and nesting. A good book on ANOVA is Pinheiro and Bates' *Mixed Effects Models in S and S-Plus* [38].

Fig 23.2 A plot of the Tukey HSD test results showing the R group differing from the other three workshops

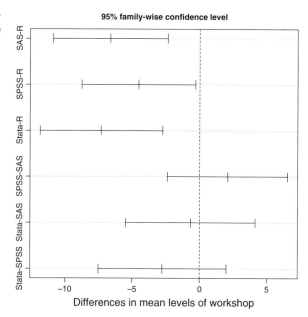

23.12 Sums of Squares

In ANOVA, SAS and SPSS provide partial (type III) sums of squares and F-tests by default. SAS also provides sequential sums of squares and F-tests by default. SPSS will provide those if you ask for them. R provides sequential ones in its built-in functions. For one-way ANOVAs or for two-way or higher ANOVAS with equal cell sizes, there is no difference between sequential and partial tests. However, in two-way or higher models that have unequal cell counts (unbalanced models), these two sums of squares lead to different F-tests and p-values.

The R community has a *very* strongly held belief that tests based on partial sums of squares can be misleading. One problem with them is that they test the main effect after supposedly partialing out significant interactions. In many circumstances, that does not make much sense. See the paper, *Exegesis on Linear Models*, by W.N. Venables, for details [39].

If you are sure you want type III sums of squares, the car package will calculate them using its Anova function. Notice that the function begins with a capital letter "A"! To run it, we must first load the car package from the library.

```
> library("car")

Attaching package: 'car'

        The    following    object(s)    are    masked    from
package:Hmisc :
        recode
```

You can see from the above message that the `car` package also contains a recode function that is now blocking access to (masking) the `recode` function in the `Hmisc` package. We could have avoided that message by detaching `Hmisc` first, but this is a good example of function masking and it does not affect the function we need. We use the `Anova` function with the `type="III"` argument to request those sums of squares and F-tests. Since our model is a one-way ANOVA, the results are identical to those from the lower-case `anova` function that is built-in to R, except that the p-values are expressed in scientific notation and the intercept (the grand mean in this case) is tested.

```
> Anova(myModel, type ="III")
Anova Table (Type III tests)

Response: posttest
             Sum Sq Df   F value     Pr (>F)
 (Intercept) 230654  1 6430.4706  < 22e>-16 ***
 workshop       875  3    8.1356  7.062e> -05 ***
 Residuals     3408 95
 -----
 Signif codes:  0 '***' 0.001 '**' 0.01 '*' 0.05 '.' 0.1 ' ' 1
```

23.13 Kruskal–Wallis Test

The non-parametric equivalent to a one-way ANOVA is the Kruskal–Wallis test. The Kruskal–Wallis test compares groups on the mean rank of a variable that is at least ordinal (e.g., low, medium, high). Its assumptions are:

- The distributions in the groups must be the same, other than a shift in location. This is often misinterpreted to mean that the distributions do not matter at all. That is not the case. They do not need to be normally distributed, but they do need to generally look alike. Otherwise the test can produce a significant result if, for example, the distributions are skewed in opposite directions but are centered in roughly the same place.
- The distributions should have the same variance. Since the test requires no particular distribution, there is no single test for this assumption. Examining a histogram is a good idea if you have at least 30 subjects. A boxplot is also helpful regardless of the number of subjects in each group.
- The observations are independent. For example, if each group contains subjects that were measured repeatedly over time or who are correlated (e.g., same family), you would want a more complex model to make the most of that information.

Here is an example that uses the `kruskal.test` function to compare the different workshops on the mean ranks of the q4 variable:

```
> kruskal.test(posttest important typo ~workshop)
        Kruskal-Wallis rank sum test

data:  posttest by workshop
```

```
Kruskal-Wallis chi-squared = 21.4448, df = 3, p-value =
8.51e-05
```

The `kruskal.test` function call above has two arguments.

1. The formula q4 ~ workshop is in the form dependent ~ independent. For details, see Sect. 8.5.2.
2. The data argument specifies the data frame to use for the formula. Alternatively, you could also use the short form of variable names by entering `attach(mydata)` before using the `kruskal.test` function and then leave the `data` argument off.

The *p*-value of 8.51e–05 or 0.0000851 is smaller than the typical cutoff of 0.05, so you would reject the hypothesis that the groups do not differ. The next question would be: which of the groups differ? The `pairwise.wilcoxon.test` function answers that question. The only arguments we use below are the measure and the factor, respectively.

```
> pairwise.wilcox.test(posttest, workshop)

        Pairwise comparisons using Wilcoxon rank sum test

data:   posttest and workshop

        R        SAS       SPSS
SAS     0.0012   -         -
SPSS    0.0061   0.4801    -
Stata   0.0023   0.5079    0.4033

P value adjustment method: holm
Warning messages:
1: In wilcox.test.default(xi, xj, ...) :
   cannot compute exact p-value with ties
```

So we see that R workshop group differs significantly from the other three and that the other three do not differ among themselves. For a description of the Holm p-value adjustment method, see Sec. 23.11 under the `pairwise.t.test` function. The median is the more popular measure to report with a non-parametric test. We can use the `aggregate` function to apply the `median` function (Table 23.1).

```
> aggregate( posttest,
+    dataframe(workshop),
+    median, na.rm=TRUE)
    workshop    x
1          R 86.0
2        SAS 78.5
3       SPSS 81.0
4      Stata 78.0
```

Example programs that demonstrate all these analyses in SAS, SPSS and R is located in Table 23.1. The R examples require installing the `Hmisc` and `car` packages before running. For details, see section 5.1.

Table 23.1 Example programs for data analysis

SAS programming statements	

```
* SAS Program for Basic Statistical Tests;
* Statistics.sas ;

LIBNAME myLib 'C:\myRfolder';
DATA  temp;
  SET myLib.mydata100;
  *pretend q2 and q1 are the same score
   measured at two times & subtract;
  myDiff=q2-q1; run;

* Basic stats in compact form;
PROC MEANS; VAR q1--posttest; RUN;

* Basic stats of every sort;
PROC UNIVARIATE; VAR q1--posttest; RUN;

* Frequencies & percents;
PROC FREQ; TABLES workshopndash; --q4; RUN;

* Chi-square;
PROC FREQ;
  TABLES workshop*gender/CHISQ; RUN;

*---Measures of association;

* Pearson correlations;
PROC CORR; VAR q1-q4; RUN;

* Spearman correlations;
PROC CORR SPEARMAN; VAR q1-q4; RUN;

* Linear regression;
PROC REG;
  MODEL q4=q1-q3;
  RUN;

*---Group comparisons;

* Independent samples t-test;
PROC TTEST;
  CLASS gender;
  VAR q1; RUN;

* Nonparametric version of above
  using Wilcoxon/Mann-Whitney test;
PROC NPAR1WAY;
  CLASS gender;
  VAR q1; RUN;

* Paired samples t-test;
PROC TTEST;
  PAIRED pretest*posttest; RUN;
```

Table 23.1 (continued)

```
* Nonparametric version of above using
  Signed Rank test;
PROC UNIVARIATE;
 VAR myDiff;
 RUN;

*Oneway Analysis of Variance (ANOVA);
PROC GLM;
   CLASS workshop;
   MODEL posttest=workshop;
   MEANS workshop / TUKEY; RUN;

   *Nonparametric version of above using
      Kruskal-Wallis test;
   PROC npar1way;
     CLASS workshop;
     VAR posttest; RUN;
```

SPSS programming
statements

```
* SPSS Program for Basic Statistical Tests.
* Statistics.sps

  CD   'C:\myRfolder'.
  GET FILE='mydata sav'.
  DATASET NAME DataSet2 WINDOW=FRONT.

* Descriptive stats in compact form.
  DESCRIPTIVES VARIABLES=q1 to posttest
    /STATISTICS=MEAN STDDEV VARIANCE
    MIN MAX SEMEAN.

 Descriptive stats of every sort.
  EXAMINE VARIABLES=q1 to posttest
    /PLOT BOXPLOT STEMLEAF NPPLOT
    /COMPARE GROUP
    /STATISTICS DESCRIPTIVES EXTREME
    /MISSING PAIRWISE.

* Frequencies and percents.
  FREQUENCIES VARIABLES=workshop TO q4.

* Chi-squared
  CROSSTABS
    /TABLES=workshop  BY gender
    /FORMAT = AVALUE TABLES
    /STATISTIC=CHISQ
    /CELLS = COUNT ROW
    /COUNT ROUND CELL.

*---Measures of association.
```

Table 23.1 (continued)

```
* Person correlations.
CORRELATIONS
 /VARIABLES =q1 TO q4

* Spearman correlations.
NONPAR CORR
 /VARIABLES=q1 to q4
 /PRINT=SPEARMAN

*  Linear regression
  REGRESSION
    /MISSING LISTWISE
    /STATISTICS COEFF OUTS R ANOVA
    /CRITERIA=PIN (.05>) POUT(.10)
    /NOORIGIN
    /DEPENDENT q4
    /METHOD=ENTER q1 q2 q3

  REGRESSION
    /DEPENDENT q4
    /METHOD=ENTER q1 q2 q3

*---Group comparisons;

*  Independent samples t-test
  T-TEST
    GROUPS = gender('m' 'f')
    /VARIABLES = q1.

* Nonparametric version of above using
* Wilcoxon/Mann-Whitney test
* SPSS requires a numeric form
* of gender for this procedure.
  AUTORECODE
    VARIABLES=gender  /INTO genderN
    /PRINT.
  NPAR TESTS
    /M-W = q1   BY genderN(1,2).

* Paired samples t-test.
  T-TEST
    PAIRS = pretest  WITH posttest (PAIRED).

* Oneway analysis of variance (ANOVA).
  UNIANOVA posttest  BY workshop
    /POSTHOC = workshop ( TUKEY )
    /PRINT = ETASQ HOMOGENEITY
    /DESIGN = workshop .

* Nonparametric version of above using
    Kruskal Wallis test.
  NPAR TESTS
    /K-W=posttest BY workshop (1 3).
```

Table 23.1 (continued)

R programming statements	

```
# R Program of Basic Statistical Tests.
# Statistics.R

setwd("/myRfolder")
load("mydata100.Rdata")
attach(mydata100)
options(linesize=64)

head(mydata100)

# --- Frequencies & Univariate Statistics ---

# The easy way using the Hmisc package.
library("Hmisc")
describe> (mydata100)

# R's build in function.
summa ry(mydata100)

# The flexible way using built-in functions.

table(workshop)
table(gender)

# Proportions of valid values.
proptable( table(workshop) )
proptable( table(gender) )

# Rounding off proportions.
round( prop.table( table(gender) ), 2 )

# Converting proportions to percents.
round( 100* ( proptable( table(gender) ) ) )

# Frequencies & Univariate
summary(mydata100)

# Means & Std Deviations
options(width=64)
sapply( mydata100 [3:8], mean,  na.rm=TRUE)
sapply( mydata100[3:8], sd,   na.rm=TRUE)

# Descriptive stats and frequencies.
# Note that you can't get frequencies on q1-q5.
summary(mydata)

# --- Crosstabulations---

# The easy way, using the gmodels package.
library("gmodels")
```

Table 23.1 (continued)

```
CrossTable(workshop, gender,
 chisq=TRUE, format="SAS")

# The flexible way using built in functions.

# Counts

myWG <- table (workshop, gender)
myWG    # Crosstabulation format.
myWGdata <- as.data. frame(myWG)
myWGdata # Summary or Aggregation format.

chisq.test(myWG)

# Row proportions.
prop.table(myWG, 1)

# Column proportions.
prop.table(myWG, 2)
# Total proportions
prop.table(myWG)

# Rounding off proportions
round( prop.table(myWG, 1), 2 )

# Row percents.
round( 100* ( prop.table(myWG, 1) ) )

# Adding Row and Column Totals.

addmargins(myWG, 1)
addmargins(myWG, 2)
prop.table( addmargins(myWG, 1:2) )

# ---Correlation & Linear Regression---

# The rcorr function from the Hmisc package

library("Hmisc")

rcorr(q1,q4)

myQs <- cbind(q1, q2, q3, q4)
rcorr(myQs, type="pearson")

# See just the P values.
options(digits=7)
myCorrs<- rcorr(myQs, type="pearson")
myCorrs$P

# See P values without scientific notation.
options(scipen=999)
myCorrs$P
```

Table 23.1 (continued)

```
# Spearman correlations using the Hmisc rcorr
function.
rcorr( cbind(q1,q2,q3,q4), type="spearman" )

# The built-in cor function.
cor( data.frame(q1, q2, q3, q4),
 method="pearson", use="pairwise")

# The built-in cor.test function
cor.test(q1, q2, use="pairwise")

# Linear regression
lm( q4 ~ q1+q2+q3, data=mydata100
)
myModel <- lm( q4 ~ q1 + q2 + q3, data=mydata100
)
myModel
summary(myModel)
anova(myModel) #Same as summary result
# Set graphics parameters for 4 plots
(optional).
par( mfrow=c(2,2),mar=c(5,4,2,1)+0.1 )

plot(myModel)

# Set graphics parameters back to default
Settings.
par( mfrow=c(1,1), mar=c(5,4,4,2)+0.1 )

myNoMissing <- na.omit(mydata100[ ,
c("q1","q2","q3","q4") ] )
myFullModel   <- lm( q4 ~ q1 + q2 + q3,
data=myNoMissing)
myReducedModel <- lm( q4 ~ q1,
data=myNoMissing)

#---Group Comparisons---

# Independent samples t-test.
t. test( q1 ~ gender, data=mydata100)

t.test ( q1[ gender=='Male'] , q1[ gender=='Female']
)
library("car")
levene.test(posttest, gender)

var.test(posttest~gender)

# Nonparametric version of above
  using Wilcoxon/Mann -Whitney test.
wilcox.test( q1 ~ gender, data=mydata100)
wilcox.test ( q1[ gender==' Male'] ,
      q1[ gender==' Female'] )
```

Table 23.1 (continued)

```
aggregate( q1, data.frame(gender),
 median, na.rm=TRUE )

# Paired samples t-test
t.test(posttest, pretest, paired=TRUE)

# Nonparametric version of above using
wilcox.test( posttest, pretest, paired=TRUE )
median(pretest)
median(posttest)

# Analysis of Variance (ANOVA).
aggregate( posttest,
 data.frame(workshop),
 mean, na.rm=TRUE )

aggregate( posttest,
 data.frame(workshop),
 var, na.rm=TRUE )

library("car")
levenetest(posttest, workshop)

myModel <- aov(posttest~workshop,
 data=mydata100)
myModel

anova(myModel)
summary(myModel) #same as anova result

# type III sums of squares
library("car")
Anova(myModel, type=" III ")

pairwisettest(posttest, workshop)

  TukeyHSD(myModel, "workshop")
  plot( TukeyHSD(myModel, "workshop") )

  # Set graphics parameters for 4 plots
  (optional).
  par( mfrow=c(2,2), mar=c(5,4,2,1)+01 )
  plot(myModel)
  # Set graphics parameters back to default
  Settings.
  par( mfrow=c(1,1), mar=c(5,4,4,2)+01 )

  #Nonparametric oneway ANOVA using
  # the Kruskal-Wallis test
  kruskaltest(posttest~workshop)
pairwise.wilcoxtest(posttest, workshop)
aggregate( posttest,
 dataframe(workshop),
 median, na.rm=TRUE )
```

Chapter 24
Conclusion

As we have seen, R differs from SAS and SPSS in many ways.

R has many features that SAS and SPSS lack such as: multiple data structures, a very wide range of variable selection methods, functions that optimize their output automatically for different data structures, data structure conversion tools, and workspace management functions. In short, R offers a complete and powerful programming environment rather than just a set of analytic procedures. Appendix D provides a summary of the many ways these packages differ.

If you are a SAS or SPSS user who has happily avoided the complexities of output management, macros, and matrix languages, R's added functionality may seem daunting to learn at first. On the other hand, R makes those added features so much easier to use than SAS or SPSS that you may find yourself more eager to expand your horizons. The added power of R and its free price make it well worth the effort.

This book has covered how R compares to SAS and SPSS, and how you can do the very same things in each. However, what we have not covered literally fills many volumes. I hope this will start you on a long and successful journey with R.

I hope to improve this book as time goes on, so if there are changes you would like to see in the next edition, please drop me a line at muenchen. bob@gmail.com. Negative comments are often the most useful, so do not worry about being critical.

Have fun working with R!

R.A. Muenchen, *R for SAS and SPSS Users*, DOI: 10.1007/978-0-387-09418-2_24,   441
© Springer Science+Business Media, LLC 2008

Appendix A
A Glossary of R Jargon

Below is a selection of common R terms defined using SAS/SPSS jargon (or plain English when possible) and R jargon. Some definitions in SAS/SPSS jargon are quite loose given the fact that they have no direct analog of some R terms. Items in *italics* are included in the glossary. Definitions in R terms are often quoted (with permission) or paraphrased from S Poetry, by Patrick Burns [40].

Table A.1 Glossary of R Jargon

	Defined in SAS/SPSS Terms	Defined in R Terms
Apply	The process of having a procedure work on variables or observations/cases. Determines whether a procedure will act as a typical procedure or as a function instead. Also a function that does that.	The process of targeting a function on rows or columns. Also a *function* that does that.
Argument	Parameter, option or setting that controls what a procedure does. Includes variables to analyze.	Input to a *function*.
Array	Multiple datasets that are linked in layers. All variables must be only one type, e.g., all numeric or all character.	A *vector* with a *dim* attribute. The dim controls the number and size of dimensions.
Assignment function	The two-key sequence, "<-", that places data or results of procedures or transformations into a variable or dataset.	The two-key sequence, "<-", that gives *names* to *objects*.
Atomic object	A variable whose values are all of one type such as all numeric or all character.	An *object* whose *components* are all of one *mode*. Modes allowed are numeric, character, logical, or complex.
Attach	The process of adding a dataset or add-on module to your path. Attaching a dataset appears to copy the variables into an area	The process of adding a *database* to your *search list*. Also a *function* that does this.

Table A.1 (continued)

	Defined in SAS/SPSS Terms	Defined in R Terms
	that lets you use them by a simple name like "gender" rather than by compound name like "mydata$gender". Done using the `attach` function.	
Attributes	Traits of a dataset like its variable names and labels.	Traits of objects such as *names*, *class*, or *dim*.
Class	An attribute of a variable or dataset that a procedure uses to change its default settings automatically. For variables, this is similar to setting the scale of a variable to help you decide what procedures it will work with.	The class attribute of an object determines which *method* of a *generic function* is used when the *object* is an *argument* in the *function* call.
Component	Like an entry in a SAS catalog. Can also be a variable in a dataset.	An item in a *list*. The *length* of a list is the number of components it has.
CRAN	The Comprehensive R Archive Network at http://cran.r-project.org/. Consists of a set of sites around the world called mirrors that provide R and its add-on packages for you to download and install.	
Data frame	A dataset.	A set of *vectors* bound together in a *list*. They can be different *modes* or *classes*, e.g., numeric and character, but they must have equal *length*.
Database	One dataset or a set of them in a library, or an add-on module.	An item on the *search list*, or something that might be. Can be an R data file or a *package*.
Dim	A variable whose values are the number of rows and columns in a dataset. It is stored in the dataset itself. Also a procedure that prints or sets these values.	The *attribute* that describes the dimensions of an array. Also the *function* that retrieves or changes that attribute.
Element	A value.	An item in an *atomic vector*.
Extractor function	A procedure that gets more results from a dataset created by another procedure.	A *function* that has *methods* that apply to *modeling objects*.
Factor	A categorical variable and its value labels. Value labels may be nothing more than "1", "2" if not assigned explicitly.	The type of *object* that represents a categorical variable. It stores its *labels* in its *levels attribute*.
Function	A procedure and/or a function. When you apply it down through cases, it is just like a procedure. But you can also apply it across rows like a function.	A program that is stored as an *object*.
Generic function	A procedure or function that has different default parameters set	A function whose behavior is determined by the *class* of one or

Table A.1 (continued)

	Defined in SAS/SPSS Terms	Defined in R Terms
	depending upon the type of data you give it.	more of its *arguments*. The class of the relevant argument(s) determines which *method* the generic function will use.
Index	The order number of a variable in a dataset, or of a value in a variable. In our practice dataset, gender is the second variable so its index is 2. Gender is mydata[,2]. The first index selects rows, the second columns. If empty, it refers to all rows/columns.	The number of a *component* in a *list* or *data frame*, or of an *element* in a *vector*.
Install	You install packages just like add-ons, just once per version.	Adding a *package* into your *library*.
Label	A procedure that creates variable labels. Also a parameter that sets value labels using the `factor` or `ordered` procedures.	A *function* from the `Hmisc` *package* that creates variable labels. Also an *argument* that sets factor labels using the `factor` or `ordered` functions.
Length	The number of observations/cases in a variable (including missing values), or the number of variables in a dataset.	A measure of *objects*. For *vectors*, it is the number of its *elements* (including NAs). For *lists* or *data frames*, it is the number of its *components*.
Levels	The values that a categorical variable can have. Actually stored as a part of the variable itself in what appears to be a very short character variable (even when the values themselves are numbers).	An attribute to a factor object that is a character *vector* of the values the *factor* can have. Also an *argument* to the `factor` and `ordered` functions that can set the levels.
Library	Where a given version of R stores its base packages and the add-on modules you have installed. Also a procedure that loads a package from the library into working memory. You must do that in every R session before using a package.	A directory containing R *packages* that is set up so that the *library* function can *attach* it. Also a *function* that *attaches* a package from the *library* onto your *search list*. You must do that in every R session before using a function in the package.
List	Like a zipped collection of datasets that you can analyze easily without unzipping.	A set of *objects* of any *class*. Its *components* can be *vectors*, *data frames*, *matrices* and even other *lists*.
Load	Bringing a dataset (or collection of datasets) from disk into memory. You must do this before you can use data in R. Also the procedure that performs that task.	Bringing a R data file into your *workspace*. Also a *function* that performs that task.

Table A.1 (continued)

	Defined in SAS/SPSS Terms	Defined in R Terms
Matrix	A dataset that must contain only one type of variable, e.g., all numeric or character. Helpful in cases where you might create a SAS/SPSS array to process repetitively.	A two-dimensional array; that is, a *vector* with a *dim* attribute of *length* 2.
Method	The analyses and/or graphs that a procedure will perform by default, that is different for different types of variables. The default settings for some procedures depend upon the scale of the variables you provide. E.g., summary (temperature) provides mean temperature, summary (gender) counts males & females.	A *function* that provides the calculation of a *generic function* for a specific *class* of object.
Mode	A variable's type such as numeric or character.	A fundamental property of an object. Can be numeric, character, logical, or complex.
Modeling function	A procedure that tests association or group differences.	A function that tests association or group differences and usually accepts a formula (e.g., $y \sim x$) and a data = *argument*.
NA	A missing value.	A missing value. See also *NaN*.
Names	Variable names. They are stored in a character variable that is part of a dataset or variable. Since R can use an *index* number instead, names are optional. Also a procedure that extracts or changes variable names.	An *attribute* of many *objects* that labels the *elements* or *components* of the object. Also the *function* that retrieves or sets this attribute.
NaN	A missing value.	Not a Number. Something that is undefined mathematically such as zero divided by zero.
NULL	An object you can use to drop variables or values. E.g., $x <-$ NULL drops the variable x.	NULL has a zero *length* and no particular *mode*.
Numeric	A variable that contains only numbers.	The *atomic* mode that represents real numbers. This contains storage *modes* double, single, and integer.
Object	A dataset, a variable, or even a procedure.	Almost everything in R. If it has a *mode*, it is an object. Includes *data frames, vectors, matrices, lists,* and *functions*.
Object-oriented programming	A style of software in which the output of a procedure depends upon the type of data you provide it. R has an object orientation, but SAS and SPSS do not.	
Option	Settings that control some aspect of your R session, such as the width of each line of output. Also a function that queries or changes the settings. See also *par*.	

Table A.1 (continued)

	Defined in SAS/SPSS Terms	Defined in R Terms
Package	An add-on module like SAS/STAT or SPSS Advanced Models.	A collection of *functions*
Par	A function that queries or sets the parameters that control some aspects of traditional graphics output, like how many graphs appear on a page.	
R	"R is a language and environment for statistical computing and graphics. It is a GNU project which is similar to the S language and environment which was developed at Bell Laboratories (formerly AT&T, now Lucent Technologies) by John Chambers and colleagues. R can be considered as a different implementation of S. There are some important differences, but much code written for S runs unaltered under R." - http://www.r-project.org/ What is R	
Replacement	When you use subscripts on the left side of an assignment to change the values in an object. E.g., setting 9 to missing: x[x = = 9] < - NA	
S	The language from which *R* evolved.	
S3, S4	Used in the r-help files to refer to different versions of *S*. The differences between them are of importance mainly to advanced programmers.	
Save	Saves the datasets you choose by name.	Saves the *objects* you request to an R data file.
Save.image	Saves all your open datasets into a single file.	A *function* that writes all *objects* in your *workspace* to a R data file.
Search list	The collection of *databases* that R will search, in order, for objects. Similar to a path for your operating system.	
S-PLUS	The commercial version of *S*. Its main difference from *R* is that it includes a graphical user interface.	
Subscript	Choosing variables or values by the order in which they appear or by their name.	The extraction or *replacement* of an object using its *index* or *name* in square [brackets].
Vector	A variable. It can exist on its own in memory or it can be part of a dataset.	A set of values or elements that have the same *mode*, i.e., an *atomic object*.
Workspace	The area of main memory where R does all its work. Data must be *loaded* into it from files and *packages* must be loaded into it from the *library* before you can use either.	

Appendix B
A Comparison of SAS and SPSS Products with R Packages and Functions

With over 1200 add-on packages, many containing multiple procedures, R can do almost everything that SAS and SPSS can do and quite a bit more. People are releasing new packages at a rapid pace and R can give you the latest count with the following two commands. The first one uses the `available.packages` function to check Internet repositories for the packages that are currently available and store them in myPackageNames. The second command determines the number of unique names.

```
> myPackageNames <- available.packages()

> length (unique( rownames(myPackageNames) ))

[1] 1449
```

So at the time of publication, there were 1449 add-on packages! If you use the `setRepositories ()` function (or Packages > Select repositories... on Windows) to add the bioinformatic packages, the number is even higher at 2242.

Table B.1 below focuses only on SAS and SPSS *products* and which of them have counterparts in R. As a result, some categories are extremely broad (e.g., regression) while others are quite narrow (e.g., conjoint). This list does not contain the hundreds of R packages that have no counterparts in the form of SAS or SPSS products. There are many important topics (e.g., mixed models, offered by all three) that are not listed because neither SAS nor SPSS sell a product focused just on that.

Much more detailed information about R packages is available organized in Task Views at http://cran.r-project.org/web/views/index.html. Another site to search by task is at http://biostat.mc.vanderbilt.edu/s/finder/finder.html. Detailed information about most R packages is available at http://www.r-project.org/, choose CRAN, then choose a mirror, then choose Packages./

Table B.1 Comparison of SAS and SPSS products to R packages

Topic	SAS Product	SPSS Product	R Package (some are package- function)
Advanced models	SAS/STAT®	SPSS Advanced Models™	stats, MASS, *many* others
Basics	SAS®	SPSS Base™	R
Conjoint analysis	SAS/STAT®: Transreg	SPSS Conjoint™	homals, psychoR, bayesm
Correspondence analysis	SAS/STAT®: Corresp	SPSS Categories™	homals, MASS, FactoMineR, ade4, PTAk, cocorresp, vegan, made4, PsychoR
Custom tables	SAS Base® Report, SQL, Tabulate	SPSS Custom Tables™	reshape
Data access	SAS/ACCESS®	SPSS Data Access Pack™	DBI, foreign, RODBC
Data mining	Enterprise Miner™	Clementine®	rattle, arules, FactoMineR
Data preparation	Various procedures	Various procedures, SPSS Data Preparation™	dprep, various functions
Exact tests	SAS/STAT®: various	SPSS Exact Tests™	coin, elrm, exactLoglinTest, exactmaxsel, exactRankTests, and as options in many others
Genetics	SAS/Genetics®, SAS/ Microarray® Solution®, JMP Genomics®	None	Bioconductor at http:// www.bioconductor.org/
Geographic information systems/ mapping	SAS/GIS®, SAS/ Graph®	SPSS Maps™ (no full GIS)	maps, mapdata, mapproj, GRASS via spgrass6, RColorBrewer, see Spatial in Task Views link above
Graphical user interface	Enterprise Guide®	SPSS Base™	JGR, R Commander, pmg, SciViews
Graphics – interactive with linked windows	SAS/INSIGHT®	None	GGobi via rggobi, iPlots, Mondrian via Rserve
Graphics – static	SAS/GRAPH®	SPSS Base™	ggplot, gplots, graphics, grid, gridBase, hexbin, lattice, plotrix, scatterplot3d, vcd, vioplot,

Table B.1 (continued)

Topic	SAS Product	SPSS Product	R Package (some are package- function)
			`geneplotter`, `Rgraphics`,
Guided analysis	SAS/LAB®	None	None
Matrix/linear algebra	SAS/IML®, SAS/STAT Studio®	SPSS Matrix™	R, `matlab`, `Matrix`, `sparseM`
Missing values imputation	SAS/STAT®: MI	SPSS Missing Values Analysis™	`Hmisc` - `aregImpute`, `EMV`, `Design` - `fit.mult.impute`, `mice`, `mitools`, `mvnmle`
Operations research	SAS/OR®	None	`glpk`, `linprog`, `LowRankQP`, TSP
Power analysis	SAS® Power and Sample Size Application, SAS/STAT: Power, GLM Power	SamplePower™	`asypow`, `powerpkg`, `pwr`, `MBESS`
Quality control	SAS/QC®	SPSS Base™	`qcc`, `spc`
Regression models	SAS/BASE®	SPSS Regression Models™	R, `Hmisc`, `Design`, `lasso`, `VGAM`, `pda`
Sampling, complex or survey	SAS/STAT®: surveymeans, etc.	SPSS Complex Samples™	`pps`, `sampling`, `sampling`, `spsurvey`, `survey`
Structural equations	SAS/STAT®: Calis	Amos™	`Sem`
Text analysis	Text Miner	SPSS Text Analysis for Surveys™, Text Miner for Clementine®	`Rstem`, `lsa`, `tm`
Time series	SAS/ETS®	SPSS Trends™ Expert Modeler	Over 40 packages that do time series are described at Task View link above under Econometrics.
Time series, automated	SAS Forecast Studio®	SPSS Trends, DecisionTime/ WhatIf™	None
Trees, decision or regression	Enterprise Miner™	SPSS Classification Trees™, AnswerTree™	`ada`, `adabag`, `BayesTree`, `boost`, `caret`, `GAMboost`, `gbev`, `gbm`, `maptree`, `mboost`, `mvpart`, `party`, `pinktoe`, `quantregForest`, `rpart`, `rpart.permutation`, `randomForest`, `randomForests`, `tree`

Appendix C
Automating Your Settings

SAS has its autoexe.sas file that exists to let you automatically set options and run SAS code. R has a similar file called .Rprofile. This file is stored in your initial working directory, which you can locate with the getwd() function.

Below is my .Rprofile. It sets options just as you would in R. See enter help (options) for many more. Let us step through it one command at a time.

First, I set the console width to 64 so my output fits training examples better. I also ask for five significant digits and tell it to mark significant results with stars. The latter is the default, but since many people prefer to turn that feature off, I included it. You would turn them off with a setting of FALSE.

```
options (width=64, digits=5, show.signif.stars=TRUE)
```

Setting the random number seed is a good idea if you want to generate numbers that are random but repeatable. That is handy for training examples in which you would like every student to see the same result. Here I set it to the number 1234.

```
set.seed (1234)
```

The setwd function sets the working directory, the place all your files will go if you don't specify a path.

```
setwd ("/myRfolder")
```

I also like to define the set of packages that I install whenever I upgrade to a new version of R. With these stored in myPackages, I can install them all with a single command. For details, see Chap. 5. This is the list of all packages used in this book.

```
myPackages <- c("car","hexbin",
  "ggplot2","gmodels","gplots", "Hmisc",
  "reshape","Rcmdr")
```

You can have R load your favorite packages automatically too. This is particularly helpful when setting up a computer to run R with a graphical user interface like R Commander.

Loading packages at startup does have some disadvantages though. It slows down your startup time, takes up memory in your workspace, and can create conflicts when different packages have functions with the same name. Therefore, you do not want to load too many. Loading packages at startup requires the use of the `local` function. The `getOption` function gets the names of the original packages to load and stores them in a character vector I named myOriginal. I then created a second character vector, myAutoLoads, containing the names of the packages I want to add to the list. I then merged them into one character vector, myBoth. Finally, I used the `options` function to change the default packages to the combined list of both the original list and my chosen packages:

```
local({
    myOriginal <- getOption("defaultPackages")
    # edit next line to be your list of favorites.
    myAutoLoads <- c("Hmisc","ggplot2")
    myBoth <- c(myOriginal,myAutoLoads)
    options(defaultPackages = myBoth)
})
```

If you want R to run any functions automatically, you create your own single functions that do the required steps. To have R run a function before all others, name it ".First". To have it run the function after all others, name it .Last. Notice that utility functions require a prefix of "`utils:: `" or R will not find them while it is starting up. The `time-stamp` function is one of those. It simply returns the time and date. The `cat` function simply prints messages.

```
.First <- function()
    {
        cat("\n                        Welcome to R!\n")
        utils::timestamp()
        cat("\n")
    }
```

You can also have R run any functions before exiting the package. As a Windows user, I would like to save my command history. Below I print a farewell message and then save the history to a file named myLatest.Rhistory.

```
.Last <- function()
    {
        graphics.off()
        cat("\n\n myCumulative.Rhistory has been saved." )
        cat("\n\n Goodbye!\n\n")
    utils::savehistory(file="myCumulative.Rhistory")
    }
```

Warning: Since these functions begin with a period, they are invisible to the ls function by default. The command ls (all.names = TRUE) will show them to you. Since they are functions, if you save a workspace that contains them, they will continue to operate whenever you load that workspace, even if you delete the .Rprofile! As usual, you can display them by typing their names and run them by adding empty parentheses to them: .First(). If you need to delete them from the workspace, rm will do it with no added arguments:

```
rm(.First,.Last).
```

Here is the .Rprofile with all commands together. You can download it with the practice data sets and programs from http://RforSASandSPSSusers.com./

```
#        Startup Settings
# Place any R commands below.
options(width=64, digits=5, show.signif.stars=TRUE)
set.seed(1234)
setwd("/myRfolder")
myPackages <- c("car",   "hexbin",
    "ggplot2","gmodels", "gplots",   "Hmisc",
    "reshape", "Rcmdr")
utils::loadhistory(file = "myCumulative.Rhistory")
# Load packages automatically below.
  local({
    myOriginal <- getOption("defaultPackages")
    # Edit next line to include your favorites.
    myAutoLoads <- c("Hmisc","ggplot2")
    myBoth <- c(myOriginal,myAutoLoads)
    options(defaultPackages = myBoth)
} )
# Things put here are done first.
.First <- function()
   {
      cat("\n                          Welcome to R!\n")
      utils::timestamp()
      cat("\n")
   }
# Things put here are done last.
.Last <- function()
   {
      graphics.off()
      cat("\n\n myCumulative.Rhistory has been saved.")
      cat("\n\n Goodbye!\n\n")
      utils::savehistory(file="myCumulative.Rhistory")
   }
```

Appendix D

Appendix D. A comparison of the major attributes of SAS and SPSS to R

	SAS and SPSS	R
Aggregating data	One pass to aggregate, another to merge (if needed, SAS only), a third to use. Few basic statistics are available.	A statement can mix both raw and aggregated values. Can aggregate on all statistics.
Choosing data	All the data for an analysis or graph must be in a single dataset.	Analyses and graphs can freely combine variables from different data frames or other structures.
Choosing observations	Uses logical conditions in IF, SELECT IF, WHERE	Uses wide variety of selection by index value, variable name, logical condition (same as when selecting variables).
Choosing variables	Uses the simple lists of variable names in the form of: x, y, z; a to z; a–z	Uses wide variety of selection by index value, variable name, logical condition (same as when selecting observations).
Controlling procedure or function	Statements such as CLASS and MODEL and options control the procedure.	You can control functions by manipulating the data's structure (its class), setting function options (arguments) and using separate *apply* and *extraction* functions.
Converting data structures to match procedure or function	In general, all procedures accept all variables; you rarely need to convert variable type.	Original data structure plus variable selection method determines structure. You commonly use conversion functions to get data into acceptable form.
Cost	Each module has its price.	R and all its packages are free.
Data size	Most procedures are limited only by hard disk size.	Most functions must fit the data into the computer's smaller random access memory.
Data structure	Rectangular dataset.	Vector, factor, data frame, matrix, list, etc.
Graphical user interface	SAS Enterprise Miner uses flowchart approach that provides audit trail and repeatability in that form. SPSS offers well developed menus that control most things. Depends upon its language for repeatability.	R has several. R Commander looks much like SPSS. It offers easy control of the basics but is not as comprehensive as either the SAS or SPSS GUIs. Uses R language for repeatability.

Appendix D. (continued)

	SAS and SPSS	R
Graphics	SAS' are easy but relatively inflexible. SPSS Graphics Production Language (GPL) is slightly ahead of R.	Traditional graphics are extremely flexible. The `ggplot2` package provides functionality very close to GPL using a similar programming style.
Help and documentation	Aimed at beginner to intermediate users.	Aimed at intermediate to advanced users.
Macro language	A separate language used mainly for repetitive tasks or adding new functionality. User-written macros run differently from built-in procedures.	R does not have a macro language as its language is flexible enough to not require one. User-written functions run the same way as built-in ones.
Managing datasets	Relies on standard operating system commands to copy, delete, etc. Standard search tools can find datasets since they are in separate files.	Uses internal environments with its own commands to copy, delete, etc. Standard search tools cannot find multiple data frames if you store them in a single file.
Matrix language	A separate language used only to add new features.	An integral part of R that you use even when selecting variables or observations.
Missing data	When data is missing, procedures use all the data they can. Some procedures offer listwise deletion as an alternative.	When data is missing, functions often provide no results by default; different functions require different missing value options.
Output management system	People rarely use output management systems for routine analyses.	People routinely get additional results by passing output through additional functions.
Publishing results	See it formatted immediately in any style you choose. Quick cut and paste to word processor maintains fonts, table status, and style. Can also export to a file.	Process output with additional procedures that route formatted output to a file. You do not see it formatted as lined tables with proportional fonts until you import it to a word processor or text formatter.
Statistical methods	SAS is slightly ahead of SPSS but both trail well behind R. SPSS can run R programs within SPSS programs.	Most new methods appear in R around five years before SAS and SPSS.
Tables	Easy to build and nicely formatted but limited in what they can display.	Can build table of the results of virtually all functions but you need to view them outside R to see them nicely formatted..
Variable labels	Built in. Used by all procedures.	Added on. Used by few procedures.

Bibliography

1. Chambers, John M. *Software for Data Analysis: Programming with R.* s.l.: Springer Science+Business Media, LLC, 2008. ISBN 978-0-387-75935-7.
2. Team, R Development Core. *R: A Language and Environment for Statistical Computing.* Vienna, Austria: http://www.R-project.org, 2007. 3-900051-07-0.
3. Roebuck, P. *The MATLAB Package.* 2006: http://lib.stat.cmu.edu/R/CRAN/doc/packages/matlab.pdf.
4. Keeling, Kellie B. and Pavur, Robert J. *A comparative study of the reliability of nine statistical software packages.* 8, May 1, 2007, Computational Statistics & Data Analysis, Vol. 51, pp. 3811–3831.
5. Bolker, Ben. [R] software comparison. *R Help Archive.* [Online] 4 16, 2007. [Cited: 3 12, 2008.]: http://finzi.psych.upenn.edu/R/Rhelp02a/archive/97802.html.
6. Frank E. Harrell, Jr. with contributions from many others. *Hmisc: Harrell Miscellaneous.* R package version 3.4-3. s.l.: http://biostat.mc.vanderbilt.edu/s/Hmisc, 2007.
7. Lemon, Jim and Grosjean, Phillipe. *prettyR: Pretty descriptive stats.* 2007. R package version 1.1-3.
8. *Bioconductor, open source software for bioinformatics.* [Online]: http://www.bioconductor.org/whatisit.
9. *Omegahat Project for Statistical Computing.* [Online]: http://www.omegahat.org/.
10. Fox, John. *car: Companion to Applied Regression.* 2007. R package version 1.2-7: http://www.r-project.org;http://socserv.socsci.mcmaster.ca/jfox/.
11. Venables, W.N., Smith, D.M., and the R Development Core Team. *An Introduction to R.* 2007. ISBN 3-900051-12-7.
12. SPSS, Inc. *SPSS Statistics-R Integration Package.* Chicago: SPSS, Inc., 2008.
13. Fox, John, Michael Ash, Theophilius Boye, Stefano Calza, Andy Chang, Philippe Grosjean, Richard Heiberger, G. Jay Kerns, Renaud Lancelot, Matthieu Lesnoff, Samir Messad, Martin Maechler, Duncan Murdoch, Erich Neuwirth, Dan Putler, Miroslav Ristic. *Rcmdr: R Commander.* 2008: http://www.r-project.org;http://socserv.socsci/mcmaster.ca/jvox/Misc/Rcmdr/.
14. Rattle: Gnome R Data Mining. *Togaware.* [Online]: http://rattle.togaware.com/.
15. Helbig, Markus and Urbanek, Simon. *JGR: Java GUI for R.* 2007. Version 1.5-8: http://cran.r-project.org/doc/packages/JGR.pdf.
16. JGR: Java GUI for R. *Dept. of Computer Oriented Statistics and Data Analysis.* [Online] 2008: http://rosuda.org/JGR/.
17. R-core members and Saikat DebRoy, Roger Bivand and others. *Foreign: Red Data Stored by Minitab, S, SAS, SPSS, Stata, Systat, dBase.* 2007. R package version 0.8-23.
18. Venables, W.N. and Ripley, B.D. *Modern Applied Statistics with S.* Fourth. New York: Springer Science+Business Media, LLC, 2002. 0-387-95457-0.
19. Spector, Phil. *Data Manipulation with R.* s.l.: Springer Business+Science, 2008. ISBN 978-0-387-74730-9.

20. Baron, Jonathan. R Site Search. *Help for R: A Language and Environment for Statistical Computing and Graphics.* [Online] [Cited: February 12, 2008.]: http://finzi.psych.upenn. edu/search.html.

21. Romain, Francois. A firefox extension for R Site Search. [Online] [Cited: 2 12, 2008.]: http://addictedtor.free.fr/rsitesearch/.

22. Wickham, Hadley. *reshape: Flexibly reshape data.* 2007. R package version 0.8.0.

23. Therneau, Terry M., Atkinson, Beth and Ripley, Brian (ported). *rpart: Recursive Partitioning.* 2007. R package version 3.1-38.

24. Lumley, Thomas. *The biglm Package.* 2006: http://cran.r-project.org/doc/packages/ biglm.pdf.

25. iPlots. *Department of Computer Oriented Statistics and Data Analysis.* [Online] 2008. [Cited: 2 11, 2008.]: http://www.rosuda.org/iplots/.

26. Swayne, D., et al. *GGobi: XGobi redesigned and extended.* Vols. In Proc. of the 33th Symposium on the Interface: Computing Science and Statistics, 2001.

27. Sarkar, Deepayan. *lattice: Lattice Graphics.* 2008. R package version 0.17-4.

28. Wickham, Hadley. *ggplot.* s.l.: http://had.co.nz/ggplot2/book.pdf, 2007.

29. Murrell, Paul. *R Graphics.* Boca Raton, FL: Chapman & Hall/CRC, 2006. ISBN 978-1-58488-486-6.

30. Cleveland, William S. *Visualizing Data.* s.l.: Hobart Press, 1993. ISBN 978-0963488404.

31. Sarkar, Deepayan. *Lattice: Multivariate Data Visualization with R.* s.l.: Springer Science+Business Media, LLC, 2008. ISBN 978-0-387-75968-5.

32. Wilkinson, Leland. *The Grammar of Graphics, Second Edition.* New York: Springer Science+Business Media, Inc, 2005. ISBN 978-0387-24544-8.

33. Friendly, Michael. *Visualizing Categorical Data.* s.l.: SAS Publishing, 2000. ISBN 978-1580256605.

34. Dan Carr, porte by Nicholas Lewin-Koh and Martin Maechler. *hexbin: Hexoganal Binning Routines.* 2006. R package version 2.3.2.

35. Warnes, Gregory R., et al. *gplots: Various R programming tools for plotting data.* R package version 2.3.2.

36. Dalgaard, Peter. *Introductory Statistics with R.* s.l.: Springer Science+Business Media, Inc., 2002. ISBN 978-0387-95475-2.

37. Warnes, Gregory R., et al. *gmodels: Various R Programming Tools for Model Fitting.* 2007. R package version 2.14.1: http://cran.r-project.org/src/contrib/PACKAGES. htm;http://www.sf.net/projects/r-gregmisc.

38. Pinheiro, Jose C. and Bates, Douglas M. *Mixed Effects Models in S and S-Plus.* New York: Springer Science+Business Media, Inc., 2000. ISBN 978-0387989570.

39. Venables, W.N. *Exegeses on Linear Models.* 1998: http://www.stats.ox.ac.uk/pub/ MASS3/Exegeses.pdf.

40. Burns, Patrick J. *S Poetry.* 1998: http://www.burns-stat.com/pages/Spoetry/ Spoetry.pdf.

41. Mitchell, Michael N. *Strategically using General Purpose Statistics Packages: A Look at Stata, SAS and SPSS.* Statistical Consulting Group, UCLA Academic Technology Services. 2007. Technical Report Series: http://www.ats.ucla.edu/stat/technical-reports/.

42. Burns, Patrick. *R Relative to Statistical Packages: Comment 1 on Technical Report Number 1 (Version 1.0) Strategically using General Purpose Statistics Packages: A Look at Stata, SAS and SPSS.* Los Angeles: UCLA Academic Technology Services, 2006. Technical Report Series: http://www.ats.ucla.edu/stat/technicalreports/.

43. Muenchen, Robert A. *R for SAS & SPSS Users.* 2006: http://RforSASandSPSSusers. com.

44. Alzola, Carlos and Harrell, Frank. *An Introduction to S and The Hmisc and Design Libraries. September 24, 2006.* http://cran.r-project.org/doc/contrib/AlzolatHarrel-Hmisc-Design-Intro.pdf/.

45. Roebuck, P. MATLAB emulation package, version 0.8-1. *MATLAB emulation package, version 0.8-1*. 2006: http://cran.r-project.org/doc/packages/matlab.pdf.

46. Lumley, Gregory R. Warnes. Includes R source code and/or documentation contributed by Ben Bolker and Thomas. *gplots: Various R programming tools for plotting data*. R package version 2.3.2.

47. Hadley Wickham, Micheal Lawrence, Duncan Temple Lang, and Deborah F Swayne. An introduction to rggobi. *R-news*, Under revision. http://ggobi.org/rggobi.

Index

- SAS operator, 103
-- SAS operator, 103
~ operator, 65
!IN SPSS keyword, 111, 133
%in% operator, 110, 111, 131,
      133, 146
%INCLUDE SAS statement, 27
.First, 455
.GlobalEnv, 266
.Last, 455
.RData file extension, 270
.Rhistory file extension, 23, 25,
      26, 272
.Rprofile
    file, 23, 16, 453
/ operator, 148
| operator, 109
: SAS operator, 101
^ R operator, 148
LAST SAS parameter, 113
+ R operator, 169
<-, 42, 50

Abline function, 369
Accuracy of R, 3
ADD FILES SPSS command, 189
Add1 function, 421
Addmargins function, 412
Aes function, 372
Aggregate function, 433
AGGREGATE SPSS procedure,
    214, 221
Aggregating, 203
AITR
    An Introduction to R, 25
ALL SPSS keyword, 104
Anova function, 6, 432
Array, 66
as.data.frame function, 231, 242, 411

as.logical function, 129
as.matrix function, 152
as.table function, 226
as.vector function, 171
Assignment operator, 42, 50
Attach, 131
Attach function, 120, 149, 264, 266
Attaching data frames, 264
Attaching files, 266
Attribute, 78
Attributes function, 262, 271, 354
Axis function, 321

Bar function, 355
Barplot function, 289
Bartlett.test function, 424
Batch processing, 27
Biglm package, 259
Bioconductor, 17
Boxplot function, 323
By function, 206, 208, 214, 324
BY processing example programs, 208
BY processing, SAS, 208
BY SAS statement, 243

c function, 57, 173
Car package, 19, 180, 433
Cases, 58
Cat function, 454
cbind function, 57
CD SPSS command, 269, 330
chisq.test function, 411
Class, 73, 82
Class function, 66, 69, 226, 262
CLASS SAS statement, 225
Cleanup.import function, 268
Coercion, 51, 56
colClasses argument, 83
Comments, in programs, 62

Complete.cases function, 167
Components, of a list, 60
Contents function, 44, 263
CONTENTS SAS procedure, 263
Conversion
    class removal, 144
    data frame to matrix, 144
    from logical vector to index
        when seleting variables, 143
    index to logical, 146
    list to separate vectors, 144
    list to vector, 144
    lists or dataframes into list, 144
    matrix to data frame, 144
    matrix to vector, 144
    variable names to indexes, 115
    vector to matrix, 144
    vectors into one long one, 144
    vectors to columns of a matrix, 144
    vectors to data frame, 144
    vectors to rows of a matrix, 144
coord_polar function, 390
Coplot function, 310
Cor function, 37, 38, 58, 416, 439
coord_flip function, 346
Correlation
    Kendall, 413
    Pearson, 434
    Spearman, 439
CRAN, 17, 29
CROSSTABS SPSS procedure, 200
Crosstabulation
    column proportions, 411
    row proportions, 411
CTABLES SPSS procedure, 200
cut2 function, 180

Data
    generating a data frame, 251
    generating continuous measures, 249
    generating factors, 246
    generating integer measures, 248
    generating numeric sequences, 245
    generating repititious patterns, 247
    generation, 251
    storage and memory considerations, 263
Data argument, 65
Data editor, 86
Data frame, 67, 81
Data function, 18
Datasets
    reshaping, example programs, 219
    adding, 186

concatenating, 186
joining, 190
listing in all packages, 19
merging, 190
stacking, 186
data.frame function, 56, 116
DATASET NAME SPSS command, 103
DELETE VARIABLES SPSS command,
    185
Deleting objects. See removing objects,
Demo function, 276
Describe function, 14, 407
DESIGN SPSS keyword, 65
Detach function, 14, 114
Dim function, 57
Directory, setting. See working directory
do.call function, 207, 215
DROP SAS statement, 185
drop1 function, 421
Duplicate observations, 210

Edit function, 79
Editor
    JGR data editor, 37, 40
    R Editor, 22, 24, 37, 173
    SAS Program Editor, 22, 24
    SPSS Syntax Editor, 22, 24
Elements, of a vector, 51
Exiting R. See quitting R
Exp function, 148
EXPLORE SPSS procedure, 406
Exporting data
    example programs, 95
Exporting data, 97
Expression function, 325
Extracting parts of objects
    getting help on, 42
Extractor functions, 67–69, 418, 444

Factor, 82, 56, 66, 266
    character, 226
    converting many variables to, 232
    converting to variables, 228
    dropping unused levels, 233
    numeric, 247
    ordered, 229
file.show function, 101
FILTER SPSS command, 124
First observation per group, 214
FIRST SPSS keyword, 214
First.variable SAS variable, 214
Fix function, 79
Foreign package, 98

FORMAT SAS procedure, 225
FORMAT SAS statement, 225
Formats, 225
Formulas, 64
FREQ SAS procedure, 200
Function, 63
    arguments, 63
    controlling with arguments, 62
    controlling with class, 65
    controlling with formulas, 64
    generic, 41, 44, 66, 67, 143, 285, 319,
        444, 445, 446
    n, SAS or SPSS, 156
    visible vs. non-visible, 67
    writing your own, 73–75
Function (drop "s"), 14

Generating data. See data generation
Generic function. See function, generic
Geom_bar function, 355
Geom_jitter function, 378
Geom_segment function, 363
GET FILE SPSS command, 113
getOption function, 454
getwd function, 61, 268, 453
getwd function, 453
ggplot function, 374
ggplot2 package, 341, 386
ggsave function, 278, 385
gl function, 246, 253, 254, 255
gmodels package, 409, 437
GOPTIONS SAS statement, 277, 290
GPL. See graphics:Graphics Production
        Language
Graphics
    comparing R's packages, 278
    density countours, 366
    devices, 277
    File> Export SPSS menu, 277
    GGobi, 274
    ggplot2 package, 274, 278
    ggplot2, example programs, 387
    Grammar of Graphics, 275–276, 341,
        342, 347, 387, 389
        example programs, 387
    grid graphics system, 307
    history recording, 278
    lattice, 310
    lattice package, 275, 310
    overview, 273
    procedures vs. systems, 277
    SAS/GRAPH, 273
    SAS/INSIGHT, 274

SPSS, 274
    traditional, 274, 281–339
    traditional graphics system, 277
    traditional, example programs, 331
Graphics Production Language, 274, 341,
        374, 458
Graphics, ggplot2
    aesthetics, defined, 342
    aspect ratio, 382
    axes, logarithmic, 381
    bar charts, 344
    bar charts with subgroups, 348
    box plots, 376
    box plots with jittered points, 378
    coordinate system, defined, 342
    density curve, 355
    dot charts, 352
    elements and parameters, summary, 386
    error bar plots, 337
    facets, defined, 342
    geoms, defined, 342
    histograms, 354
    labels, 353
    linear fit by groups, 373
    multiple plots on a page, 382
    normal QQ plots, 299
    overview of qplot and ggplot, 342
    pie charts, 347
    plot symbols, setting by groups, 372
    plots by group or level, 309
    point display variations, 361
    presummarized data, 351
    scales, defined, 342
    scatterplot matrix, 374
    scatterplot with density contours, 366
    scatterplots, 361
    scatterplots faceted by group, 374
    scatterplots with fit lines, 367
    scatterplots with jitter, 363
    scatterplots with large datasets, 364
    scatterplots with reference lines, 368
    statistics, defined, 342
    strip plots, 360
    titles, 353
Graphics, traditional
    abline function, 307, 308, 309, 369
    adding titles, labels, colors, legends, 288
    adj parameter, 327
    arrows function, 327
    ask parameter, 291
    axis function, 321
    bar plots, 337
    barplot, 281

Graphics, traditional (cont.)
    barplots of counts with subgroups,
        281
    barplots of means, 286
    box function, 327
    box plots, 322
    cex parameter, 327
    col parameter, 327
    coplots, 309
    demonstration plot, 147
    density, 366
    dot charts, 352
    dual-axes plots, 320
    error bar plots, 380
    family parameter, 326
    font parameter, 327
    formula and symbol display, 324
    graphics parameters, 325
    grid function, 350
    histograms, 293
    histograms overlaid, 297
    identifying points, 311
    interaction plots, 324
    jitter, 304
    las parameter, 327
    linear fit, 308
    linear fit by group, 308
    lines function, 294
    lty parameter, 327
    lwd argument, 327
    main sub title argument, 324
    main title argument, 326
    mar parameter, 326
    mfcol parameter, 326
    mfrow parameter, 326
    mosaic plot, 275, 286
    mtext function, 328
    multiple plots on a page, 290
    new parameter, 326
    normal QQ plots, 299
    options and elements table, 325
    par function, 297
    pch parameter, 327
    pie charts, 347
    ps parameter, 326
    scatterplot matrices, 318
    scatterplots, 303
    scatterplots with confidence and
        prediction intervals, 312
    scatterplots with confidence ellipse, 311
    scatterplots with jitter, 304
    scatterplots with large datasets, 305
    scatterplots with lines, 307
    spine plot, 348
    srt parameter, 327
    strip charts, 301
    text function, 321
    types of point displays, 304
    usr parameter, 289, 326
    xlab argument, 326
    xlog parameter, 326
    ylab argument, 326
    ylog parameter, 326
Grep function, 110, 111, 132

Hat function, 325, 328, 338
Head function, 18, 214, 264, 290
Help
    examples, 42
    extracting and replacing elements, 42
    for datasets, 45
    for generic functions, 44
    for packages, 44
    help files, 41
    running help examples, 32
    searching, 46
    via mailing lists, 45
    via web searches, 46
    vignettes, 47
help.search function, 42
help.start function, 41
Hexbin package, 307
Hist function, 297
Hmisc package, 13, 43, 45, 95, 180, 219, 239,
        263, 326

ID variable. See row names
Identify function, 301, 303, 304
IML SAS product, 6
Importing data
    from SAS, 95
    from SAS, example programs, 95
    from SPSS, 96
    from SPSS, example programs, 96
IN SAS operator, 111, 133
INCLUDE SPSS command, 27
Inputting text files. See reading text files
install.packages function, 13, 16, 33
Installation
    of packages, 15
    of R, 12
interaction.plot function, 324
is.na function, 166

JGR user interface, 36

KEEP SAS statement, 185

Keywords, SPSS, 51
kruskal.test function, 432, 433

Label function, 239
LABEL SAS statement, 239
Lapply function, 154, 169, 207
    compared to do.call, 207, 215
Last observation per group, 214
LAST SPSS keyword, 214
last.variable SAS variable, 214
Lattice package, 275, 277, 290, 307, 341
Layout function, 290
Length function, 155, 157
LENGTH SAS statement, 95
levene.test function, 424
LIBNAME SAS statement, 269
Library function, 13, 14
Library, for SAS formats, 225
Lines function, 294, 308, 316, 319
List, 41, 45, 59
    created by lm, 68
    created in your function, 66
List function, 59, 75, 92, 145, 287
lm function
lm function, 5, 64, 68, 262, 308, 314,
    417, 418
Load function, 23, 25, 26, 61, 261, 270
Loadhistory function, 26, 272
Loading a package, 14, 33
Local function, 454
Log function, 147
log10 function, 381
Logical comparisons
    to missing values, 89, 129, 159,
    165, 166,
Logical operators, 159
Long dataset format, 190
Lowess function, 308, 367
ls function, 60, 61, 261, 262, 263,
    264, 455
ls.str function, 263, 271

Macro. See function
Macro language, 5, 90, 458
Macro substitution, 90–91, 92
Mailing lists, 45, 103
Mapply function, 157
Masking functions, 432
MASS
    Modern Applied Statistics in S, 45, 403
MATCH FILES SPSS procedure, 214
MATLAB, 2, 451
Matlines function, 313, 316

Matrix, 67, 68
    character, 68
    numeric, 68
Matrix language, 5, 6, 441, 458
Matrix, SPSS product, 5
Mean function, 42, 58, 63, 67, 68, 147, 152,
    195, 197, 204, 286
MEANS SAS procedure, 153
Median function, 408, 427, 433
Merge function, 191, 192, 193, 199
Methods function, 66, 319, 441
Missing values. See NA
    example programs to assign, 169
MISSOVER SAS option, 89
Mode, 21, 27, 51, 57, 66, 68, 86, 144, 196, 200,
    205, 300, 326, 418, 446
Model object, 68, 70
MODEL SAS statement, 65
Mosaic function, 286
mtext function, 321, 328

N function, SAS and SPSS, 155, 156, 158
NA, missing value, 30, 52, 53, 54, 56, 63, 79,
    80, 82, 146, 153, 167
na.omit function, 167, 421
na.rm argument, 63, 153
Names function, 56, 70, 106, 115, 127,
    128, 132, 134, 172, 174, 262, 298,
    317, 415
Naming objects, 60
ncol function, 107
NODUPKEY SAS option, 210
Noint SAS option, 65
Not run, 43
nrow function, 127
NULL object, 185
NVALID SPSS function, 155

Object, 61
Objects function, 68, 261
Observations, 56
    converting to variables, 232
    selecting, 42, 103, 123–140, 295
    selecting all, 124
    selecting by index number, 124
    selecting by row name, 127
    selecting by string search, 132
    selecting in SAS and SPSS, 123
    selecting using logic, 108–110
    selecting with subset function, 114
ODS. See Output Delivery System
OMS. See Output Management System
Operators, mathematical, 148

Options, 486
    scipen, 404, 415
Options
    digits, 403, 415
    number of siginificant digits, 403, 453
    SAS, linesize, 403, 437
    SAS, probsig, 403
    setting automatically, 444
    SPSS, small, 5, 222, 403
    SPSS, width, 50
    width, 50
Options function, 50
Order function, 221, 222
Ordered factor, 225, 229
Ordinal data, 225, 229
Output Delivery System, SAS, 5, 68, 417
Output management, 5, 6, 68, 417, 441
Output Management System, SPSS, 5, 68,
    417, 458
Packages, 18
Pairs function, 67, 319, 320
Pairwise.t.test function, 429, 433
Pairwise.wilcoxon.test function, 433
par function, 290, 302, 307, 322, 326, 341
Paste function, 108, 176
Plot function, 5, 67, 281
plotmeans function, 324
Points function, 294
Predict function, 315, 316
prettyR package, 14, 156
Print function, 27, 31, 51, 69, 70, 74, 116, 135,
    145, 262, 383
    as related to SPSS, 31, 52
PROC SORT NODUPKEY, 212
Procedures, SAS or SPSS, 73
Production Facility, SPSS, 28
Programmability Extension, SPSS, 23
Programming syntax, 24
prop.table function, 407, 411, 412
PRX SAS function, 110, 132

qplot function, 341, 342, 343
qq.plot function, 300
qqnorm function, 301
quit function, 23, 26, 32
    in SPSS-R programs, 28
Quitting R, 26

R Commander user interface, 33
R, running
    from JGR interface, 36
    from R Commander, 33, 34, 35
    from Rattle interface, 34

from within SPSS, 27
in batch mode, 27
in standard R interface, 31, 343
programs that include programs, 31
Rattle package, 34
Rattle user interface, 34
rbind function, 187
    used with do.call, 207
rcorr function, 413, 414
read.fwf function, 89, 93
read.table function, 79, 81, 82, 89, 165,
    166, 226
Reading data from keyboard, 86
Reading data within a program
    example programs, 85
Reading text files
    2 records per case
        example programs, 94
    comma separated values, 80
    delimited, 79
    example programs, 80
    one record per case, 87
    skipping columns, 82
    two or more records per case, 92
    within a program, 84
Recode function, 180, 432
Records, 55
Regression, linear, 31, 64, 68
Regular expressions, 110, 112, 132,
    133, 262
Removing objects, 185, 216
Rename function, 171, 199
Renaming
    columns, 37, 175
    rows, 177
    variables, 37, 174
    variables, example programs, 177
rep function, 247, 254
Replacing parts of objects
    getting help on, 43
Repositories, 17
Reshape package, 201, 217
rm function, 61, 267, 268, 313
    regarding First. and Last., 455
Round function, 407, 412
row names, 56
rownames function, 449
row.names argument, 89, 177, 190
row.names function, 56, 127, 128, 132,
    134, 317
rpart *function*, 180
rpart *package*, 180
Running R, *See* R, running

Sample function, 248
Sapply function, 155, 182, 231, 408
sasxport.get function, 95
SAS/IML SAS Product, 6
Save function, 270, 278, 385
save.image function, 22, 26, 150, 270
savehistory function, 272
Saving
    data and functions, 22, 24, 26
    history (journal), 271
    output, 271
    program and output, 22, 24, 26
    programs, 271
    workspace, 269
Scale function, 198
Scan function, 87
Scientific notation, 403
sd function, 155
Search, 14
Search function, 14, 264
Search path, 264
SELECT IF SPSS Command, 124
Selecting observations
    example programs, 135
seq function, 246
SET SAS statement, 186
SET SPSS command, 404
setRepositories() function, 449
setwd function, 61, 269, 453
sink UNIX command, 2, 25
slashes, in filenames, 61
Sorting
    data, 221
    example programs, 223
Source function, 25, 27, 31
    as related to SPSS, 31
SPLIT FILE processing example
    programs, 208
SPLIT FILE SPSS command, 204
Split function, 186
split.screen function, 290
spss.get function, 96
spsspivottable.Display function, 31
SPSS-R Integration Package, 28, 29
sqrt function, 148
Statements, SAS, 51
Statistics
    nested, 65
    analysis of covariance, 65
    Analysis of variance, 427
        post hoc tests, 436
        specifying interactions, 430
        sums of squares, 431

Chi-squared test, 408
Correlations, 413
Crosstabulation, 438
Descriptive, 404
Example programs, 433, 434
Kruskal-Wallis test, 432
Mann-Whitney U test, 425
Models, comparing, 421
Overview, 481
percentages, 407, 412
Predictions on new data, 422
Proportions, 406, 411
Regression, linear, 417
    interaction, 65
    partial tests, 419
    QQ plot, 421
    residual plot, 421
    residuals-leverage plot, 420, 421
    scale-location plot, 421
    t-tests on parameters, 419
    without intercept, 65
    diagnostics, 420
    sequential tests, 420
t-tests
    for independent groups, 422
    paired, 424
Variance, equality testing, 424
Wilcoxon rank sum test, 425
Wilcoxon signed rank test for paired
    groups, 426
Step function, 421
stepAIC function, 421
str function, 262, 263
stringsAsFactors argument, 56
stripchart function, 301, 303
subset function, 114
Summarizing, 157
Summary function, 406, 407
    from within SPSS, 31
    regarding variable labels, 239
SUMMARY SAS procedure, 201, 221

t.test function, 424, 426
Table function, 58, 81, 282, 286
Table object, 65
TABULATE SAS procedure, 200
Tail function, 18, 262
Tapply function, 196
TEMPORARY SPSS command, 124
Text files, viewing, 87
Text files, reading. See reading text files
Text function, 317, 325
textConnection function, 84, 85

Timestamp function, 454
TO SPSS keyword, 100
Transcript of code and results, 26
Transform function, 148
Tree function, 180
TukeyHSD function, 430
type. *See* mode

Unclass function, 70, 144
Uninstalling
    packages, 16
    R, 16
UNIVARIATE SAS procedure, 406
Unique function, 449
Unlist function, 144
Unload packages. *See* detach function
update.packages function, 15
Updating packages, 15
USE ALL SPSS command, 124

VALUE LABEL SPSS command, 150
Value labels, 225
    example programs, 234
var function, 408, 428
var.test function, 424
VARIABLE LABELS SPSS command, 239
VARIABLE LEVEL SPSS command, 225
Variables, 65
    converting to observations, 217
    dropping, 185
    keeping, 185
    labels, 239
    recoding, 180
    renaming, 56, 171
    selecting, 103

selecting all variables, 104
selecting by column name, 125107
selecting by index number, 104
selecting by list index, 115
selecting by simple name, 113
selecting by string search, 110
selecting in SAS and SPSS, 103
selecting to save in new dataset, 116
selecting using $ notation, 112
selecting using logic, 108
selecting with subset function, 114
selection example programs, 116
Vector, 51
    character, 68
    character or string, 51
    numeric, 63, 73

Where function, 130
WHERE SAS statement, 103, 123
Which function, 162
Wide dataset format, 217
wilcox.test function, 426, 427
with function, 51, 84, 114, 120, 147, 196, 208,
        404
within function, 147
Working directory, 61, 272
Workspace
    managing, 261
    minimizing, 268
    saving, 269
write.foreign function, 98
write.table function, 97
writing text files, 99

Z score, 198

Printed in the United States of America